# Cambrian–Ordovician lingulate brachiopods from Scandinavia, Kazakhstan, and South Ural Mountains

LEONID POPOV AND LARS E. HOLMER

Popov, L. & Holmer, L.E. 1994 09 15: Cambrian–Ordovician lingulate brachiopods from Scandinavia, Kazakhstan, and South Ural Mountains. *Fossils and Strata*, No. 35, pp. 1–156. Oslo. ISSN 0300-9491. ISBN 82-00-37651-6.

Lingulate brachiopods are described from the Upper Cambrian – Lower Ordovician (Tremadoc–Arenig) of Scandinavia (Sweden, Denmark, and Norway), South Ural Mountains, northeastern Central Kazakhstan, and the southern Kendyktas Range in southern Kazakhstan. The faunas comprise a total of 56 species, of which 20 are new; these are assigned to 10 genera, of which the lingulids *Agalatassia* and *Keskentassia*, the siphonotretid *Siphonotretella*, and the acrotretids *Galinella*, *Longipegma*, *Ottenbyella*, *Akmolina*, *Mamatia*, *Sasyksoria*, and *Otariella* are new. The new Subfamily Elliptoglossinae is proposed. The poorly known Cambrian–Ordovician stratigraphy of the South Urals, northeastern Central Kazakhstan, and the southern Kendyktas Range is reviewed. Many sequences in these areas that were previously referred to the Upper Cambrian and Tremadoc can now be correlated with the lower Arenig Hunneberg Stage in Baltoscandia. Three main types of faunal assemblages can be distinguished: (1) the *Broeggeria* assemblage; (2) several microbrachiopod assemblages; and (3) the *Leptembolon–Thysanotos* assemblage. The *Broeggeria* assemblage is distributed world-wide in the Tremadoc of the southern Kendyktas Range, Scandinavia, Belgium, Great Britain, Canada, and Argentina, while the *Leptembolon–Thysanotos* assemblage is confined to the Arenig of an area surrounding the East European platform, including northern Estonia, Poland, Germany, Bohemia, Serbia, and the South Urals. The microbrachiopod assemblages are known mainly from the Upper Cambrian – Arenig of Scandinavia, South Ural Mountains, northeastern Central Kazakhstan, and the southern Kendyktas Range. □*Brachiopoda, Lingulata, Lingulida, Siphonotretida, Acrotretida, new subfamily Elliptoglossinae, new genera, AGALATASSIA, KESKENTASSIA, SIPHONOTRETELLA, GALINELLA, LONGIPEGMA, OTTENBYELLA, AKMOLINA, MAMATIA, SASYKSORIA, OTARIELLA, Upper Cambrian, Lower Ordovician, Tremadoc, Arenig, biostratigraphy, palaeogeography, Scandinavia, South Ural Mountains, northeastern Central Kazakhstan, southern Kendyktas Range, southern Kazakhstan.*

*Leonid Popov, VSEGEI, Srednij Pr, St. Petersburg 199 026, Russia; Lars E. Holmer, Institute of Earth Sciences – Historical Geology & Palaeontology, Norbyvägen 22, S-752 36 Uppsala, Sweden; 25th February, 1993; revised 13th January, 1994.*

# Contents

# Introduction

This study deals with Upper Cambrian – Lower Ordovician lingulate ('phosphatic inarticulate') brachiopods from Scandinavia (Sweden, Denmark, and Norway), South Ural Mountains, northeastern Central Kazakhstan, and the southern Kendyktas Range, southern Kazakhstan (Fig. 1).

In Baltoscandia, most previous studies of Ordovician lingulates (e.g., Gorjansky 1969; Biernat 1973; Bednarczyk & Biernat 1978; Holmer 1986, 1989) have been restricted mainly to faunas from the Arenig Series and upwards. The comprehensive study by Popov *et al.* (1989), of brachiopods from the Middle Cambrian to Tremadoc Obolus sandstone of the northern East Baltic, and a study of the lingulate faunas of correlative Swedish sequences (Puura & Holmer 1993) are exceptions. In Sweden and Norway, our work has concentrated on Tremadoc faunas, mainly from six key sections. The Scandinavian sequences provide an important basis for correlations with the South Urals and Kazakhstan in view of their more continuous and tectonically undisturbed sequences that are rich in conodonts and graptolites.

By contrast, the stratigraphy and complex tectonic structure of the South Urals, northeastern Central Kazakhstan, and the southern Kendyktas Range, southern Kazakhstan, are poorly understood. The rich and diverse lingulate faunas of these districts have previously received only cursory attention (Lermontova & Razumovskij 1933; Nikitin 1956; Koneva *et al.* 1990); they occur mostly in scattered, tectonically disturbed outcrops. Most of our faunas come from sequences that were previously considered to be of Tremadoc age, but our studies now indicate that the majority of the lingulate asssemblages from the South Urals and Kazakhstan can be correlated with the lower Arenig Hunneberg Stage in Baltoscandia (Fig. 2). This correlation is based mainly on the affinities of the lingulate faunas themselves but is confirmed occasionally by the occurrence of conodonts.

The Late Cambrian assemblages from northeastern Central Kazakhstan contain some of the earliest known taxa of acrotretid families, which otherwise diversified mainly from Ordovician times. These Cambrian assemblages also illustrate the major changes in composition of lingulate faunas across the Cambrian–Ordovician boundary.

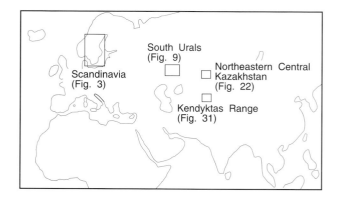

*Fig. 1.* Location of investigated areas with Upper Cambrian – Lower Ordovician lingulate faunas.

*Acknowledgements.* – We are grateful to Mikhail K. Apollonov, Igor F. Nikitin and Damir Tsay (Alma-Ata, Kazakhstan) for their assistance during field work in Kazakhstan during 1978–1981 as well as for providing important information on the geology and stratigraphy of the region.

The main lingulate collections from the South Urals were kindly given to us by Viktor G. Korynevskij (Miass), who also provided valuable geological data. We are grateful to Kirill S. Ivanov (Ekaterinburg) for material and for providing information on the age of localities in the South Urals. Svetlana V. Dubinina (Moskow), Tatjana Yu. Tolmacheva (St. Petersburg), and Aidar M. Zhylkaidarov (Alma-Ata) gave important information on conodont assemblages from Kazakhstan and in the South Urals.

Valdar Jaanusson and Jan Bergström (Stockholm) kindly helped during our work with the collections in the Swedish Museum of Natural History. David Bruton (Oslo) was most helpful during the field work in the Oslo region and allowed access to the collections in the Palaeontological Museum. Anita Löfgren, Kristina Lindholm, Kent Larsson and Per Ahlberg (Lund) helped in obtaining material and information relating to Scania. Walter Kegel Christensen (Copenhagen) arranged the loan of Danish material. Frederick J. Collier, Rex Doescher, and Richard Grant (Washington D.C.) assisted by arranging facilities to study material stored in the U.S. National Museum. Robin M. Cocks (London) and Michael G. Bassett (Cardiff) gave access to collections in the Natural History Museum and the National Museum of Wales, respectively.

The manuscript has benefitted from comments on both the language and the scientific content by Stefan Bengtson (Uppsala), John S. Peel (Uppsala) and Michael G. Bassett (Cardiff).

This work has been supported by grants (to LEH) from the Swedish Natural Science Research Council (NFR). LP gratefully acknowledges a one-year NFR visiting-scientist grant and a grant from the Swedish Institute that have enabled him to work extensively at the Institute of Earth Sciences – Historical Geology and Palaeontology, Uppsala University.

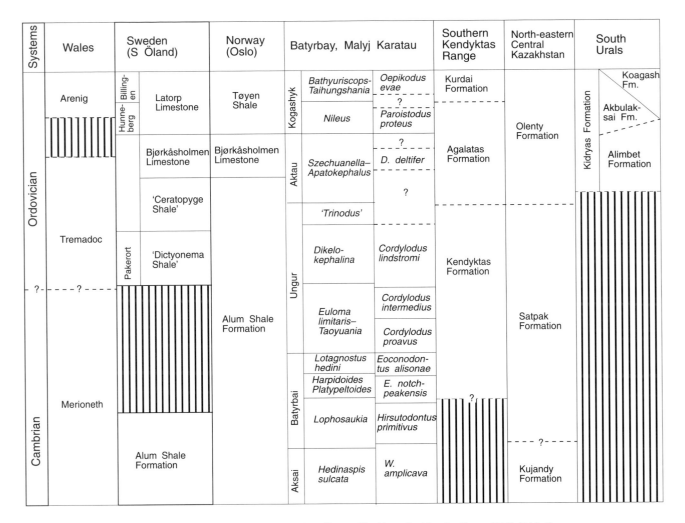

| Systems | Wales | Sweden (S Öland) | | Norway (Oslo) | Batyrbay, Malyj Karatau | | | Southern Kendyktas Range | North-eastern Central Kazakhstan | South Urals |
|---|---|---|---|---|---|---|---|---|---|---|
| Ordovician | Arenig | Hunneberg · Billingen | Latorp Limestone | Tøyen Shale | Kogashyk | Bathyuriscops-Taihungshania | Oepikodus evae | Kurdai Formation | | Koagash Fm. |
| | | | | | | | ? | | Olenty Formation | Akbulak-sai Fm. |
| | | | | | | Nileus | Paroistodus proteus | | | |
| | | | Bjørkåsholmen Limestone | Bjørkåsholmen Limestone | Aktau | | ? | Agalatas Formation | | Alimbet Formation |
| | | | | | | Szechuanella–Apatokephalus | D. deltifer | | | |
| | | | 'Ceratopyge Shale' | | | | ? | | | |
| | Tremadoc | Pakerort | | | Ungur | 'Trinodus' | | | | |
| | | | 'Dictyonema Shale' | | | Dikelo-kephalina | Cordylodus lindstromi | Kendyktas Formation | | |
| | ? ----- ? ----- | | | Alum Shale Formation | | Euloma limitaris–Taoyuania | Cordylodus intermedius | | Satpak Formation | |
| | | | | | | | Cordylodus proavus | | | |
| | | | | | Batyrbai | Lotagnostus hedini | Eoconodon-tus alisonae | | | |
| | | | | | | Harpidoides Platypeltoides | E. notch-peakensis | | | |
| Cambrian | Merioneth | | | | | Lophosaukia | Hirsutodontus primitivus | ? | | |
| | | | Alum Shale Formation | | | | | | ? ----- | |
| | | | | | Aksai | Hedinaspis sulcata | W. amplicava | | Kujandy Formation | |

*Fig. 2.* Correlation of Upper Cambrian – Lower Ordovician successions discussed in this study. After Apollonov (1991, Table 1).

# Upper Cambrian – Lower Ordovician stratigraphy and lingulate faunas

There is as yet no formally established stratotype for the Cambrian–Ordovician boundary (e.g., Norford 1991). For the sake of standard reference and convenience the base of the Ordovician is taken here as coincident with the lower boundary of the *Cordylodus lindstromi* Biozone; this level appears to be roughly correlative with the base of the Tremadoc Series in Wales (e.g., Fortey *et al.* 1991; Fig. 2 herein). The base of the Dictyonema Shale (and top of the *Acerocare* Biozone) in Sweden and Norway is also close to this level (Bruton *et al.* 1988), which has been taken traditionally as the base of the Ordovician in Baltoscandia. In Estonia, the first appearance of *Rhabdinopora*[*Dictyonema*]-bearing grapto-litic shales is apparently at a somewhat lower level, the upper part of the *Cordylodus proavus* Biozone (Kaljo *et al.* 1988).

Lower Ordovician is here used as a synonym for the Oeland Baltoscandian Regional Series (Kaljo, Rõõmusoks & Männil 1958).

A general legend to all the geological maps and strati-graphical columns discussed below is given in Fig. 4. The distribution and abundance of lingulates in the examined samples is given in Appendix 1.

## Scandinavia

In Norway and Sweden, our work was restricted to the lower Tremadoc Dictyonema and Ceratopyge shales (belonging to the upper part of the Alum Shale Formation in the sense of Andersson *et al.* 1985 and Owen *et al.* 1990) and the upper Tremadoc Bjørkåsholmen Limestone (term introduced by Owen *et al.* 1990 to replace the Ceratopyge Limestone). Lingulates from this interval were studied in samples ob-tained mainly at: (1) the Ottenby section on the island of Öland, (2) Flagabro and (3) Fågelsång sections in Scania, (4) Stora backor and (5) Mossebo sections in Västergötland, and (6) Bjørkåsholmen section, Oslo region (Fig. 3).

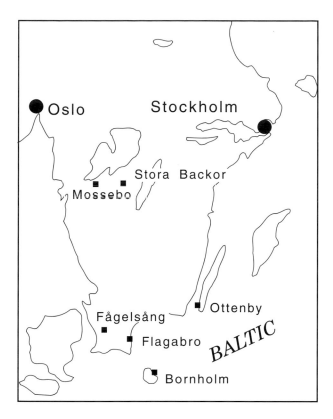

*Fig. 3.* Localities of material investigated from Scandinavia.

The stratigraphy of the Cambrian and Ordovician sequences in Sweden was summarised by Martinsson (1974; see also Mens *et al.* 1987) and Jaanusson (1982a), respectively; Bruton & Owen (1982; and references therein) provided a general summary of the Ordovician of the Oslo region, Norway.

## Öland

Most of the material examined from the Bjørkåsholmen Limestone is from the Ottenby section (Fig. 5), where the Dictyonema and Ceratopyge shales are not particularly well exposed; a number of specimens from the latter two units come from other localities on Öland. The thickness of the Dictyonema and Ceratopyge shales varies considerably over the island, with a general reduction towards the north, where the Dictyonema Shale and the Bjørkåsholmen Limestone are generally absent (e.g., Jaanusson & Mutvei 1982).

At Borgholm, the glauconitic Ceratopyge Shale is extremely rich in lingulates. Walcott (1902, 1908, 1912; his locality 310d, which perhaps corresponds to the Furuhäll section; see Bagnoli *et al.* 1988 for a detailed description of this section) described and recorded an assemblage including *Broeggeria salteri* (Holl, 1865), *Lingulella ferruginea* Salter, 1867 [=*L. antiquissima* (Jeremejew, 1856)], *L. lepis* Salter, 1866 [=*L. antiquissima*], *Acrotreta seebachi* (Walcott, 1902) [=?*Eurytreta* cf. *bisecta* (Matthew, 1901)], and *Acrothele borgholmensis* Walcott, 1908 [=*Orbithele ceratopygarum* (Brøgger, 1882)].

Walcott (1902, 1912) also described *Acrotreta conula* [=?*Ottenbyella carinata*], which was reported as coming from an unnamed Upper Cambrian locality (no. 310a) on Öland within the *Olenus truncatus* Biozone. However, examination of the slab bearing the type specimens and the associated brachiopods (including *Broeggeria salteri*; USNM 35267; Walcott 1912, Pl. 75:2) establishes that it comes from the Ceratopyge Shale; the lithology of the shale is identical to that of the rest of his Borgholm material but quite unlike that of the Upper Cambrian (see also Waern 1952, p. 235).

The Cambrian–Ordovician sequence in the Bödahamn core on Öland was described by Waern (1952), who recorded *Broeggeria salteri*, *Acrotreta conula* Walcott, 1912 [=?*Ot-*

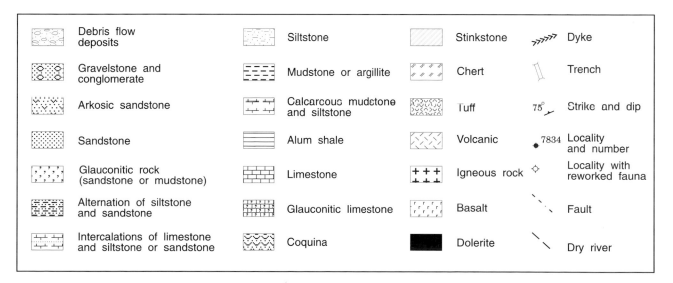

*Fig. 4.* Legend for geological maps and stratigraphical columns.

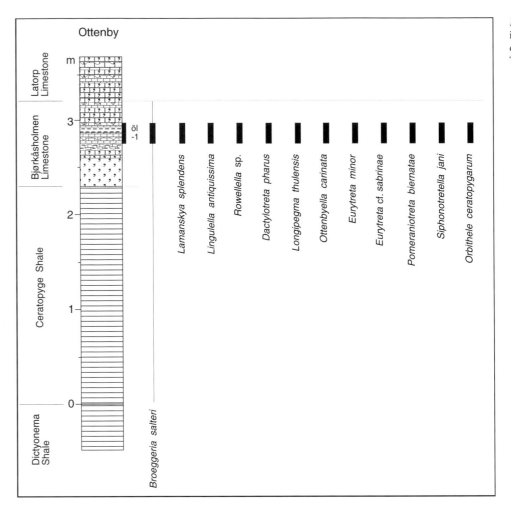

*Fig.* 5. Coastal section at Ottenby, island of Öland, showing distribution of lingulates. After Tjernvik (1956, Figs. 17, 19).

tenbyella carinata], and *Acrotreta* cf. *oelandica* Westergård, 1909 [=?*O. carinata*] from the Ceratopyge Shale (interval 30.90– 40.88 m), following directly above the Obolus conglomerate in the core.

*Ottenby.* – The coastal section at Ottenby is near the southern point of Öland, west of Ottenby Kungsgård (Fig. 3). This section exposes a sequence, somewhat more than 5 m thick, from the Ceratopyge Shale up to the Volkhov Stage Lanna Limestone (Fig. 5). It has been described by various authors since the end of the 19th Century (Tullberg 1882; see account by Moberg & Segerberg 1906, Walcott 1912, and Tjernvik 1956). Tjernvik (1956, p. 145, Figs. 17, 19) gives the most recent detailed description (see also Jaanusson & Mutvei 1982, p. 10, Fig. 3). The black Dictyonema Shale is not exposed, but in the nearby Ottenby core it is 7.8 m thick (Westergård 1944).

The Ceratopyge Shale is also developed as a black alum shale, in which glauconite is absent. The lower boundary of the unit is not exposed, but according to Westergård (1944, p. 5) it is 2.3 m thick in the Ottenby core. *Broeggeria salteri* is the only lingulate recorded by Tjernvik (1956) from the

Ceratopyge Shale at this locality, but Westergård (1909, p. 76) described *Acrotreta oelandica* [=?*Ottenbyella carinata*] from this level (Fig. 5).

Directly above the Ceratopyge Shale follows a 30 cm thick unit of glauconitic clay, which lacks recognizable fossils other than fragments of *Broeggeria* (Fig. 5). The overlying Bjørkåsholmen Limestone is about 60 cm thick, consisting of grey limestone that is glauconitic in its upper and lower part, intercalated with mudstone. It is particularly rich in fossils at around 20 cm above the base (sample Öl-1; collected by Moberg & Segerberg; Fig. 5 herein; see also Appendix 1K), and most of Moberg & Segerberg's (1906) lingulate fauna came from this level including the following taxa: *Lingula? producta* [=*Lingulella antiquissima*], *Lingula? ordovicensis* [=*L. antiquissima*], *Lamanskya splendens*, *Acrotreta* [=*Ottenbyella*] *carinata*, *Acrotreta circularis* [=*O. carinata*], *Acrothele barbata* [=*Orbithele ceratopygarum* (Brøgger, 1882)]. This interval also contains *Rowellella* sp., *Dactylotreta pharus* sp. nov., *Longipegma thulensis* gen. et sp. nov., *Eurytreta minor* Biernat, 1973, *E.* cf. *sabrinae* (Callaway, 1877), *Pomeraniotreta biernatae* Bednarczyk, 1986, and *Siphonotretella jani* gen. et sp. nov.

Above the Bjørkåsholmen Limestone follow the Hunneberg–Billingen Latorp Limestone and Volkhov Lanna Limestone; this sequence is also extremely rich in lingulates (Holmer, unpublished).

## Scania

Bergström (1982) gives a general introduction to the Ordovician in Scania. As noted by him, the break associated usually with the Cambrian–Ordovician boundary seems to be lacking in some parts of Scania.

The lingulate fauna from the Dictyonema Shale and the Bjørkåsholmen Limestone was studied by Moberg & Segerberg (1906) and Westergård (1909). Most of the described fauna comes from the Flagabro and Fågelsång districts (Fig. 3), but Westergård (1909) also described *Lingula?* [=*Elliptoglossa] linguae* and *Obolus? inflatus* [=?*Broeggeria salteri*] from Bjørkåsholmen Limestone at the Jerrestad section (Bergström 1982, p. 195, Fig. 7), eastern Scania.

*Flagabro.* – The section and core at Flagabro (Fig. 3) was described by Tjernvik (1958; see also Bergström 1982, p. 192, Fig. 4). The Dictyonema Shale in the core is 11.20 m thick, consisting of black pyritic and baritic alum shales with only a single bed of anthraconite (bituminous limestone; sometimes also called 'stinkstone' or 'orsten') close to the base; *Broeggeria salteri* was recorded by Tjernvik from throughout this interval along with *Lingulella* sp. and '*Acrotreta* sp'. In the core, the Ceratopye Shale is 26 cm thick, consisting of alum shales with pyrite and phosphorite, and a thin bed of anthraconite close to the base; no lingulates were listed from this unit by Tjernvik (1956). The Bjørkåsholmen Limestone is 32 cm thick, consisting of dark, very hard limestone with only a minor amount of glauconite; only *Lingulella producta* (Moberg & Segerberg) [=*L. antiquissima*] was recorded.

For the present work, one sample (Sk-1; kindly made available by Dietmar Andres, Berlin; see also Appendix 1K) from the exposure of Bjørkåsholmen Limestone along the Flagabro rivulet (see Tjernvik 1958, Fig. 6) was investigated. It contained *Eurytreta minor, E.* cf. *sabrinae, Pomeraniotreta biernatae, Dactylotreta pharus* sp. nov., *Ottenbyella carinata*, and *Siphonotretella jani* gen. et sp. nov.

The matrix of Bjørkåsholmen Limestone at this and many other localities in Scania is partly silicified, and the lingulate specimens had to be isolated by digestion in HF.

*Fågelsång.* – This classic area situated east of Lund (Fig. 3) includes scattered exposures of Palaeozoic rocks ranging in age from Cambrian to Silurian, described by Moberg (1910) for the excursion of the International Geological Congress (see also Regnell 1960).

The Upper Cambrian – Middle Ordovician sequence in the Fågelsång core was described by Hede (1951), who recorded but did not illustrate the rich brachiopod fauna from this interval.

The Dictyonema Shale in the core is 5.3 m thick and identical in lithology to that at Flagabro; the brachiopod fauna is dominated by *Broeggeria salteri, Lingulella antiquissima* and poorly preserved acrotretids.

Moberg & Segerberg (1906) described *Lingula? corrugata* [=?*Lingulella antiquissima*], *Lingula?* [=*Ralfia?*] *bryographtorum*, and *Obolella* (*Acrotreta?*) *sagittalis* Salter, 1866 [=?*Eurytreta* cf. *bisecta* (Matthew, 1901)] from this interval in the Fågelsång area.

The Ceratopyge Shale is 49 cm thick in the core, comprising alum shales in the lower part and a dark-grey shale in the upper part, with numerous phosphoritic nodules. Hede (1951) records a lingulate fauna identical to that of the underlying unit.

In the core, the Bjørkåsholmen Limestone is 106 cm thick, consisting mainly of a dark, compact, hard limestone that is usually siliceous. Hede (1951) records a lingulate fauna similar to that of the underlying units.

Moberg & Segerberg (1906) described *Lingula? producta* [=?*Lingulella antiquissima*], *Lingula? ordovicensis* [=?*L. antiquissima*], *Acrotreta* [=*Ottenbyella*] *carinata*, and *Acrotreta circularis* [=*O. carinata*] from the Bjørkåsholmen Limestone at Fågelsång. Two samples (Sk-2 and 3; collected by Moberg & Segerberg; see also Appendix 1K) were used to obtain the etched material examined for this study. The first (Sk-2; etched with acetic acid) yielded *Eurytreta minor, E.* sp. a, Acrotretidae gen. et sp. nov. a, and *Siphonotretella jani* gen. et sp. nov. The second (Sk-3; etched with HF) yielded *Elliptoglossa linguae, Pomeraniotreta biernatae, Eurytreta minor, Longipegma thulensis* gen. et sp. nov., *Rowellella* sp., *Siphonotretella jani* gen. et sp. nov., and *Lingulella antiquissima*.

## Bornholm

The Lower Palaeozoic sequence on Bornholm is much reduced by comparison with other parts of Baltoscandia, and the total thickness of the Ordovician is only around 20 m (e.g., Poulsen 1966).

The Upper Cambrian is developed as alum shales, about 21 m thick; the only recorded lingulate from this interval is *Broeggeria salteri* (Poulsen 1923; Poulsen 1966). The fauna of the overlying Dictyonema Shale, 2.5 m thick, was studied by Poulsen (1922); it includes *Acrotreta sagittalis* Salter var. *lata* [=?*Eurytreta bisecta*], *Broeggeria salteri* and *Lingulella lepis* [=*L. antiquissima*]. The Dictyonema Shale on Bornholm was not sampled for the present work, but the type material of Poulsen (1922) was examined.

The Bjørkåsholmen Limestone is absent on the island, where the Volkhov Skelbro Limestone follows directly on the Dictyonema Shale (Poulsen 1966).

## Västergötland

For a summary of the Ordovician sequence in this region, see Jaanusson (1982b). As noted by him, the Tremadoc is poorly

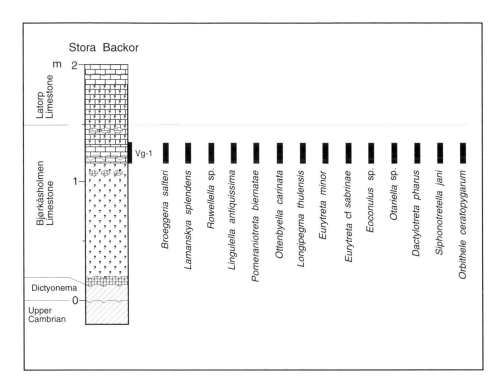

*Fig. 6.* Section in the abandoned quarry at Stora Backor, Mösseberg, Västergötland, showing distribution of lingulates. After Tjernvik (1956, Fig. 13).

developed: the Dictyonema Shale is only up to a metre thick and known only from Hunneberg and southern Falbygden; the Ceratopyge Shale does not seem to be developed at all; and the upper Tremadoc Bjørkåsholmen Limestone is found only in southern Falbygden (e.g., at Stora Backor), Kinnekulle, and on Hunneberg (e.g., Mossebo), where it reaches a maximum thickness of 1.5 m.

*Stora Backor.* – This old abandoned quarry, situated on the northwestern slopes of Mösseberg (Fig. 3), exposes a sequence from the Olenid shales to the Volkhov Lanna Limestone; it was described in detail by Tjernvik (1956, p. 139, Fig. 13).

The Dictyonema Shale is represented only by a 10 cm thick bed of anthraconite containing dendroid graptolites. Above follows a glauconitic mudstone, about 1 m thick, with some lenses of hard limestone. The overlying main Bjørkåsholmen Limestone is about 35 cm thick.

A bulk sample (Vg-1; Fig. 6; Appendix 1K) from the Bjørkåsholmen Limestone, 1.1–1.4 m above the base, was investigated for this study; it yielded *Lingulella antiquissima, Rowellella* sp., *Lamanskya splendens, Orbithele ceratopygarum, Siphonotretella jani* gen. et sp. nov., *Longipegma thulensis* gen. et sp. nov., *Eurytreta minor, E.* cf. *sabrinae, Pomeraniotreta biernatae, Eoconulus* sp., *Otariella* sp., *Dactylotreta pharus* sp. nov., *Ottenbyella carinata.*

*Mossebo.* – This locality on the northeastern slopes of Hunneberg (Fig. 3) was described in detail by Tjernvik (1956, p. 118, Fig. 5).

The Dictyonema Shale is absent, and the Bjørkåsholmen Limestone, which is only about 40 cm thick, sits directly on the Upper Cambrian; Tjernvik (1956) recorded a rich trilobite fauna including the common species for this interval in Sweden. The investigated bulk sample (Vg-2; Fig. 7; Appendix 1K) from the Bjørkåsholmen Limestone was collected by Torsten Tjernvik; it yielded *Lingulella antiquissima, Laman-*

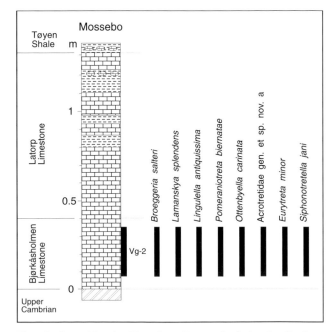

*Fig. 7.* Section at Mossebo, Hunneberg, Västergötland, showing distribution of lingulates. After Tjernvik (1956, Fig. 5).

*skya splendens, Siphonotretella jani* gen. et sp. nov. *Pomeraniotreta biernatae, Eurytreta minor,* and Acrotretidae gen. et sp. nov. a.

## Oslo region

Henningsmoen (1982, p. 92) gives an historical review of research on the Ordovician sequence in this region. Owen *et al.* (1990) revised the succession in terms of modern lithostratigraphy, introducing many new units.

The stratigraphy of the Cambrian Ordovician boundary interval was studied in detail by Bruton *et al.* (1982, 1988), and the boundary (traditionally taken at the top of the *Acerocare* Biozone), within the Alum Shale Formation (in the sense of Owen *et al.* 1990), does not seem to be marked by any hiatus.

The Lower Ordovician, upper part of the Alum Shale [=Dictyonema Shale] is poor in lingulates, and mostly indeterminable obolids and acrotretids have been recorded (e.g., Bruton *et al.* 1982, Fig. 4). Brøgger (1882) described *Broeggeria salteri* from this interval.

The uppermost part of the Alum Shale Formation, above the so-called 'Platypeltoides incipiens Limestone' (see Owen *et al.* 1990, p. 7) corresponds to the Ceratopyge Shale in other parts of Scandinavia. In the road section near Rortunet Shopping centre, this interval is about 7 m thick (Owen *et al.* 1990, p. 7, Fig. 2). The lingulate fauna from the Ceratopyge Shale was studied most recently by Gjessing (1976), who recorded a fauna including *Broeggeria salteri, Lingulella lepis* [=*L. antiquissima*], and *Orbithele ceratopygarum* from several localities in the Oslo region. In addition, Brøgger (1882) and Walcott (1912) recorded *Acrotreta sagittalis* [=?*Eurytreta* cf. *bisecta*].

All the lingulate species from the Ceratopyge Shale were also recorded from the overlying Bjørkåsholmen Limestone by Brøgger (1882) and Walcott (1912). In addition, Gjessing (1976; see also Fjelldal 1966) recorded *Conotreta* sp. [=*Dactylotreta pharus* sp. nov.], *Eurytreta minor, Spondylotreta* sp. [=*Ottenbyella carinata*], and *Torynelasma rossicum* Gorjansky [=*Biernatia circularis* sp. nov.]. Walcott (1902) also described *Acrotreta seebachi* [=?*Eurytreta* cf. *bisecta*] from this unit.

Most of the material examined for this study comes from the Bjørkåsholmen Limestone at the type locality. The specimens of *Biernatia circularis* sp. nov., however, come from the same unit in the Teigen–Stablum core from the Eiker–Sandsvaer district (Gjessing 1976).

*Bjørkåsholmen (stratotype).* – This beach section is close to Slemmestad in the Oslo–Asker district (Figs. 3, 8); the exposed sequence was described recently by Owen *et al.* (1990, p. 8, Fig. 3). The Bjørkåsholmen Limestone is 1.2 m thick, consisting of nodular limestone with intercalations of shale in the lower part, and two bedded, dark, compact limestone units, separated by a parting of shale in the upper part (Fig. 8).

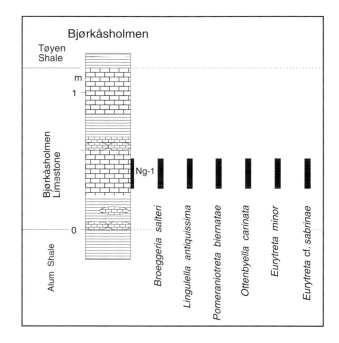

*Fig. 8.* Section at Bjørkåsholmen, Oslo region, showing distribution of lingulates. After Owen *et al.* (1990, Fig. 3).

For the present work, a bulk sample (Ng-1; collected by Holm; Fig. 8; Appendix 1K) from the middle, bedded limestone unit of Bjørkåsholmen Limestone was used; it yielded *Broeggeria salteri, Lingulella antiquissima, Pomeraniotreta biernatae, Eurytreta minor,* and *E.* cf. *sabrinae.*

## South Urals

The Lower Ordovician rocks in the South Ural Mountains are exposed mostly within the so-called 'Sakmara Belt' (Varganov *et al.* 1973, Fig. 2; Korynevskij 1989, Fig. 1) situated to the south and southeast of the Sakmara River (Fig. 9). The sequence was recently described by Korynevskij (1983, 1989, 1992), who summarized the existing information on the stratigraphy and introduced a revised lithostratigraphic classification that is mainly followed here. Unfortunately, Korynevskij's (1989) work is poorly available and printed only in 200 copies; thus, the most important part of the stratigraphical and geological information concerning the described faunas is reviewed and summarized below.

In the western part of the Sakmara Belt, the Lower Ordovician comprises clastic sequences of late Tremadoc to early Arenig age. The lower, mostly argillaceous part of the sequence, to the south of the Ural River is referred to the Alimbet Formation ('Koktugai beds' *in* Keller & Rozman 1961). The upper unit of arkosic glauconitiferous sandstones, with interlayers of argillites and limestones, is referred to the Akbulaksai Formation. Towards the southeast, the clastics are replaced by the Koagash Formation, consisting mainly of volcanics.

*Fig. 9.* Map of the Aktyubinsk district, South Urals (Fig. 1) with main investigated localities. 1 – Tyrmantau Ridge, 2 – Alimbet Farm, 3 – Alimbet River, 4 – Akbulaksai section, 5 – Bolshaya and Malaya Kayala rivers, 6 – Karabutak River, 7 – Kosistek River, 8 – Koagash River, 9 – Medes River.

The Kidryas Formation (introduced by Lermontova & Rozomovskij 1933) is used here for the entire Lower Ordovician sequence at the Tyrmantau Ridge; it is a lateral equivalent of the Alimbet, Akbulaksai, and Koagash formations (Fig. 2).

For the present study, nine main sections were studied in the southern part of the Sakmara Belt (Fig. 9): (1) Tyrmantau Ridge, (2) Alimbet Farm, (3) Alimbet River, (4) Akbulaksai River, (5) Bolshaya and Malaya Kayala rivers, (6) Karabutak River, (7) Kosistek River, (8) Koagash River, and (9) Medes River.

In this area, lingulate brachiopods were first recorded by Antipov & Meglitsky as early as 1858, but the only subsequent work dealing with the rich early Ordovician fauna was published by Lermontova & Razumovskij (1933). The Lower Ordovician sequences in the South Urals locally also contain abundant and diverse assemblages of articulate brachiopods (Andreeva *in* Markowskij 1960), trilobites (Balashova 1961), as well as molluscs and conodonts, but the latter two groups are relatively rare.

The complicated nature of the Lower Palaeozoic geology of this district and the imperfect knowledge of the distribution of the mostly endemic fossils have made it difficult to understand the relationships between major lithostratigraphic units; moreover, the stratigraphic terminology of the Lower Ordovician rocks varies considerably (cf., e.g., Keller & Rozman 1961; Vagranov *et al.* 1973; Korynevskij 1989).

According to the traditional view, the Upper Cambrian to Lower Ordovician (lower Arenig) sequence in the South Urals is subdivided into four regional stages: the Khmeliev (of assumed Late Cambrian age), the Kidryas (early Tremadoc), the Kolnabuk (late Tremadoc), and the Koagash (Arenig) stages. However, the supposed Late Cambrian age of the Khmelev Stage is based mainly on a few poorly preserved endemic species of trilobites, as well as the problematic brachiopod *Billingsella* [=*Protambonites*] *akbulakensis* Andreeva. The Kidryas Stage was considered to be of early Tremadoc age mainly because of the presence of obolids that were compared erroneously with the lingulate assemblage from the 'Obolus' beds in the East Baltic (Lermontova *in* Lermontova & Razumovskij 1933; Vagranov *et al.* 1973).

Our studies of the lingulate assemblages and their stratigraphic distribution have led to some reinterpretation of the stratigraphic and structural relationships of the main regional and litostratigraphic units within the South Urals. Thus, we find no evidence for the presence of Late Cambrian or early Tremadoc faunas (Fig. 2). The oldest lithostratigraphic units belong to the Alimbet Formation ('Koktugai beds' in Keller & Rozman 1961) and lower part of the Kidryas Formation. The trilobite assemblage described (Balashova 1961) or listed (Vagranov *et al.* 1973; Korynevskij 1983; 1989) from this stratigraphic interval contains some diagnostic taxa of the *Ceratopyge* fauna of Baltoscandia (including *Shumardia oelandica*, *Ceratopyge forficula*, and *Orometopus elatifrons*), as well as some brachiopods which are known from the Ceratopyge Shale and Bjørkåsholmen Limestone in the Lower Ordovician sequences of Sweden and Norway.

The trilobite faunas from the South Urals have not been revised since Balashova (1961), and no conodont biozonation is available. The upper part of the Akbulaksai Formation is characterized by an endemic assemblage of articulate brachiopods including *Altorthis kinderlensis* Andreeva, *Alimbella armata* Andreeva, *Medessia uralica* Andreeva, and *Protambonites lermontovae* (Andreeva). In the lower part of the Akbulaksai Formation ('Sarytugai beds' of Keller & Rozman 1961), a *Ceratopyge* assemblage of trilobites is also recorded, but in the upper part of the sequence there are only a few trilobite taxa, which lack stratigraphic significance. The lingulate assemblage associated with this fauna includes some diagnostic species, such as *Thysanotos siluricus*, *Leptembolon lingulaeformis*, *Eurytreta chabakovi* (Lermontova), *Acrotreta korynevskii* Holmer & Popov, *Siphonobolus uralensis* (Lermontova), and *Orbithele ceratopygarum*.

In Estonia and Ingria, *Thysanotos siluricus* is associated only with conodonts of the *Paroistodus proteus* Biozone (Viira 1966; Mägi & Viira 1976). In the South Urals, a few samples with *T. siluricus* have yielded conodonts, and they are also invariably from the *P. proteus* Biozone, and thus it is apparent that the local *T. siluricus* Biozone here correlates in part with the Hunneberg Stage (lowermost Arenig).

The *Thysanotos siluricus* assemblage also appears near the top of the Alimbet Formation, together with the last occurrence of a *Ceratopyge* trilobite assemblage (Figs. 11, 13, 18). Conodonts have not been found at this stratigraphic level, but it is assumed here that the level of the first appearance of the *T. siluricus* assemblage in this sequence is also close to the base of the Arenig.

Our study of the lingulate assemblages from the Koagash Formation at the type section shows that they include taxa characteristic of both the Hunneberg and Billingen stages in Baltoscandia, and this correlation is also confirmed by available data on the distribution of conodonts. The lower (Hunneberg) part of the Koagash Formation appears to be a correlative of the upper part of the Kidryas and Akbulaksai formations. The Koagash Formation contains a distinctive assemblage of lingulate and articulate brachiopods, such as *Ferrobolus fragilis* sp. nov., *Acrotreta korynevskii*, *Mamatia retracta* (Popov), *Otariella intermedia* gen. et sp. nov., *Eoconulus primus* sp. nov., *Orbithele ceratopygarum*, *Clarkella supina*, and *Tetralobula latens*; the formation is correlated with the Olenty Formation of northeastern Central Kazakhstan (Fig. 2).

## Tyrmantau Ridge

This is the type area for the Kidryas Formation, introduced by Lermontova & Razumovskij (1933, p. 188) for a clastic sequence exposed about 0.5 km to the south of Kidryassovo village.

The assemblage listed by Lermontova & Razumovskij includes the lingulate brachiopods *Acrothyra* [=*Eurytreta*] *chabakovi* and *Obolus* [=*Palaeoglossa*?] *razumovskii*, but most of the brachiopod specimens described in their paper were collected about 9 km to the northwest, from an isolated locality (Lermontova & Razumovskij 1933, p. 190) to the east of the Novo-Dmitrievka village. The geology and stratigraphy of this area were discussed briefly by Voinova (*in* Voinova *et al.* 1941), Leonenok (1955), and Keller & Rozman (1961). The most detailed description of this section was given by Korynevskij (1983, 1989), who also reviewed the stratigraphy and discussed the faunal assemblages.

In the area, the lower and upper contacts of the Lower Ordovician units are invariably faulted, and the original stratigraphic relationships are unknown. According to Korynevskij's (1983, 1989) description, the Kidryas Formation forms a homocline structure within a fault-bounded belt. However, the character of the lithology and the distribution of fossils suggest to us that the sequence is repeated, and the four lower units (about 465 m thick) in Korynevskij's description appear to represent equivalents of the lower Arenig in the upper part of the main section. The contact between units 4 and 5 is unexposed; in view of the differences in dip and strike, it is presumably tectonic (Fig. 10).

The following beds crop out at Tyrmantau (Fig. 11; Appendix 1A).

KIDRYAS FORMATION:

Unit 1. – More than 35 m of argillites.
Unit 2. – 70 m of arkosic and quartzose, cross-bedded sandstones.
Unit 3. – 70 m of argillites with nodules of argillaceous limestones. According to Korynevskij (1983), the nodules contain the trilobites *Ceratopyge* sp., *Dikelokephalina* sp., and *Geragnostus*. sp. (sample B-768-5). It seems that the trilobites *Apatokephalus arduus* and *Macropyge foliacea*, which were listed by Vagranov *et al.* (1973, p. 16) as coming from the Tyrmantau Ridge, were collected from this unit.
Unit 4. – 40 m of fine-grained, arkosic, glauconitic sandstones, with interbeds of siltstones and argillites, and few layers of coarse-grained

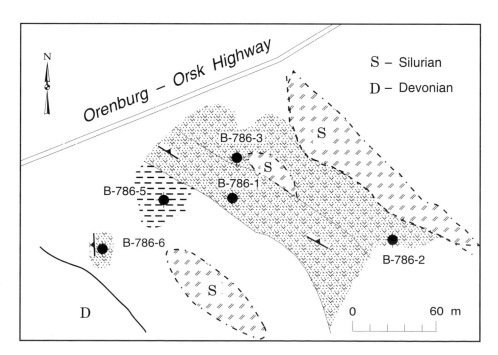

*Fig. 10.* Schematic geological map of the Lower Palaeozoic exposures at the Tyrmantau Ridge, showing the position of the main sampled localities. After V.G. Korynevskij, personal communication, 1991.

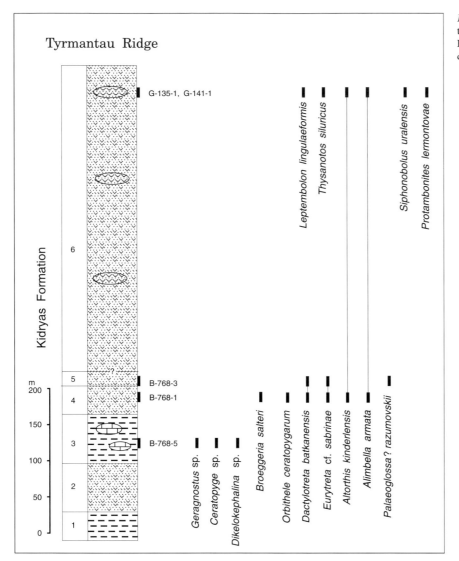

*Fig. 11.* Generalized stratigraphical column through the Kidryas Formation at Tyrmantau Ridge, showing the levels sampled, and distribution of brachiopods and trilobites.

arkosic sandstones. This unit is equivalent to 'Unit H' of Lermontova & Razumovskij's (1933) description. Nodules of calcareous mudstone, near the top of this unit (sample B-768-1), contain *Broeggeria salteri*, *Dactylotreta batkanensis* sp. nov., *Eurytreta* cf. *sabrinae*, and *Orbithele ceratopygarum*.

Unit 5. – 22 m of coarse to medium-grained, arkosic sandstones with ripple marks and interlayers of fine-grained sandstone and argillite. This unit is equivalent to Unit I of Lermontova & Razumovskij (1933). The assemblage from this unit (sample B-768-3) includes *Palaeoglossa? razumovskii*, *Dactylotreta batkanensis* sp. nov., and *Eurytreta* cf. *sabrinae*.

Unit 6. – More than 130 m of arkosic sandstones with interlayers of fine-grained sandstones, siltstones and argillites ('Unit K' in Lermontova & Razumovskij 1933). The lower boundary is not exposed in this section, and according to Korynevskij (1983) it is presumably marked by a fault. The thickness of the unit increases to up to 430 m in the area between the Pismenka and Issakul rivers, to the north of Kidryassovo village, where it conformably overlies 'Unit 5' (Khvorov *in* Korynevskij 1983, p. 4). Lens-shaped coquina beds within the unit contain the brachiopods *Altorthis kinderlensis*, *Alimbella armata*, *Protambonites lermontovae*, *Thysanotos siluricus*, *Leptembolon lingulaeformis*, and *Siphonobolus uralensis* (samples G-135-1; G-141-1). A similar assemblage of articulate brachiopods, including *Altorthis kinderlensis*, *Medessia uralica*, and

*Protambonites lermontovae*, was found at one isolated locality (sample B-768-6; Fig. 10), which was referred to the basal unit of the Kidryas Formation by Korynevskij (1983). However, the lithology and brachiopod assemblages suggest that it is part of a separate tectonic block of Lower Ordovician rocks belonging to the upper part of the Kidryas Formation.

Tyrmantau Ridge is also the type area for the Kidryas Regional Stage ('Horizon'), which is traditionally correlated with the Pakerort Stage of the East Baltic (Vagranov *et al.* 1973; Apollonov 1991; Fig. 2, herein). As noted above, this correlation is not supported here. The lingulate assemblage from the lower part of the Kidryas Formation (Units 4 and 5) has affinities to that of the Ceratopyge Shale and Bjørkåsholmen Limestone in Baltoscandia. The trilobites from Unit 3 (listed by Vagranov *et al.* 1973) also suggest a Tremadoc age for the lower part of the Kidryas Formation. In the upper part of the formation, this fauna is replaced by an assemblage with *Thysanotos siluricus* and some endemic articulate brachiopods indicating an early Arenig (Hunneberg) age.

## Alimbet Farm

This locality, in the northern part of the Aktyubinsk district (Fig. 9), is on the right side of the Alimbet River, to the east and southeast of Alimbet Farm (Fig. 12). It is the type locality for the Alimbet and Akbulaksai formations (Fig. 2), and the exposed sequence has been described by Keller & Rosman (1961), Leonenok (1955), Varganov *et al.* (1973), and Korynevskij (1989); Balashova (1961) described the trilobite fauna.

The lower part of the sequence was studied in the core C-50, about 1.4 km to the southwest of the Alimbet Farm (Fig. 12). To the east, Lower Ordovician rocks have a faulted contact with Middle Ordovician basalts and lower Silurian cherts. According to Korynevskij (1989, p. 27, Figs. 7, 8), the following sequence is present (Figs. 13–14; Appendix 1B):

ALIMBET FORMATION:

[The base of the formation is not exposed.]
Unit 1. – More than 900 m of siltstones with numerous nodules of argillaceous limestones; interbeds of fine-grained sandstones, and siltstones occur in the lower part of the unit. Sample B-675 yielded a fauna including the trilobites *Geragnostus* aff. *crassus*, *Leiagnostus* aff. *peltatus*, *Hospes* sp., *Dikelokephalina* sp., *Procyclopyge* sp., *Shumardia* cf. *pusilla*, *Glaphurus* sp., *Parapilekia* sp., *Apatokephalus* sp., *Euloma* sp., *Megistaspis* sp., *Harpides* sp., *Ceratopyge forficula*, *Orometopus* sp., *Niobella* sp., *Promegalaspides kasachstanensis*, *Alimbetaspis* sp., *Macropyge* sp., *Niobe* sp., and *Acerocarina* sp., as well as the graptolites *Rhabdinopora uralense* and *Anisograptus* sp.
A diverse assemblage of trilobites was found by Korynevskij in nodules of argillaceous limestone close to the top of the unit (sample B-578-1) including *Asaphellus alimbeticus*, *Promegalaspides kaschstanensis*, *Geragnostus* (*Micragnostus*) sp., *Protopliomerops* (*Parapilekia*) sp., *Shumardia pusilla*, *Platypeltoides* sp., *Ceratopyge* sp., *Leiagnostus* sp., *Parabolinella* sp., *Tersella*? sp., and *Alimbetaspis* sp. The associated assemblage of brachiopods includes *Altorthis kinderlensis*, *Medessia uralica*, *Eurytreta chabakovi*, *Siphonobolus uralensis*, and *Thysanotos siluricus*.

AKBULAKSAI FORMATION:

Unit 2. – 3 m of glauconitic sandstones and siltstones.
Unit 3. – 10–12 m of basalts.
Unit 4. – 45 m of siltstones with nodules of argillaceous limestones and interbeds of siltstones, fine-grained sandstone, and mudstone. The nodules of limestone contain (samples B-578-2; 381-N) the trilobites *Alimbetaspis* sp., *Dikelokephalina* sp., and *Nyaya* sp., as well as the brachiopods *Thysanotos siluricus*, *Eurytreta chabakovi*, and *Siphonobolus uralensis*. Interbeds of sandstones from the upper part of the unit (sample G-39-2) yielded the trilobites *Dikelokephalina* sp. and *Alimbetaspis* sp. and the brachiopods *Altorthis kinderlensis* and *Leptembolon lingulaeformis*.
Unit 5. – 55 m of cross-bedded sandstones with intercalations of siltstones and gravelstones.
Unit 6. – 25 m of siltstones with numerous intercalations of fine-grained, arkosic sandstones with glauconite, and nodules of silty limestone. The nodules contain (sample B-578-4) the trilobites *Alimbetaspis kelleri*, *Dikelokephalina* sp., *Nileus* sp., and *Euloma* sp. as well as the brachiopods *Altorthis kindelensis*, *Alimbella* cf. *armata*, *Eurytreta chabakovi*, *Semitreta*? aff. *magna*, and *Leptembolon lingulaeformis*.
Unit 7. – 75 m of fine-grained arkosic sandstones and siltstones.
Unit 8. – 25 m of siltstones with rare intercalations of siltstones and nodules of argillaceous limestone.
Unit 9. – 9 m of siltstones with a bed of fine-grained sandstone in the middle part of the unit, where the sample B-578-5, yielded the trilobites *Kainella*

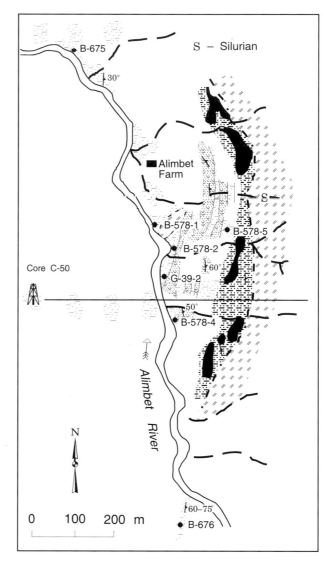

*Fig. 12.* Schematic geological map of the area around the Alimbet Farm, on the right side of the Alimbet River, showing the distribution of the Lower Palaeozoic and position of the main sampled localities. The location of the profile in Fig. 14 is indicated by the horizontal line from Core C-50 to the east. After Korynevskij (1989, Fig. 4).

sp., and *Geragnostus* sp., and the brachiopods *Altorthis kinderlensis*, *Leptembolon lingulaeformis*, and *Siphonobolus uralensis*.
Unit 10–11. – 18 m of siltstones and fine-grained, arkosic sandstones.
Unit 12. – 50 m of siltstones. In the southern part of the area, at the left side of the Alimbet River (sample B-676), an assemblage containing the trilobites *Alimbetaspis kelleri*, *Geragnostus* (*Micragnostus*) sp., *Niobella* sp., *Shumardia* sp., and *Nyaya* sp., as well as the brachiopods *Altorthis kinderlensis*, *Leptembolon lingulaeformis*, and *Eurytreta chabakovi* was collected.
Unit 13. – 13 m of arkosic sandstones.
Unit 14. – 30 m of siltstones.
Unit 15. – 180 m of alternating sandstones and siltstones, with argillaceous intercalations, containing numerous trace fossils and fragments of indeterminate lingulate brachiopods.
Unit 16. – 16 m of conglomerates with sandy matrix, and intercalations of sandstones and siltstones.

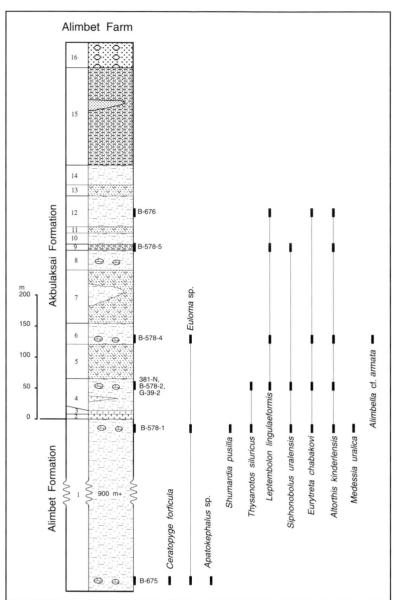

*Fig. 13.* Generalized stratigraphical column through the Alimbet and Akbulaksai formations at Alimbet Farm, showing distribution of selected species of trilobites and brachiopods. After Korynevskij (1989, Fig. 7).

*Fig. 14.* Generalized profile (location indicated in Fig. 12) through the Alimbet and Akbulaksai formations at Alimbet Farm. Numbered units as in Fig. 13. After Korynevskij (1989, Fig. 6).

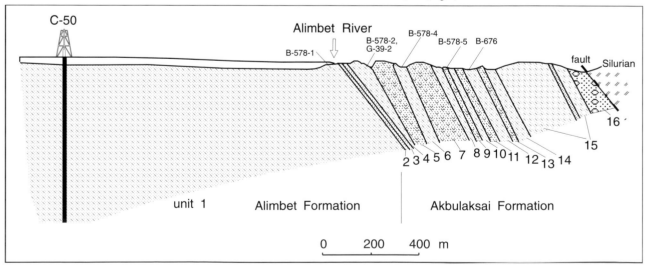

*Fig. 15.* Schematic geological map of the area around the upper reaches of the Alimbet River, showing the distribution of the Lower Palaeozoic and position of the main sampled localities. After Korynevskij (1989, Fig. 8).

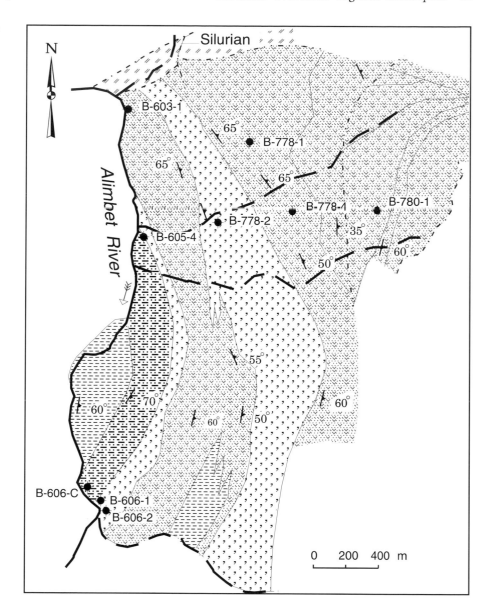

The total thickness of the Akbulaksai Formation in this section is about 556 m.

The succession of assemblages in the Alimbet Farm section is comparable with that at the Tyrmantau Ridge. In both sections, the first appearance of the *Leptembolon–Thysanotos* assemblage is associated with endemic articulate brachiopods, such as *Altorthis kinderlensis*, *Protambonites lermontovae*, *Medessia uralica* and *Alimbella armata*.

## Alimbet River

Along the upper reaches of the Alimbet River, about 2.25 km to the south of the southern exposures at Alimbet Farm, Ordovician deposits are exposed in another area along the Alimbet River (Fig. 15). The Lower Ordovician sequence described below is based on studies of a series of excavations on the left side of the river, as well as numerous scattered outcrops on the right side of the river. The following description of the sequence (Fig. 16; Appendix 1C) is adopted from Korynevskij (1989, p. 32).

Unit 1. 300 m of argillites, fine-grained sandstones, and siltstones with intercalations of cross-bedded, arkosic and quartzose sandstones.

Unit 2. 130 m of alternating siltstones and sandstones (glauconitic, arkosic, and quartzose) with nodular beds of conglomerate and gravelstone. The best exposures of this unit are situated at the southern part of the area along the right side of the Alimbet River. *Leptembolon lingulaeformis* is present in the lower part of the unit (sample B-606, about 13 m above the base). The brachiopod and trilobite assemblages higher up in the unit (sample B-606-1; about 27 m below the top) includes *Geragnostus*

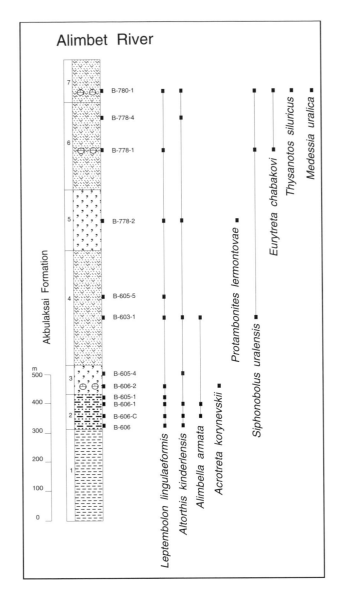

*Fig. 16.* Generalized stratigraphical column through the Akbulaksai Formation along the upper reaches of the Alimbet River, showing the distribution of selected species of trilobites and brachiopods. After Korynevskij (1989, Fig. 13).

(*Micragnostus*) sp., *Dikelokephalina* sp., *Nyaya* sp., *Leptembolon lingulaeformis*, *Altorthis kinderlensis*, and *Alimbella armata*.

Unit 3. 100 m of glauconitic siltstones, with rare interbeds of arkosic and quartzose sandstones with glauconite. This unit contains rare trilobites, such as *Acerocarina* sp. and *Dikelokephalina* sp., and the brachiopod *Altorthis kinderlensis*. Numerous specimens of *Acrotreta korynevskii* were found in limestone nodules (sample B-606-2). The thickness of the unit varies significantly from 40 m in the north to 100 m in the south of the area.

Unit 4. 400 m of cross-bedded, arkosic and quartzose sandstones with rare interbeds of gravelstone and siltstone. The assemblage includes the brachiopods (sample B-603-1) *Altorthis kinderlensis*, *Alimbella armata*, *Leptembolon lingulaeformis*, and *Siphonobolus uralensis*.

Unit 5. 180 m of fine-grained, glauconitic sandstones, with interbeds of siltstones. The recorded fossils include (sample B-778-2) the brachiopods *Altorthis kinderlensis* and *Protambonites lermontovae*.

Unit 6. 300 m of coarse-grained, arkosic and quartzose sandstones, with interbeds of siltstones in the lower part of the unit. Nodules of silty glauconitic limestones within siltstones in the lower part of the unit (sample B-778-1) contain the trilobites *Acerocarina* sp., and *Dikelokephalina* sp., and the lingulates *Leptembolon lingulaeformis*, *Eurytreta chabakovi*, *Semitreta*? aff. *magna*, and *Siphonobolus uralensis*. *Altorthis kinderlensis* was found near the top of the unit, within a bed of coarse-grained, glauconitic, arkosic sandstone (sample B-778-4).

Unit 7. More than 150 m of fine-grained, arkosic sandstones with glauconite and argillites. The trilobites *Acerocarina* sp. and *Geragnostus* sp. were found in beds of argillites, near the base of the unit. Nodules of silty, argillaceous limestones in the upper part of the unit (sample B-780-1) contain the brachiopods *Altorthis kinderlensis*, *Alimbella armata*, *Leptembolon lingulaeformis*, *Thysanotos siluricus*, *Semitreta*? aff. *magna*, and *Siphonobolus uralensis*.

## Akbulaksai River

This section along at the upper reaches of the Akbulaksai River (Fig. 17) has been described several times (e.g., Vagranov *et al.* 1973; Korynevskij 1989), but with widely different interpretations. The most detailed description of the Lower Palaeozoic rocks was given by Korynevskij (1989, p. 37, Fig. 9), based on a series of trenches excavated along the left side of the river. According to him, the section comprises a gradual sequence of clastic rocks. Our study indicates that the Lower Ordovician rocks in this area may form as many as three separate blocks with faulted contacts (Figs. 17–18; Appendix 1D).

The first of these (Unit 1; Figs. 17–18) occurs in the southwestern part of the area, where the Akbulaksai Formation is represented by a 230 m thick unit consisting of fine-grained, arkosic, glauconitic sandstones with intercalations of coarse-grained, quartzose, glauconitic, cross-bedded sandstones, siltstones, and argillites. Rare specimens and fragments of *Leptembolon lingulaeformis* were found at the base of the unit (sample B-607-5). About 120 m above the base (sample B-607-6), *Altorthis kinderlensis*, *Alimbella armata*, and *Leptembolon lingulaeformis* were found; *A. kinderlensis*, *L. lingulaeformis*, and the trilobite *Dikelokephalina* sp. were also collected from a sample (B-607-1) higher up in the unit.

Along the eastern fault, towards the northwest, this unit is in contact with the second block (Units 2–3 in Figs. 17–18), about 120 m thick, of fine-grained sandstones alternating with siltstones with nodules of argillaceous limestones. The trilobites *Ceratopyge* sp., *Platypeltoides* sp., *Shumardia* sp., *Promegalaspides* sp., *Dikelokephalina* sp., and *Apatokephalus* sp. occur in nodules at the base of the unit (sample B-607-2). In the upper part of the unit, *Altorthis kinderlensis*, *Thysanotos siluricus* (sample G-42-3), *Eurytreta chabakovi*, and *Leptembolon lingulaeformis* (sample B-607-6-1) appears. This part of the sequence belongs to the Alimbet Formation.

A similar assemblage with early Arenig brachiopods was found also in the overlying beds of fine-grained, arkosic sandstones and siltstones, some 300 m thick (Unit 3; Figs. 17–18); it includes *Altorthis kinderlensis*, *Medessia uralica*, and *Protambonites lermontovae* (sample B-607-7). About

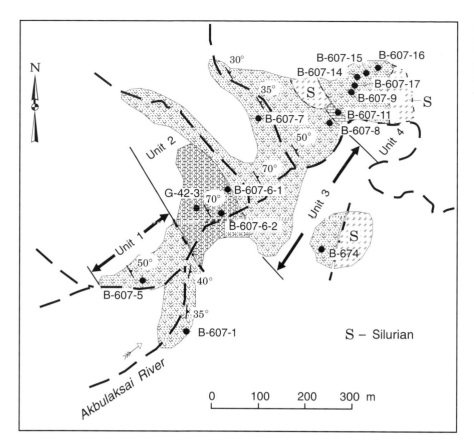

*Fig. 17.* Schematic geological map of the area around the Akbulaksai River, showing the distribution of the Lower Palaeozoic and position of the main sampled localities. Location of Units 1–4 in Fig. 18 indicated. After Korynevskij (1989, Fig. 9).

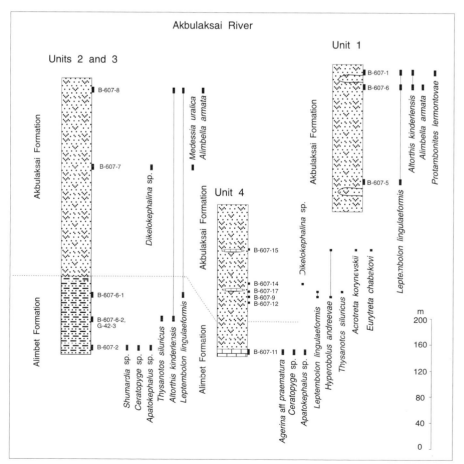

*Fig. 18.* Three generalized stratigraphical columns (location of Units 1–4 indicated in Fig. 17) through the Akbulaksai Formation around the Akbulaksai River, showing distribution of selected species of trilobites and brachiopods. After Korynevskij (1989, Fig. 9).

100 m above this sample, near the top of the unit, the trilobites *Dolgeuloma* sp. and *Dikelokephalina* sp. and the brachiopods *Altorthis kinderlensis, Alimbella armata, Protambonites lermontovae,* and *Leptembolon lingulaeformis* (sample B-607-8) were found. This part of the sequence belongs to the Akbulaksai Formation.

The third faulted block (Unit 4; Figs. 17–18) is present in the northeastern part of the area, where a similar sequence is exposed. In the lower exposed part, the Alimbet Formation consists of bedded limestones with glauconite, containing the trilobites *Homagnostus kasachstanicus, Ceratopyge* sp., *Apatokephalus* sp., *Varvia* sp., *Agerina* aff. *praematura,* and *Nyaya* sp. (sample B-608-11). Towards the top, arkosic sandstones, glauconitic sandstones, argillites, and siltstones yielded only the graptolite *Rhabdinopora* sp. The lower part of the overlying Akbulaksai Formation consists of arkosic sandstones, siltstones, argillites containing the brachiopods *Protambonites akbulakensis* and *Hyperobolus andreevae* sp. nov. (samples B-607-12; B-607-9). Higher up, nodules of sandy limestones within the arkosic sandstones contain the brachiopods *Leptembolon lingulaeformis* and *Thysanotos siluricus* (sample B-607-17). A similar type of nodules in the overlying interval yielded the brachiopods *Protambonites akbulakensis,* juvenile *Hyperobolus andreevae* sp. nov., *Eurytreta chabakovi,* and *Acrotreta korynevskii,* as well as the trilobites *Geragnostus* (*Micragnosus*) sp. and *Koldinioidea* sp. (sample 607-15).

According to Vagranov *et al.* (1973), the upper part of the Alimbet River section belongs within the Upper Cambrian, based mainly on the presence of the problematic *'Billingsella'* and some rare, poorly preserved trilobites that were assigned to the Upper Cambrian genus *Aphelaspis* (Ancygin *et al.* 1977). This proposal is not supported here; *'Billingsella' akbulakensis* Andreeva differs markedly from *Billingsella* in having well-developed dental plates and is here assigned to *Protambonites* Havlíček.

## Bolshaya and Malaya Kayala rivers

These two localities are situated on both sides of the Bolshaya Kayala River, near the ruins of Molokanskij village (Fig. 9). The geology of this area was described recently by Korynevskij (1992, p. 131, Fig. 2), who also recorded numerous brachiopods.

At both localities, lingulates occur in nodules of argillaceous limestone, occurring within a bed of shale near the top of the exposed Lower Ordovician sequence (Appendix 1E).

Sample G-104-1 was taken from the left side of the Bolshaya Kayala River, about 250 m east of the village ruins; it yielded only the acrotretid *Eurytreta* cf. *sabrinae,* but Korynevskij also recorded the trilobites *Ceratopyge* sp., *Shumardia* sp., and *Acerocarina* sp.

About 60–70 m to the west, a sample (G-104) from the same unit yielded the articulate brachiopods *Protambonites akbulakensis* and the conodont *Protopanderodus* sp.

Sample G-113 was taken from the right side of the Malaya Kayala River, northwest of the village ruins, about 250 m west of the previous sample (G-104). The recorded lingulate assemblage includes *Leptembolon lingulaeformis, Eurytreta chabakovi, Orbithele ceratopygarum,* and *Siphonobolus uralensis,* and the trilobites *Apatokephalus* sp., *Geragnostus* sp. *Leiagnostus* sp., *Orometopus* sp., and *Pliomerops* sp.

## Medes River

At this locality, Ordovician rocks are exposed along the left side of the river, within a triangle formed by the valleys of of the Kolymbay River to the west, Medes River to the east, and the Kosagash Rivulet to the northeast (Fig. 9). Ancygin *et al.* (1977) and Ivanov & Puchkov (1984) have described the geology and stratigraphy of the Lower Palaeozoic strata within this area According to Ancygin *et al.* (1977, p. 184, Fig. 19), there is a continuous succession from the Upper Cambrian to the Lower Ordovician. According to Ivanov & Puchkov (1984), however, the presence of Silurian and Devonian cherts and serpentinites within an area of Ordovician exposures, suggest that the geology is more complicated and in need of further study.

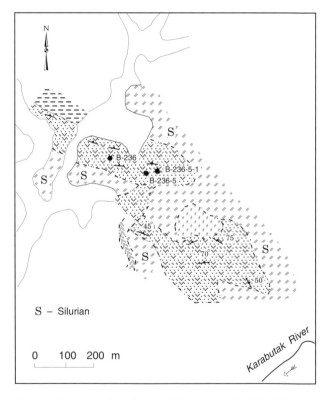

*Fig. 19.* Schematic geological map of the area on the right side of the Karabutak River, showing the distribution of the Lower Palaeozoic and position of the main sampled localities. After V.G. Korynevskij, personal communication, 1991.

Sample G-212 was taken from the left side of the Medes River about 1120 m upstream from the mouth of the Kolymbay Rivulet. The Lower Ordovician strata at this locality are exposed in a tectonic window surrounded by Silurian cherts. Argillites with nodules of argillaceous limestones yielded *Eurytreta chabakovi* and juvenile specimens of *Thysanotos siluricus*. A complete specimen of *T. siluricus* was found in this area by Korynevskij (sample B-237-1), but the exact position of this locality is uncertain.

## Karabutak River

The Lower Ordovician rocks at this locality are exposed mainly in a small area on the right side the river (Fig 19; Appendix 1E). According to V.G. Korynevskij (personal communication, 1992), these strata form two separate blocks, with faulted contacts, surrounded by Silurian cherts and volcanics. The sequence in the southeastern block is strongly disturbed tectonically. The northwestern block consists of arkosic sandstones, about 50 m thick, with overlying argillites and nodules of argillaceous limestone, about 70 m thick. The lower unit contains a lower Arenig (Hunneberg) brachiopod assemblage comprising *Leptembolon lingulaeformis* (sample B-236), *Altorthis kinderlensis*, *Thysanotos siluricus*, and *Alimbella armata* (samples B-236-5; B-236-5-1; B-236).

A small collection of early Arenig brachiopods was obtained by K.S. Ivanov (Ekaterinburg) from the right side of the Karabutak River, about 1.1 km to the south of the mouth of the the Shyrshaly–Sai Rivulet (his sample 1163), from a lens-like body of argillaceous limestone, about 30 cm thick, situated within volcanic rocks. The lingulates include *Mamatia prisca*, *Otariella intermedia* sp. nov., Siphonotretidae gen. et sp. indet., and a few fragmentary valves of *Thysanotos siluricus*. The associated conodont assemblage contains *Paroistodus proteus*, *Drepanodus sublongibasis*, and *Oistodus inaequalis*, all indicating the presence of the *Paroistodus proteus* Biozone.

## Kosistek River

The Lower Ordovician beds here are exposed in a small block on the right side of the river, to the north of the Kargaily storage-lake (Figs. 9, 20; Appendix 1E).

At Kosistek, lingulate brachiopods were found only in an isolated exposure with mudstone and nodules of an argillaceous limestone, about 100 m to the northwest of the river (Fig. 20; sample G-30-18). The assemblage includes *Acrotreta korynevskii*, *Thysanotos siluricus*, *Orbithele ceratopygarum*, *Siphonobolus uralensis*, and *Otariella intermedia* gen. sp. nov.

According to Korynevskij (1989, p. 58), a small trench through a lens-like bed of glauconitic limestones (Fig. 20; samples G-30-1; G-30-6), about 150 m to the west of sample G-30-18, yielded trilobites and conodonts indicating the *Drepanoistodus deltifer* Biozone.

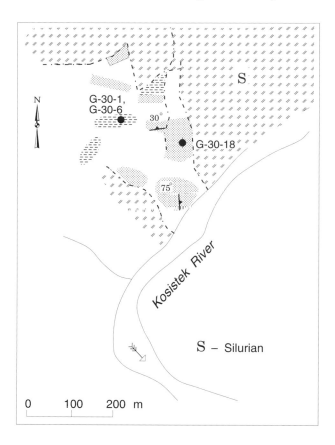

*Fig. 20.* Schematic geological map of the area on the left side of the Kosistek River, showing the distribution of the Lower Palaeozoic and position of the main sampled localities. After V.G. Korynevskij, personal communication, 1991.

## Koagash River

At this locality, the Lower Ordovician is exposed mainly on the right side of the river, southwest of the ruins of Kyzylzhol village, and to the east and southeast of Karagansai Mountain (Fig. 21; Appendix 1F). This is the type locality of the Koagash Formation, established by Leonenok (1955).

The general geology of this area has been discussed in several previous publications (Vagranov *et al.* 1973; Ivanov & Puchkov 1984), but a detailed account of the Lower Palaeozoic succession is still lacking. In view of the complex nature of the Ordovician sequence at Koagash, we can only give brief information concerning the main sampled units, and the inferred stratigraphy of the exposed rocks. Our data are based mainly on V.G. Korynevskij and K.S. Ivanov (personal communications, 1991) as well as our own results from the study of the collections from the area.

Lower Ordovician clastic rocks crop out in the northern part of the area; they were studied by Korynevskij in a series of trenches between the site of core C-360 and the river (Fig. 21). To the west of the excavated section, Silurian and Devonian black cherts of the Sakmara Formation are thrust on Lower Ordovician rocks. According to Korynevskij, the following section is present in the trenches.

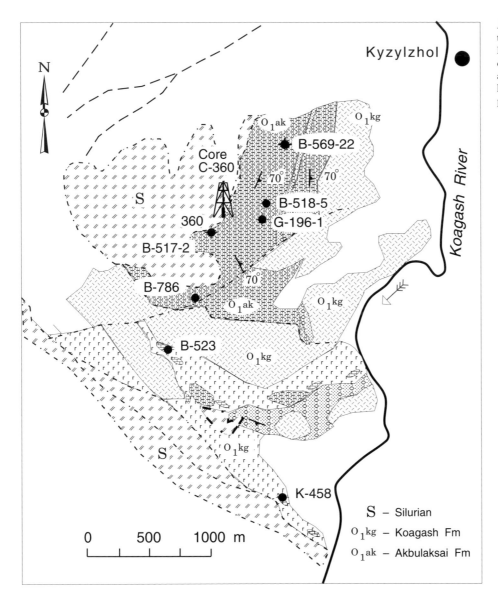

*Fig. 21.* Schematic geological map of the area on the right side of the Koagash River, showing the distribution of the Lower Palaeozoic and position of the main sampled localities. After V.G. Korynevskij, personal communication, 1991.

## AKBULAKSAI FORMATION:

Unit 1. 80 m of alternating argillites and fine-grained sandstones with nodules of argillaceous limestone. The nodules contain the brachiopods *Leptembolon lingulaeformis*, and *Altorthis kinderlensis* (samples B-517-2; B-517-5). About 100 m to the south of the site of core C-360, a lens of limestone, about 4 m thick, was excavated in a trench (sample 360). The lens contained the trilobites *Homagnostus kasachstanicus*, *Apatokephalus karabutakensis*, *Niobe emarginula*, *Glaphurella insolita*, *Pseudosphaerexochus taiketkensis*, and *Lichakephalina schilicta*, along with conodonts of the *Paroistodus proteus* Biozone. Brachiopods collected by Ivanov from the same sample (360) includes *Lamanskya splendens* (see Holmer 1993) and *Mamatia retracta*.

In the upper 35 m of the unit, the trilobites *Ceratopyge* sp., *Parapilekia* sp., *Acerocarina* sp., *Orometopus* sp., and *Euloma* sp. were found together with the articulate brachiopod *Altorthis kinderlensis*.

Unit 2. 85 m of argillites, with rare, thin interlayers of fine-grained sandstones. Only rare specimens of *Apatokephalus* sp. and *Eurytreta chabakovi* were found.

Unit 3. 85 m of fine-grained, glauconitic sandstones.

Unit 4. 100 m of coarse to fine-grained, glauconitic sandstones with interbeds of siltstones and argillites. The lower 25 m are composed of coarse-grained, cross-bedded, arkosic sandstones; fine-grained, parallel-bedded sandstones predominate in the upper part.

Unit 5. 95 m of glauconitic siltstones, containing numerous nodules of argillaceous limestone. About 35 m above the base, a brachiopod assemblage with *Orbithele ceratopygarum*, *Siphonobolus uralensis*, *Protambonites* cf. *lermontovae*, *Althorthis kinderlensis*, and *Medessia* cf. *uralica* was found (sample G-196-1). About 10 m above this sample, the brachiopods *A. kinderlensis*, *M. uralica* and *Leptembolon lingulaeformis* were recorded (sample B-518-5), and a similar assemblage is known also from about 30 m below the top of the unit. This unit can be traced about 600 m to the south and southwest from this section, where it is exposed on the southern flank of an anticline; the brachiopod assemblage from this sample (B-786) contains *A. kinderlensis*, *Alimbella armata*, *L. lingulaeformis*, *Thysanotos siluricus*, and *Acrotreta korynevskii*.

The total thickness of the Akbulaksai Formation exposed here is more than 445 m; according to Korynevskij, it is covered by the volcanic rocks of the Koagash Formation, the lower boundary of which is marked by conglomerates and coarse-grained sandstones.

The volcanics are also exposed in a separate, faulted block in the southern part of the area. To the north, this block is in contact with the Akbulaksai Formation. The Koagash Formation consists mainly of basalts with lenses of limestone, and interbeds of tuffs, volcanoclastic conglomerates, and sandstones. According to Korynevskij, the total thickness of the Koagash Formation is here more than 2,000 m. The proposed Lower Arenig age of the sequence exposed in the block is based on fossils from the limestones lenses within the volcanics.

The faunas from the faulted block originate mainly from two samples (K-458; B-523; Fig. 21). The first was taken from a lens of limestone on the right side of the river (Fig. 21). The assemblage includes *Mamatia retracta*, *Otariella intermedia* gen. et sp. nov., *Siphonotrella* sp., and *Ferrobolus fragilis* sp. nov. The associated conodonts (K.S. Ivanov, personal communication, 1986) includes *Acodus erectus*, *Paroistodus* cf. *proteus*, and *Oistodus acuminatus* indicating the *Drepanoistodus proteus* Biozone.

The second sample was taken from a locality about 1.5 km to the northwest of the first locality (Fig. 21) and comes from a lens of a coquina limestone, containing numerous silicified valves of the articulate brachiopods *Clarkella supina* and *Tetralobula latens*. The brachiopod assemblage includes *Eoconulus primus* sp. nov., *Acrotreta korynevskii*, and *Orbithele ceratopygarum*. The associated conodonts (identified by S.V. Dubinina; V.G. Korynevskij, personal communication 1991) suggests a Billingen (*Prioniodus elegans* or *Oepikodus evae* biozones) age.

## Northeastern Central Kazakhstan

Upper Cambrian trilobites and lingulate brachiopods were discovered in the northeast of Central Kazakhstan by Borukaev in 1931 and were described later by Lermontova (1954). Recent knowledge of the Upper Cambrian to Lower Ordovician geology and stratigraphy in this area is based mainly on work by Ivshin (1956, 1962, 1972) and Nikitin (1956, 1972). These authors introduced the main lithostratigraphic units, such as the Kujandy, Satpak and Olenty formations, as used here. Upper Cambrian and Lower Ordovician bio-

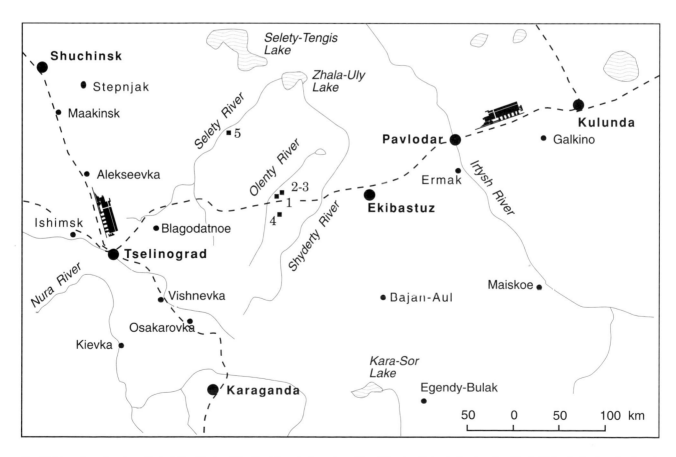

*Fig. 22.* Map of northeastern Central Kazakhstan (Fig. 1) with main investigated localities. 1 – Kujandy Section, 2 – Aksak–Kujandy Section, 3 – Satpak Syncline, 4 – Sasyksor Lake, 5 – Selety River.

stratigraphy in this area has long been based mainly on trilobites and articulate brachiopods; graptolites, lingulate brachiopods and conodonts have been studied only during the last decade and are still known inadequately.

The Late Cambrian and Lower Ordovician lingulate faunas used for this study were sampled during several field seasons betwen 1978 and 1981. The following five main outcrop areas were studied (Fig. 22): (1) Kujandy Section, (2) Aksak–Kujandy Section, (3) Satpak Syncline, (4) Sasyksor Lake, and (5) Selety River. The following description includes new data on the geology and Lower Palaeozoic stratigraphy of these localities, mainly based on field observations and information kindly supplied by M.K. Apollonov, I.F. Nikitin, and D. Tsay (Alma-Ata).

## Kujandy Section

The Kujandy and Satpak formations are exposed on the right side of the Olenty River, in a series of trenches on the northwestern slope of Kujandy Mountain, about 1.5 km from the summit (Figs. 23–25; Appendix 1G).

The lower contact of the Kujandy Formation is faulted, and the Satpak Formation is overlain unconformably by

tuffs and volcanoclastic sandstones belonging to the Olenty Formation. The section was selected by Ivshin (1972) as the stratotype for the Kujandy Formation and the Shiderty Stage.

Field work by M.K. Apollonov and others shows that part of the section that Ivshin (1972) referred to the Upper Cambrian Lermontov Horizon (unit 3 in Ivshin 1972) contains a Tremadoc assemblage of graptolites, together with the trilobite *Shumardia* (sample 8118; Fig. 23). It was also discovered that the rich assemblages of Upper Cambrian trilobites from the Lermontov and Bala–Shiderty horizons listed by Ivshin (1972; units 3 and 4, respectively) were obtained from large olistoliths of Upper Cambrian limestone within the clastic Satpak Formation. Moreover, at least two faults define a small block of Upper Cambrian rocks within the Satpak Formation. The uppermost Upper Cambrian to Lower Ordovician (Tremadoc) age of the Satpak Formation is confirmed by graptolite assemblages found by Tsay (1983) at several levels within this section.

A diverse trilobite assemblage was described from the lower part of the section by Ivshin (1956, 1962); the trilobites listed by Ivshin (1972) from the upper part of the section have never been published formally and are mainly *nomina*

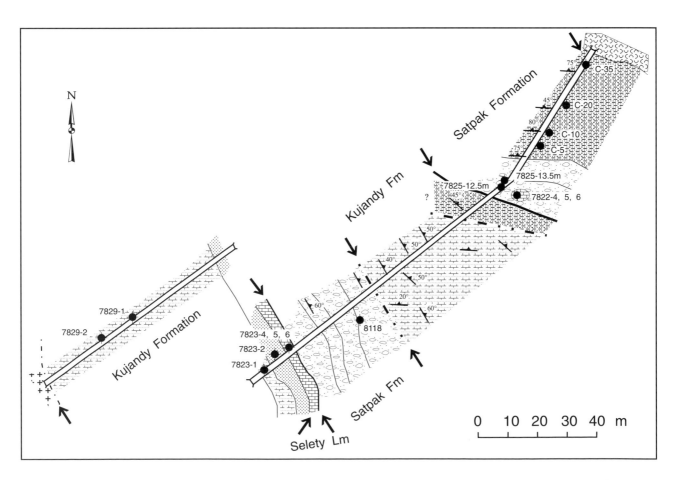

*Fig. 23.* Schematic geological map of the area around the excavated trenches in the Kujandy Section, showing the distribution of the Lower Palaeozoic and position of the main sampled localities.

*Fig. 24.* The trench at the Kujandy Section. O.P. Kovalevskij (left side of the photograph) is standing on a large olistolith of Upper Cambrian limestone (sample 7822) within the Upper Cambrian to Lower Ordovician Satpak Formation (debris-flow deposit). The bed of limestone at the top of the Upper Cambrian Kujandy Formation (samples 7823-4, 7823-5, 7823-6) is near the middle of the trench.

*nuda.* The Upper Cambrian brachiopods from this section were studied by Nikitin (1956); several acrotretid brachiopods were described by Koneva *et al.* (1990) and two bradoriid species by Melnikova (1990).

The sequence exposed in the excavated trenches is described below (Figs. 23, 25):

### KUJANDY FORMATION:

Unit 1. More than 50 m of fine to medium-grained sandstones, with lenses of a sandy, coquina limestone, about 2–10 cm thick, composed of disarticulated valves of *Billingsella*. This unit is exposed in several trenches, 0–6 m from the southwestern end of the western trench (Fig 23); the lower contact is faulted. Recorded fossils comprise mainly the articulate brachiopod *Billingsella fluctuosa* (samples 7829-1; 7829-2; Fig. 21).

Unit 2. Eight m of coarse to medium-grained sandstones with lenses of bioclastic, coquina-like, sandy limestone, about 15–20 cm thick. The bioclastic material represents mainly pelmatozoan columnals and disarticulated brachiopod valves. Brachiopods include *Billingsella alena* and *Huenella biplicata* (sample 7823-2)

Unit 3. A 2.5 m thick unit of limestone; in the lower half metre it is composed of bioclastic limestone, alternating with layers of aphanitic limestone. The upper 2 m consists of layers of bioclastic, coquina-like, limestones, made up of trilobite fragments and pelmatozoan columnals (sample 7823-4), light grey, trilobite coquina (sample 7823-5), as well as aphanitic limestones (sample 7823-6). Rich faunas include the lingulates *Experilingula* sp., *Fossuliella konevae*, *Dysoristus orientalis*, *Zhanatella rotunda* Koneva, *Treptotreta bella*, *Quadrisonia simplex*, and *Eoscaphelasma satpakensis*, the conodonts *Westergaardodina bicuspidata* and *Furnishina furnishi*, and the problematic *Utahphospha kazakhstanica.*

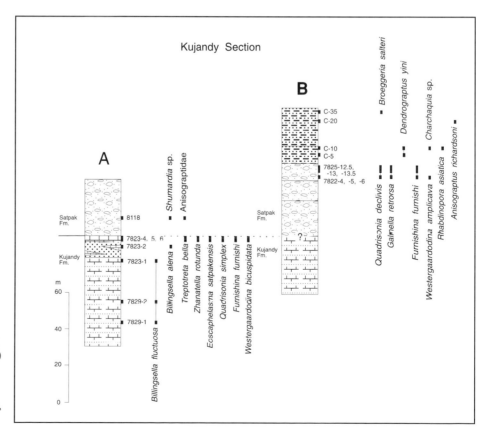

*Fig. 25.* Two generalized stratigraphical columns (locations indicated in Fig. 23) through the Kujandy, Satpak and Olenty formations at the Kujandy Section, showing distribution of selected species of conodonts, trilobites, brachiopods and graptolites.

In the second, eastern trench, the Kujandy Formation forms a small isolated olistolith (possibly displaced into the basin), exposed within the interval of 45–82 m from the southwestern end of the trench (Fig. 23). It consists of medium-grained, calcareous sandstones containing the articulate brachiopod *Billingsella lingulaeformis.*

### Satpak Formation

Unit 4. More than 30 m of debris-flow deposits with alternating layers of sandstones and siltstones and with olistoliths of argillaceous limestone. The trilobite *Shumardia* sp. and graptolites of the Family Anisograptidae were found within the unit (sample 8118).

   The Satpak Formation is also exposed within the interval from 82 m and upwards to the northeastern end of the trench (Fig. 23). The following sequence was found:

Unit 5–6. 30 m of sandstones and siltstones.

Unit 7. 10 m of debris-flow deposits, consisting of medium- to fine-grained sandstones with olistoliths and large allochtonous blocks of limestone coquina that contains a diverse assamblage of Late Cambrian trilobites, lingulates and conodonts (samples 7825-12.5; 7825-13.0; 7825-13.5; 7822-4; 7822-5; 7822-6) including *Hedinaspis regalis, Charchakia horoni, Parabolinites* sp, *Broeggeria* sp., *Galinella retrorsa, Quadrisonia declivis, Ferrobolus concavus, Coelocerodontus bicostatus, Furnishina furnishi, F. primitiva, F.* cf. *ovata, F.* sp., *Prooneotodus gallatini, P. rotundatus, Phakelodus tenulis, Westergaardodina bicuspidata, W. amplicava, W. moessebergensis,* and *W. calex.*

Unit 8. 10 m of thin bedded sandstones and siltstones, with rare nodules of calcareous siltstones. The graptolite *Dendrograptus yini* was found about 5 m above the base (sample C-5). The assemblage from the top of the unit (sample C-10) includes the graptolite *Rhabdinopora asiatica* and the trilobites *Bienvillia tetragonalis, Charchaquia* sp. and *Geragnostus* sp.

Unit 9. More than 20 m of siltstones and argillites with the graptolite *Anisograptus richardsoni* (sample C-20). *Broeggeria salteri* has been found in the upper part (sample C-35).

### Aksak–Kujandy Section

This section is on the right side of the Olenty River, about 3.5 km southwest of the Aksak–Kujandy Mountain (Figs. 26–27; Appendix 1H). The Kujandy Formation and the Satpak and Olenty formations form an anticline, which is overturned to the northeast. The description of the sequence is based mainly on exposures in a trench about 340 m long (Fig. 26).

Only a few previous studies have been made of the faunas from this section. Assemblages described in previous publications include articulate brachiopods of the Olenty Formation (Nikitin 1956), graptolites from the Satpak Formation (Tsay 1983), and a few bradoriids (Melnikova 1990) from the Upper Cambrian part of the section. The oldest rocks crop out in the trench about 132 m from its northeastern margin. The following sequence (Fig. 27) is exposed, 132 m to 173 m from the northeastern end of the trench (Fig. 26).

### Kujandy Formation:

Unit 1. More than 5 m of fine-grained, calcareous sandstones and siltstones, and nodular limestone. Some of the limestones beds are relatively fossiliferous; sample 7827 (taken about 132 m from the northeastern end of the trench) contains the trilobites *Lotagnostus* sp., *Boschekulia glabra, Parabolinites* sp., *Harpidoides* sp., *Pseudagnostus* sp., and *Parakoldinia* sp. [identified by Ivshin in Tsay 1983]; the graptolites *Dendrograptus* sp., and *Callograptus staufferi;* the conodonts *Furnishina furnishi, Westergaardodina amplicava, W. bicuspidata, Prooneotodus gallatini, P. rotundatus, Phakelodus tenuis, Prosagittodontus* sp., and *Problematoconites perforatus;* the lingulates *Broeggeria* sp., *Fossuliella konevae*

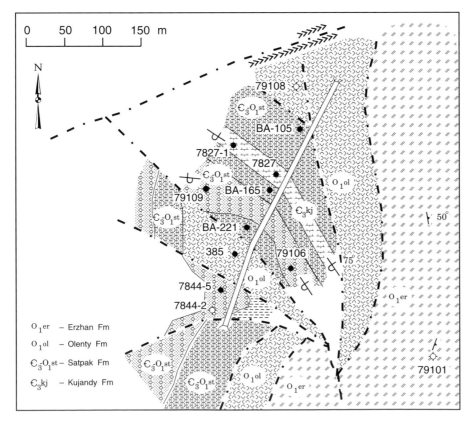

*Fig. 26.* Schematic geological map of the Aksak–Kujandy Section around the excavated trench, showing the distribution of the Lower Palaeozoic and position of the main sampled localities.

*Fig. 27.* Two generalized stratigraphical columns (locations indicated in Fig. 23) through the Kujandy, Satpak, and Olenty formations in the Aksak–Kujandy Section, showing the distribution of selected species of conodonts, trilobites, brachiopods and graptolites.

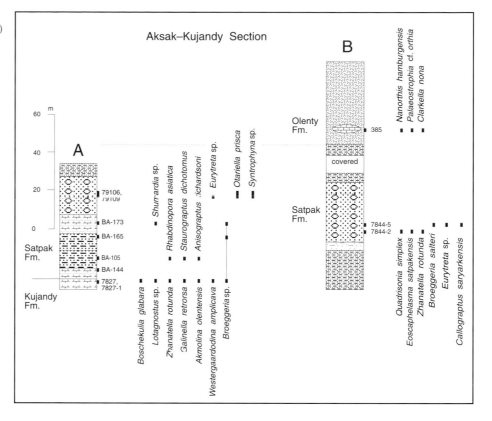

sp. nov., *Galinella retrorsa*, *Akmolina olentensis* sp. nov., *Quadrisonia declivis*, and *Zhanatella rotunda*; the bradoriid *Monasterium ivshini*; and hyolithelmintes and sponge spicules. The interval can be correlated with the *Westergardodina amplicava* Biozone in the Batyrbay section at Malyj Karatau (southern Kazakhstan). A similar assemblage was also recovered from this unit, about 70 m northeast of the trench (sample 7827-1).

SATPAK FORMATION:

Unit 2. 6 m of fine-grained, calcareous sandstones containing the graptolite *Callograptus olentensis* (sample BA-144).

Unit 3. 9 m of siltstones, yielding the graptolites *Rhabdinopora asiaticum*, *Staurograptus dichotomus*, and *Anisograptus richardsoni* in the lower part of the unit (sample BA-150). The same asemblage was also found at the northeastern flank of the anticline, about 105 m from the northeastern end of the trench (sample BA-105).

Unit 4. 9 m of thin bedded siltstones, with rare interbeds of fine-grained sandstone containing the graptolite *Callograptus* sp. and rare lingulates such as *Broeggeria salteri* and *Eurytreta* sp. (sample BA-165).

Unit 5. 10 m of calcareous siltstones yielding the lingulates *Broeggeria salteri* and *Lingulella* sp. as well as unidentifiable fragments of dendroid graptolites (sample BA-173).

Unit 6. 20 m of debris-flow deposits, with fine- to medium-grained, calcareous sandstones, including thin layers of sandstones and siltstones about 12–14 m above the base of the unit.

Olistoliths of Upper Cambrian sandy limestones and coquina-like limestones are included within the unit to the southeast and to the northwest of the trench. A diverse assemblage of articulate brachiopods was found in the olistoliths, about 75–80 m to the southwest of the trench (sample 79106); including *Syntrophina* sp.

A few specimens of the lingulates *Otariella prisca* gen. et sp. nov. and *Broeggeria* sp. were also found to the northwest of the trench (sample 79109).

Unit 7. 7 m of siltstone with interbeds of sandstones in the lower 3 m.

About 207 m from the northeastern end of the trench, the axis of the syncline is exposed. On the southwestern flank of this fold, about 207–225 m from the north eastern end of the trench, sandstones and siltstones are exposed that appear to represent equivalents of units 6 and 7 of the sequence described above. In a sample (BA-221) 221 m from the northeastern end of the trench, Tsay (1983, p. 95) recorded the graptolite *Staurograptus dichotomus*.

The Olenty Formation is exposed in the trench 225–281 m from its northeastern end; it consists of tuffs with lenses of limestone. The contact with the Satpak Formations is faulted. The limestone lenses (sample 385) contain articulate brachiopods described by Nikitin (1956).

Olistoliths of Upper Cambrian limestones are also present in the Olenty Formation, containing (sample 79108) the lingulates *Schizambon* sp. and *Treptotreta bella*, as well as the conodont *Westergaardodina* sp.

The Satpak Formation is also exposed in the southwestern part of the trench, about 281–340 m from its northeastern end, where a sample (7844-5) contained the graptolite *Callograptus saryarkensis* and the lingulates *Broeggeria salteri*, *Lingulella* sp., and *Eurytreta* sp.

A sample (7844-2) from an olistolith of bioclastic limestones within the Satpak Limestone yielded various Late Cambrian trilobites, conodonts, and lingulates including *Fossuliella konevae*, *Zhanatella rotunda*, *Quadrisonia simplex*, and *Eoscaphelasma satpakensis*.

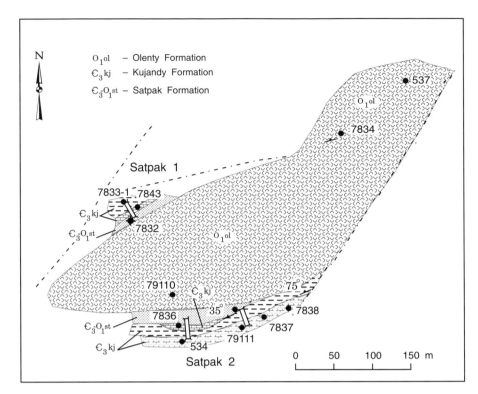

*Fig. 28.* Schematic geological map of the area around the excavated trenches in the Satpak Syncline, showing the distribution of the Lower Palaeozoic and position of the main sampled localities.

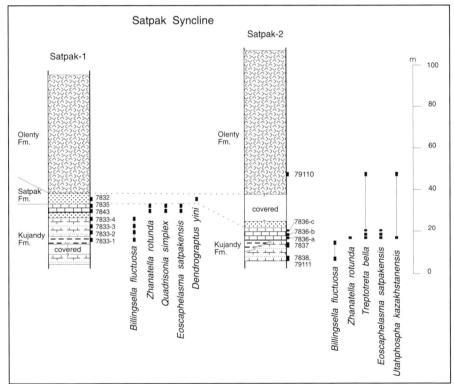

*Fig. 29.* Two generalized stratigraphical columns (locations indicated in Fig. 28) through the Kujandy, Satpak, and Olenty formations in the Satpak Syncline, showing distribution of selected species of trilobites, brachiopods and graptolites.

## Satpak Syncline

This section is situated on the right side of the Olenty River, about 3.5–4.0 km northwest of the Aksak–Kujandy Mountains (Fig. 28; Appendix 1I). The Kujandy Formation is exposed in a series of excavated trenches along the northwest

(Satpak-1 section) and southeast flanks (Satpak-2 section) of the syncline (Fig. 29[1]). The base of the Kujandy Formation is invariably faulted.

Only few previous studies have been made of the Late Cambrian to Early Ordovician faunas within the area; Nikitin (1956) described the articulate brachiopods, and the Late Cambrian trilobites were studied by Ivshin (1956, 1962). Some species of Late Cambrian lingulate brachiopods from the Selety Horizon were studied by Koneva *et al.* 1990. The graptolite *Dendrograptus yini* was discovered by Tsay (1983) within the Satpak Formation at the top of the sequence.

The tuffs and volcanoclastic sandstones of Olenty Formation are separated from the underlying rocks by a slight unconformity. A detailed description of the Upper Cambrian to the Lower Ordovician sequence within the area was given by Nikitin (1956, p. 119; 1972, p. 34).

## Sasyksor Lake

This locality is about 3 km to the east of Sasyksor Lake and 1 km to the west of the Karaultobe Mountains (Fig. 22; Appendix 1H). The Upper Cambrian and Lower Ordovician geology of this area was described briefly by Nikitin (1956; 1972, p. 163, Fig. 47).

The Olenty Formation is exposed in a series of trenches. At the eastern end of one of them a sample (601) from a lens of a coquina-like limestone, within a sequence of tuffs and volcanoclastic sandstones, yielded articulate brachiopods (see Nikitin 1956) and the lingulate brachiopods *Ferrobolus*

*fragilis* sp. nov., *Mamatia retracta*, *Sasyksoria rugosa* gen. et sp. nov., *Conotreta shidertensis* sp. nov., and *Eoconulus primus* sp. nov.

## Selety River

The Selety Limestone is exposed on the right side of the Selety River about 15 km south of Bestobe and about 4–5 km upstream from Izvestkovyj village (Figs. 22, 30; Appendix 1H). A detailed description of the sequence was given by Nikitin (1956, p. 117, Fig. 19).

The lingulate brachiopods *Mirilingula* sp., *Broeggeria salt eri*, and *Quadrisonia declivis* were collected from layers of nodular, argillaceous limestone, about 5–10 cm thick, in the upper part of the section (Unit 7 *in* Nikitin 1956; sample 325), in association with the conodont *Westergaardodina amplicava*.

## Southern Kendyktas Range

Tremadoc to early Arenig lingulate brachiopods are known from both sides of the Agalatas River and from the quarry on Keskentas Ridge in the southern Kendyktas Range (Figs. 31–35; Appendix 1J). The most comprehensive information on the geology and stratigraphy of this area was published by

*Fig. 30.* Exposure of the Upper Cambrian Selety Limestone (sample 325) on the right side of the Selety River to the south of Bestobe.

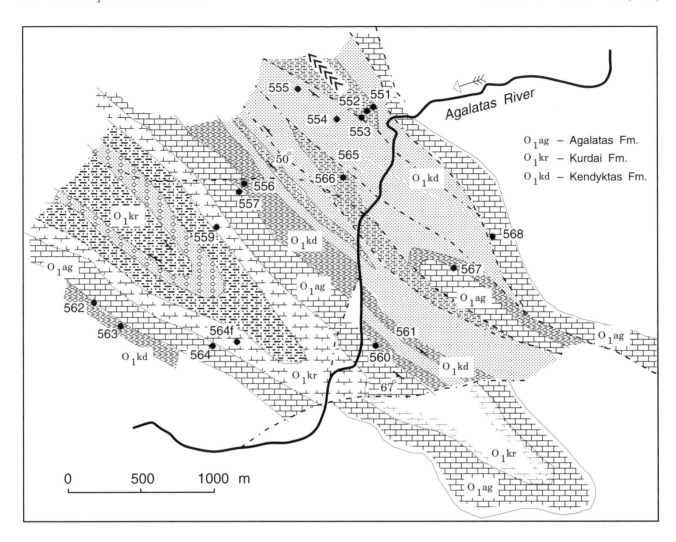

*Fig. 31.* Schematic geological map of the area around Keskentas Ridge in the southern Kendyktas Range, showing the distribution of the Lower Palaeozoic and position of the main sampled localities.

Keller & Rukavishnikova (1961). The trilobites were described by Lissagor (1961), articulate brachiopods by Rukavishnikova (1961), and graptolites by Zima (1976).

New data on the geology and stratigraphy presented here were kindly supplied by M.K. Apollonov, I.F. Nikitin, and D. Tsay (Alma-Ata).

The main lithostratigraphic units were established by Keller & Rukavishnikova (1961). They are: (1) the siliciclastic Tremadoc Kendyktas Formation, up to 350 m thick; (2) the mainly carbonate upper Tremadoc to lower Arenig Agalatas Formation, about 400 m thick; and (3) the mainly clastic lower Arenig Kurdai Formation, more than 470 m thick (Fig. 2).

The lower boundary of the Kendyktas Formation is invariably faulted; the lower part of the formation consists of unfossiliferous sandstones more than 60 m thick. Above this follow fine-grained sandstones, siltstones and argillites, yielding numerous brachiopods, trilobites, and graptolites (Fig. 31).

According to Lissagor (1961), the trilobite assemblage from the Kendyktas Formation includes *Harpides rugosus*, *Hysterolenus oblongus*, and *Bicornipyge bicornis*. Apollonov (1991) noted that the two latter species are present only in the upper part of the formation, while the lowermost assemblage includes *Hysterolenus*, an unnamed species of *Shumardia*, *Harpides*, *Euloma* (*Proteuloma*), as well as *Araiopleura stephani* and *Niobe hompharayi*. According to Zima (1977, p. 41), the graptolites *Rhabdinopora socialis*, *Anisograptus richardsoni*, *Staurograptus dichotomus*, and *Aletograptus* sp. have been recorded. The only brachiopods found in the Kendyktas Formation belong to the *Broeggeria* assemblage, including *Broeggeria salteri*, *Lingulella antiquissima*, *Elliptoglossa linguae*, and *Eurytreta sabrinae* (Figs. 31, 35).

The boundary between the Tremadoc Kendyktas and the Tremadoc–lower Arenig Agalatas formations is well exposed on the northeastern wall of the limestone quarry on the Keskentas Ridge (Fig. 34) and along the road into the quarry on the right side of the Agalatas River.

*Fig. 32.* View of the Agalatas River, with scattered exposures of the Kendyktas Formation.

*Fig. 33.* Exposure of the Kendyktas Formation on the right side of the Agalatas River (sample 552).

*Fig. 34.* Transition between the Kendyktas and Agalatas formations in the quarry on the northeastern slope of the Keskentas Ridge (samples 556, 557).

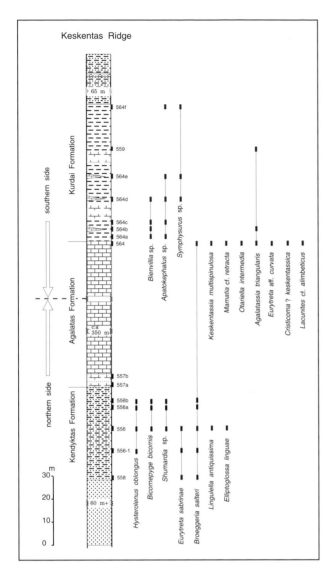

*Fig. 35.* Generalized stratigraphical column through the Kendyktas, Agalatas and Kurdai formations in the Satpak Syncline, showing the distribution of selected species of trilobites and brachiopods.

On the 'northern side' (Fig. 35) of the Kenskentas Ridge, the Kendyktas Formation consists of bedded siltstones with intercalations of fine-grained sandstones and argillites. The first lingulate assemblage occurs near the base of the unit (about 40–45 m below the top), where *Broeggeria salteri* has been found (sample 558). About 25 m below the top (sample 556-1), *Hysterolenus oblongus* and *Broeggeria salteri* were found, and about 7 m above this level (sample 556) a more diverse assemblage with *Broeggeria salteri*, *Lingulella antiquissima*, and *Elliptoglossa linguae*, and the trilobites *Bicornipyge bicornis*, *Hysterolenus oblongus*, and *Shumardia* sp. was obtained. Similar assemblages were also found, about 6–7 m (sample 556a) and 5 m (sample 556b) below the top of the unit.

The lower part of the overlying Agalatas Formation consists of sandstones and siltstones with some intercalations of limestones (Fig. 35); it is poor in fossils and has yielded only some articulate brachiopods, including *Nanorthis* and *Syntrophinella* (sample 557b).

Another section through the transitional beds between the Kendyktas and Agalatas formations is exposed on the left side of the Agalatas River, along the northeastern slope of the ridge (Fig. 31), where the Agalatas Formation yielded only *Lingulella antiquissima* (sample 560).

On the 'southern side' (Fig. 35) of the Kenskentas Ridge, the transition between the Agalatas and Kurdai formations is well exposed along the deep valley that cuts through the flank of the Keskentas Ridge, about 550 m northwest of the Agalatas River. Here, the Agalatas Formation consists of dark, bedded, pyritic limestone. A rich assemblage of lingulate brachiopods was found in the upper 50 cm of the unit (sample 564), including *Aglatassia triangularis* gen. et sp. nov., *Broeggeria salteri*, *Keskentassia multispinulosa* gen. et sp. nov., *Schizambon* sp., *Eurytreta* aff. *curvata*, *Eurytreta*? sp., *Mamatia* cf. *retracta*, *Cristicoma*? *keskentassica* sp. nov., *Otariella intermedia* gen. et sp. nov., and *Lacunites* cf. *alimbeticus*. The associated fauna includes trilobites, bradoriids, gastropods, and the conodonts *Drepanoistodus* aff. *basiovalis*, *Scandodus* sp., and *Drepanoistodus parallelus*, indicating the *Paroistodus proteus* Biozone.

According to Keller (1961) and Rukavishnikova (1961), a brachiopod assemblage including *Nanorthis multicostata*, *Syntrophynella typica*, and *Tetralobula plana* was found at Keskentas Ridge in a unit with alternating layers of sandstones, siltstones, and oolitic limestones at about the transition between the Kendyktas and Agalatas formations. Lissagor (1961) described the trilobites from the Agalatas Formation. There is, however, no exact information on the distribution of the trilobites within the formation.

The Agalatas Formation is overlain conformably by the mainly clastic Kurdai Formation, consisting of argillites, argillaceous limestone, siltstones, and sandstones; it has yielded only *Agalatassia triangularis* gen. et sp. nov. and numerous trilobites, including *Bienvillia*, *Apatokephalus*, *Orometopus*, *Symphysurus*, and *Geragnostus*. The total thickness of the formation is more than 145 m.

# Stratigraphic and geographic distribution of Upper Cambrian – Lower Ordovician lingulate assemblages

Lingulate and articulate brachiopods are not considered usually to be important in long-distance correlation of Cambrian–Ordovician boundary beds. Probably this is due partly to the poor state of knowledge of Late Cambrian and Tremadoc brachiopods.

In this study, three types of lingulate assemblages have been distinguished in the Upper Cambrian – Lower Ordovician of Baltoscandia, the South Urals, northeastern Central Kazakhstan, and the southern Kendyktas Range: (1) *Broeggeria* assemblage; (2) Upper Cambrian – Arenig microbrachiopod assemblages (in the sense of Wright & McClean 1991); and (3) *Leptembolon–Thysanotos* assemblage (Figs. 36–37).

## *BROEGGERIA assemblage*

The most distinctive and widely distributed element of the lingulate faunas is the *Broeggeria* assemblage, of which the most distinctive taxa are *Broeggeria salteri*, *Elliptoglossa lingua*, *Lingulella antiquissima* (senior synonym of *L. nicholsoni*), and *Eurytreta bisecta* or *E. sabrinae* (Figs. 36–37).

Although these species have somewhat different stratigraphic ranges and may be found partly in other benthic assemblages, they form a fairly constant association which is distributed widely during the Tremadoc, in black-shale lithofacies. The assemblage is probably controlled strongly by environmental factors, and it seems that *Broeggeria* and its associated forms may represent resistant eurytopic organisms that were well adapted to dysaerobic enviroments.

The *Broeggeria* assemblage is recorded from the Dictyonema Shale and Ceratopyge Shale of Baltoscandia, the Tremadoc Series of Wales and Shropshire (e.g., Owens *et al.* 1982), and the McLeod Brook Formation of Cape Breton, Nova Scotia (e.g., Walcott 1912; Owens *et al.* 1982), and is described here from the Satpak Formation and Kendyktas Formation of Kazakhstan. *Broeggeria salteri* also occurs in Belgium (Schmidt & Geukens 1958) and Argentina (e.g., Harrington 1938), but these faunas are still poorly known.

The fauna associated with *Broeggeria* usually includes sponges and a trilobite assemblage with *Hysterolenus* or *Bienvillia*.

## *Microbrachiopod assemblages*

Some acrotretid lineages, including *Eurytreta*, *Eoscaphelasma–Otariella–Eoconulus*, and several ephippelasmatides all appear to have evolved fairly rapidly during the Tremadoc and early Arenig; there is a succession of assemblages with diagnostic taxa that may be of use for biostratigraphic control. However, the published, pre-Arenig record of the majority of these acrotretids is extremely limited, and thus it is difficult to make wider comparisons outside of the study areas.

*Upper Cambrian.* – The oldest microbrachiopod assemblage in our collections comes from the Kujandy Formation of northeastern Central Kazakhstan; it contains the lingulids *Fossuliella konevae* sp. nov., *Zhanatella rotunda*, and *Dysoristus orientalis* sp. nov. and the acrotretids *Quadrisonia simplex*, *Treptotreta bella*, and *Eoscaphelasma satpakensis* (Fig.

37). This assemblage is associated with the trilobites of the local *Irvingella* Biozone (Ivshin 1972).

The majority of the lingulate taxa in this assemblage are unknown outside northeastern Central Kazakhstan. Only *Zhanatella rotunda* is recorded also from the Malyj Karatau Range, southern Kazakhstan, within the interval from the *Pseudagnostus curtarae* to the *Eurudagnostus ovaliformis* local biozone.

A somewhat younger assemblage consisting of *Broeggeria* sp., *Fossuliella konevae* sp. nov., *Zhanatella rotunda*, *Ferrobolus concavus* sp. nov., *Quadrisonia declivis*, *Galinella retrorsa*, and *Akmolina olentensis* gen. et sp. nov. has been found in olistoliths of Upper Cambrian limestones in the Satpak formation and *in situ* at the Aksak–Kujandy Section (Fig. 37); the assemblage is associated with the paraconodont *Prooneotodus? rotundatus*.

*Quadrisonia declivis* is also known from the Upper Cambrian of the Malyj Karatau Range within the interval corresponding to the local *Hedinaspis sulcata* Biozone. *Quadrisonia* appears to be closely related to the lineage from which *Eurytreta* was derived, at about the level of the first occurrence of *Cordylodus* (Koneva & Popov *in* Koneva *et al.* 1990).

*Tremadoc.* – The microbrachiopod assemblage from the upper Tremadoc Bjørkåsholmen Limestone in Scandinavia is characterised by a mixture of taxa, representing relics of Cambrian lineages as well as the earliest known representatives of Ordovician lineages of lingulids and acrotretids.

The taxa include *Broeggeria salteri*, *Lamanskya splendens*, *Elliptoglossa linguae*, *Lingulella antiquissima*, *Rowellella* sp., *Dactylotreta pharus* sp. nov., *Eurytreta minor*, *Eurytreta* cf. *sabrinae*, *Pomeraniotreta biernatae*, *Biernatia circularis* sp. nov., *Otariella* sp., and *Orbithele ceratopygarum* (Fig. 36).

The species of *Rowellella* and *Biernatia* are the earliest known, and *Otariella* sp. is an early eoconulid stock with close affinity to the Late Cambrian scaphelasmatides.

In Baltoscandia, *Broeggeria salteri*, *Lamanskya splendens*, and *Orbithele ceratopygarum* are known also from the Hunneberg Stage (Fig. 36).

*Arenig.* – Early Arenig lingulate assemblages from Baltoscandia have not been closely studied. In the South Urals, the early Arenig microbrachiopods *Ferrobolus fragilis* sp. nov., *Mamatia retracta*, *Acrotreta korynevskii*, *Otariella prisca* gen. et sp. nov., and *Orbithele ceratopygarum* were found together with conodonts of the *Paroistodus proteus* Biozone (Fig. 36). In the Koagash section, the early eoconulid *Otariella prisca* is replaced by *Eoconulus primus* (Fig. 36); the associated conodont assemblage at this level contains *Oistodus lanceolatus*, suggesting the *Prioniodus elegans* or *Oepikodus evae* biozones. According to our unpublished observations, *Eoconulus* appears near the base of the Billingen Stage at about the same stratigraphic level in Baltoscandia. In the South Urals, this assemblage is associated with *Clarkella supina* Nikitin and *Tetralobula latens* Nikitin.

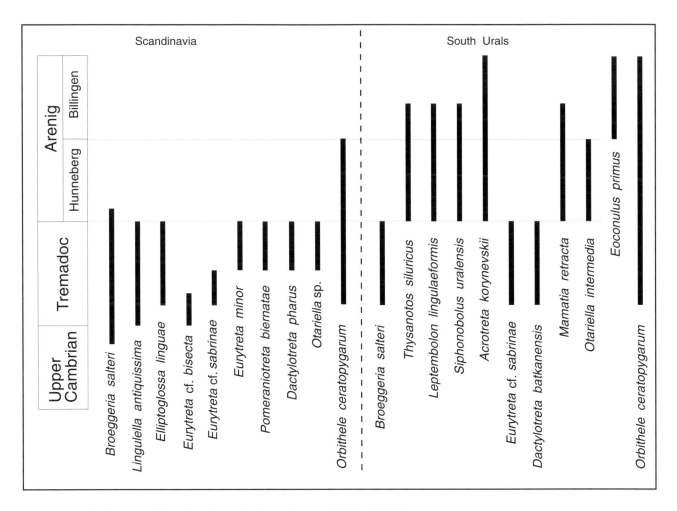

*Fig. 36.* Stratigraphical ranges of selected lingulate species in Scandinavia and the South Urals.

A similar assemblage of lingulate microbrachiopods occurs in the Olenty Formation of northeastern Central Kazakhstan, where *Ferrobolus fragilis*, *Mamatia retracta*, and *Eoconulus primus* also occur; this assemblage is also associated with *Clarkella supina* and *Tetralobula latens*. We take these data as evidence for an early Arenig age of the upper part of the Olenty Formation, which was considered previously to be of late Tremadoc age (Apollonov 1991).

The contemporaneous assemblage from the top of the Agalatas Formation in the southern Kendyktas Range includes *Aglatassia triangularis* sp. nov, *Broeggeria salteri*, *Keskentassia multispinulosa* gen. et sp. nov., *Schizambon* sp., *Eurytreta* aff. *curvata*, *Mamatia* cf. *retracta* , *Cristicoma*? *keskentassica* sp. nov., and *Otariella intermedia* gen. et sp. nov. The associated fauna includes conodonts of the *Paroistodus proteus* Biozone and a diverse assemblage of trilobites (Apollonov 1991).

Only *Otariella intermedia* and *Mamatia retracta* are known from the correlative sequences in the South Urals and northeastern Central Kazakhstan; the other lingulate taxa are unknown outside of southern Kendyktas Range.

## The LEPTEMBOLON–THYSANOTOS *assemblage*

One important problem in understanding Lower Ordovician lingulate biostratigraphy is related to the age and stratigraphic range of the *Leptembolon lingulaeformis– Thysanotos siluricus* assemblage. Apart from *Thysanotos* and *Leptembolon*, this assemblage usually includes *Acrotreta* and the siphonotretid genera *Schizambon* or *Siphonobolus*, and sometimes *Orbithele*.

The *Leptembolon–Thysanotos* assemblage has been recorded from a wide geographic area surrounding the East European platform, including northern Estonia (Gorjansky 1969), the Holy Cross Mountains in Poland (Bednarczyk 1964, 1971), Germany (Sdzuy 1955), Bohemia (Havlíček 1982), southeastern Serbia (Havlíček 1982), as well as from the South Urals. The assemblages in all these districts are not associated usually with good zonal conodonts or graptolites; the presence of *Thysanotos siluricus*, in itself, has usually been regarded as evidence for a Tremadoc age. However, this is mainly a result of the erroneous interpretation of the glauconitic sands of northern Estonia as Tremadoc in age.

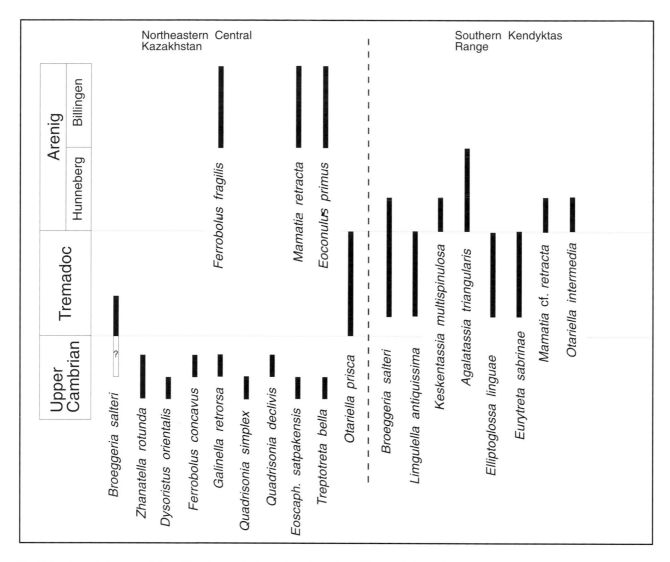

*Fig. 37.* Stratigraphical ranges of selected lingulate species in northern Central Kazakhstan and the Kendyktas Range.

In the South Urals, the *Leptembolon–Thysanotos* assemblage appears near the top of the Alimbet Formation and also in the overlying Akbulaksai Formation and in the upper part of the Kidryas Formation. These parts of the sequence were correlated previously with the Tremadoc Pakerort Stage in Baltoscandia (Vagranov *et al.* 1973), but they are here considered to belong to the Arenig (Fig. 36).

# Remarks on palaeogeography

Current reconstructions of the palaeogeographic distribution of Late Cambrian and Tremadoc benthic faunas are based mostly on trilobites. According to a recent analysis by Shergold (1988) of the distribution of trilobite biofacies, the late Cambrian and early Tremadoc benthic faunas of Baltoscandia belong to the large, cold-water, Baltic province together with the olenid-dominated faunas of Wales, eastern North America, Argentina, and Bolivia. With the exception of Baltoscandia, all these areas were closely connected to Gondwana, and during this interval the *Broeggeria* assemblage occurred widely (Fig. 38).

A significant enviromental change took place in Baltoscandia a short time before the end of Tremadoc, when clastic and black mud sedimentation was replaced by carbonates and glauconite-rich clastic deposits containing a *Ceratopyge* assemblage of trilobites and a distinctive assemblage of microbrachiopods (Fig. 38). At about the same time the *Ceratopyge* biofacies appeared also in the Urals, on the opposite side of the East European platform (Fig. 38). The late Tremadoc and Arenig benthic faunas of Baltoscandia have only a few elements in common with those of the Mediterranean areas.

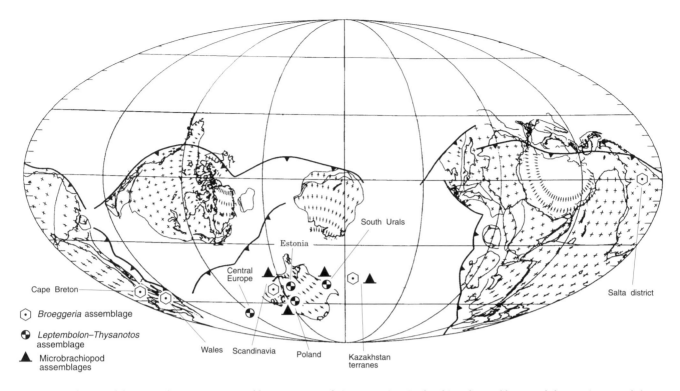

*Fig. 38.* Distribution of the Tremadoc *Broeggeria* assemblage, Upper Cambrian – Arenig microbrachiopod assemblages, and the Arenig *Leptembolon–Thysanotos* assemblage. Plate-tectonic reconstruction after Scotese & McKerrow (1991, Fig. 2).

The *Leptembolon–Thysanotos* assemblage appears to have been distributed mainly across the East European platform; during the early Ordovician, northern Estonia, the Holy Cross Mountains, and the South Urals were parts of this region (Fig. 38); *Thysanotos siluricus* and its associated taxa are unknown from the Mediterranean areas, with the exception of Bohemia and Bavaria. However, the trilobite assemblages from the Lemnitz Shale of Bavaria and the Mílina Formation of Bohemia contain numerous genera that are known also from the Lower Ordovician of Baltoscandia (Mergl 1986). This suggests that there was some faunal exchange between these three areas during the early Arenig and that only at the end of the Arenig did the Bohemian benthic assemblages became part of the Mediterranean faunas.

## Systematic palaeontology

Most of the illustrated and described material comes from Scandinavia, Kazakhstan, and the South Urals, but specimens from Siberia, Estonia, Great Britain, and North America are figured and described for comparative purposes.

We have worked with faunas etched by means of weak acid from limestone sequences as well as with material from argillaceous rocks that are usually distorted in varying degrees and sometimes preserved only as internal moulds. For obvious reasons it has been extremely difficult to make a detailed comparison between lingulid and acrotretid species preserved in these two different ways; in particular, this has been a problem in understanding the numerous species of *Lingulella* and *Eurytreta* described and discussed below.

In the following, partial synonymies are provided, listing only the key works describing and illustrating the discussed taxa, as well as those containing more complete synonymies.

The illustrated and/or discussed material is deposited in the following institutes: Swedish Museum of Natural History, Stockholm (RM Br); Palaeontological Institute, Lund (LO); Swedish Geological Survey, Uppsala (SGU); Palaeontological Museum, Uppsala (PMU); Paleontologisk Museum, Oslo (PMO); Geologisk Museum, Copenhagen (MMH); British Museum of Natural History, London (BM B), National Museum of Wales, Cardiff (NMW); Palaeontological Department, Institute of Geological Sciences, London (IGS GSM); the Department of Geological Sciences, University of Birmingham, Birmingham (BU); United

States National Museum, Washington D.C. (USNM); Institute of Palaeozoology, Warszawa (Bp); Institute of Geological Sciences, Warszawa (ING PAN); Institute of Geology, Tallinn (GT Br); Geological Museum, Institute of Geology, Alma-Ata (GA); Central Scientific Research Geologic Exploration Museum, St. Petersburg (CNIGR); Palaeontological Institute, Moscow (PIN).

Measurements (in millimetres if not stated otherwise) are as follows (Figs. 39–40): W, L, H = width, length, height of valve; Lmax = maximum length of apsaconical ventral valve; WI, LI = width, maximum length of pseudointerarea; WG = width of median groove or pedicle groove; ML = median length of pseudointerarea; WM1, LM1 = width, length of cardinal muscle field; WV, LV = width, length of visceral area; LS, BS = length, point of origin, of dorsal median septum; PHS = position of maximum height of dorsal median septum; M = position of maximum height of ventral valve; Fp, Fa = point of origin, length of pedicle foramen. OR = observed range, *S* = standard deviation, *N* = number of measurements, MAX, MIN = maximum, minimum value.

The terminology of the lingulate brachiopods used here mainly follows that of Rowell (1965), Koneva (1986), and Holmer (1989). The term *median tongue* is introduced to denote the narrow anterior extension of the dorsal visceral field of many lingulids (Fig. 40). The term *emarginature* was introduced by Koneva (1986) to denote the deep, circular indentation in the ventral umbo, formed by the pedicle groove (Fig. 67G, U).

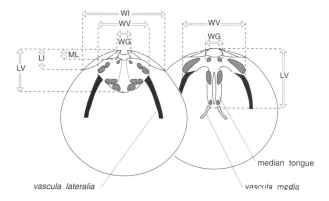

*Fig. 40.* Schematic drawings of dorsal and ventral views of *Zhanatella rotunda* Koneva, showing locations of measurements.

# Class Lingulata Gorjansky & Popov, 1985

# Order Lingulida Waagen, 1885

# Superfamily Linguloidea Menke, 1828

# Family Obolidae King, 1828

# Subfamily Obolinae King, 1828

# Genus *Experilingula* Koneva & Popov, 1983

*Type and only species.* – Original designation; *Experilingula divulgata* Koneva & Popov, 1983, p. 113; Upper Cambrian (*Pseudagnostus pseudangustilobus* – *Micragnostus mutabilis* biozones); Malyj Karatau Range, southern Kazakhstan.

*Diagnosis.* – Anteromedian parts of ventral propareas deflected strongly dorsally, to form triangular opening; pedicle groove deep, bounded laterally by ridges, and bisected medially by narrow furrow. Flexure lines well developed in both valves. Ventral visceral field not extending to midlength. Dorsal pseudointerarea anacline, with concave median groove, elevated above valve floor. Dorsal visceral field small, with narrow median tongue extending to midlength, and bisected by low median ridge; dorsal *vascula lateralia* peripheral, arcuate; dorsal *vascula media* long, moderately divergent.

*Ocurrence.* – Upper Cambrian, Kazakhstan.

## *Experilingula* cf. *divulgata* Koneva & Popov, 1983
Fig. 41

*Synonymy.* – □cf. 1983 *Experilingula divulgata* sp. nov. Koneva & Popov, p. 115, Pls. 26:1–13; 27:1–14.

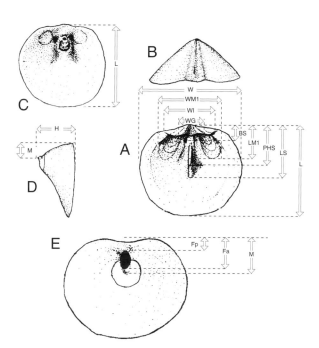

*Fig. 39.* Schematic drawings showing locations of measurements. □A–D. *Treptotreta bella* sp. nov. □E. *Eoscaphelasma satpakensis* Koneva, Popov & Ushatinskaya.

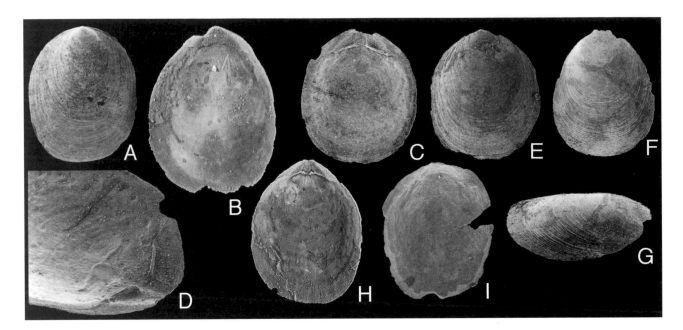

*Fig. 41. Experilingula* cf. *divulgata* Koneva & Popov; Kujandy Formation; northeastern Central Kazakhstan. □A. Dorsal exterior; Kujandy Section (sample 7823-4); RM Br 136229; ×7.5; □B. Dorsal interior; Kujandy Section (sample 7823-4); RM Br 136230; ×15. □C. Ventral interior; Kujandy Section (sample 7823-4) RM Br 136231; ×12. □D. Lateral view of C; ×24. □E. Dorsal exterior; Kujandy Section (sample 7823-4); RM Br 136232; ×13.5; □F. Ventral exterior; Kujandy Section (sample 7823-4); RM Br 136233; ×12. □G. Lateral view of E; ×13.5. □H. Ventral interior; Kujandy Section (sample 7823-4); RM Br 136234; ×15. □I. Dorsal interior; olistolith within Erzhan Formation; Aksak–Kujandy Section (sample 79101); RM Br 136235; ×18.

*Material.* – Figured. Ventral valves: RM Br 136231 (L 2.84, W 2.46, LI 0.6, WI 1.64, WG 0.16); RM Br 136233; RM Br 136234 (L 3.8, W 2.52, LI 0.56, WI 1.72, WG 0.12, LV 1.36, WV 1.56). Dorsal valves: RM Br 136229 (L 4.8, W 3.9); RM Br 136230 (L 3.04, W 2.4, LI 0.48, WI 1.47, WG 0.84, LS 1.84); RM Br 136232; RM Br 136235. Total of 4 ventral and 8 dorsal valves.

*Description.* – The shell is subequally biconvex and elongate oval. The ventral valve is evenly convex, about 1.2–1.5 times as long as wide, with a high, slightly apsacline pseudointerarea. The pedicle groove is narrow, with subparallel margins bounded by ridges, and bisected medially by a fine furrow (Fig. 41D). The anteromedian parts of the propareas are deflected strongly dorsally, forming a subtriangular pedicle opening (Fig. 41D). The flexure lines are well defined. The ventral visceral field is small, not extending to mid-length, and slightly elevated anteriorly. The visceral area is bisected by a pair of slightly divergent, median furrows (Fig. 41C, D, H).

The dorsal valve is evenly convex, about 1.2–1.5 times as long as wide, with a slightly anacline pseudointerarea. The median groove is well defined. The propareas are raised above the valve floor, bisected by flexure lines. The dorsal interior has a poorly defined visceral area, with a narrow median tongue extending slightly anterior to mid-length. A thin median ridge bisects the anterior part of the visceral area (Fig. 41B, I).

*Discussion.* – In our collections, *E.* cf. *divulgata* is represented mostly by juvenile specimens. They are closely comparable with the type species in the general shape of the shell and pseudointerareas. However, the anterior border of the dorsal visceral area and dorsal median ridge are less well defined as compared with the type material from Malyj Karatau.

The median furrows, bisecting the visceral area in this and other lingulids (see, e.g., Fig. 59A–C) probably corresponds to the scar of the pedicle nerve as present in valves of Recent lingulids (e.g., Holmer 1991).

*Occurrence.* – It occurs in the Kujandy Formation at the Kujandy Section (Appendix 1G), and in olistoliths within the Erzhan Formation at the Aksak–Kujandy Section (Appendix 1H).

## Genus *Hyperobolus* Havlíček, 1982

*Type species.* – Original designation by Havlíček (1982, p. 15); *Lingula feistmanteli* Barrande, 1879; Třenice Formation; Krušná hora, Bohemia.

*Diagnosis.* – Shell thick, subequally biconvex to dorsi-biconvex, subtriangular to subpentagonal; ornamented by fine rugellae. Pseudointerareas of both valves orthocline, highly raised above valve floor. Pedicle groove narrow and deep. Ventral visceral field thickened, rhomboidal, not extending

*Fig. 42. Hyperobolus andreevae* sp. nov.; Akbulaksai Formation; Akbulaksai River. □A. Holotype; ventral exterior (sample B-607-9); RM Br 135932; ×2. □B. Dorsal exterior (sample B-607-9); RM Br 135933; ×2. □C. Ventral internal mould (sample B-607-9); RM Br 135934; ×2. □D. Latex cast of C; ×2. □E. Dorsal internal mould (sample B-607-9); RM Br 135935; ×2. □F. Latex cast of ventral external mould (sample B-607-9); RM Br 135936; ×2. □G. Dorsal view of juvenile complete shell (sample B-607-15); RM Br 135949; ×20. □H. Lateral view of G; ×22. □I. Posterior view of G; ×32.5. □J. Dorsal umbo and ventral pseudointerarea of G; ×65. □K. Ornamentation of larval and postlarval shell of G; ×130. □L. Ornamentation of the shell of G; ×325. □M. Ornamentation of the larval shell of G; ×680. □N. Exterior of dorsal valve (sample B-607-15); RM Br 135938; ×14. □O. Posterior view of M; ×32.5. □P. Ventral interior of juvenile (sample B-607-15); RM Br 135937; ×100. □Q. Ventral interior of juvenile (sample B-607-15); RM Br 135950; ×32.

to mid-length, bisected by median groove; ventral *vascula lateralia* submedial, slightly divergent in posterior half. Dorsal visceral field with wide median tongue, extending to or slightly anterior to mid-length; anterior lateral muscle scars large, placed close to central scars; dorsal *vascula lateralia* submarginal, slightly divergent; *vascula media* short, widely divergent.

*Species included. – Lingula feistmanteli* Barrande, 1879; *Hyperobolus andreevae* sp. nov.

*Occurrence.* – Lower Ordovician (Arenig); Bohemia, South Ural Mountains.

## *Hyperobolus andreevae* sp. nov.

Fig. 42

*Name.* – In honour of Olga N. Andreeva.

*Holotype.* – Fig. 42A; RM Br 135932; ventral valve (L 31.4, W 30.7); Akbulaksai Formation (sample B-607-9); Akbulaksai River.

*Paratypes.* – Figured. Complete shell: RM Br 135949. Ventral valves: RM Br 135934 (L 19.7, LI 2.8, WI 13.5); RM Br 135936; RM Br 135937; RM Br 135950. Dorsal valves: RM Br 135933 (L 23.7); RM Br 135935 (L 32.7); RM Br 135938. Total of 4 complete shells, 20 ventral and 18 dorsal valves.

*Diagnosis.* – Shell thin, dorsi-biconvex, with fine rugellae. Ventral visceral area subtriangular, not extending to mid-length, lacking median depression. Dorsal visceral field extending slightly anterior to mid-length.

*Description.* – The shell is close to circular, flattened, and only slightly dorsi-biconvex. The ventral valve is only slightly longer than the dorsal, and has a widely triangular, orthocline pseudointerarea, about 21% as long as wide, extending about 14% of the valve length. The pedicle groove is deep and narrow; the propareas are slightly elevated above the floor of the valve, and bisected by flexure lines (Figs. 42C, D, P, Q). The ventral visceral area is weakly impressed, subtriangular, and does not extend to mid-length (Fig. 42B). The dorsal valve has a gently convex lateral profile, with maximum height placed somewhat posterior to the centre. The dorsal pseudointerarea has a wide, shallow median groove and narrow, flattened propareas that are slightly elevated above the valve floor. The dorsal visceral field is weakly impressed, situated in the posterior third of the valve. The central muscle scars are large, elongate oval, situated slightly posterior to mid-length, near the end of the wide median tongue. In a few valves, a pair of small, elongated anterior lateral muscle scars may be observed, anterior to the central muscle scars.

The larval and postlarval shell is finely pitted, with rounded pits, up to 7 μm across (Fig. 42K–M). The postlarval ornamentation has fine rugellae and widely spaced growth lamellae (about 8–10 per mm), superposed on fine, poorly defined, irregular radial plications (Fig. 42F, N–O).

*Discussion. – H. andreevae* is closely comparable with the type species in the shape and size of the shell; it differs mainly in the following three characters: (1) The shell is comparatively thinner; (2) the shape of the ventral visceral field is subtriangular and lacks a median depression; and (3) the dorsal visceral field is comparatively shorter, extending only slightly anterior to the centre of the valve. The finely pitted microornament of the larval and postlarval shell has not been observed on the type species.

*Occurrence.* – At the type locality only (Fig. 18; Appendix 1D).

## Genus *Thysanotos* Mickwitz, 1896

*Synonymy. –* □1896 *Obolus* (*Thysanotos*) subgen. nov. Mickwitz, p. 130. □1908 *Obolus* (*Mickwitzella*) subgen. nov. Walcott, p. 70. □1912 *Obolus* (*Mickwitzella*) – Walcott, p. 434. □1965 *Thysanotos* Mickwitz – Rowell, p. H266. □1969 *Thysanotos* Mickwitz – Gorjansky, p. 33. □1982 *Thysanotos* Mickwitz – Havlíček, p. 23. □1982 *Thysanobolus* gen. nov. Havlíček, p. 21.

*Type and only species.* – Original designation by Mickwitz (1896, p. 130); *Obolus siluricus* Eichwald, 1840; Leetse beds; northern Estonia.

*Diagnosis. –* Shell dorsi-biconvex, flattened. Ventral pseudointerarea orthocline with narrow, triangular, deep pedicle groove; flexure lines well defined. Ventral interior with subtriangular visceral field, slightly raised anteriorly, not extending to mid-length; ventral *vascula lateralia* submarginal, widely divergent in the proximal parts. Dorsal pseudointerarea low, anacline, raised above valve floor, with well-defined median groove and flexure lines. Dorsal interior with well-defined visceral area; dorsal median tongue bounded laterally by low ridges, and extending anterior to mid-length; dorsal central muscle scars large, elongate suboval, divided by groove; dorsal *vascula lateralia* submarginal; *vascula media* long, divergent. Shell surface covered with high, widely spaced rugellae, superposed on concentric lamellae with marginal spines; larval and postlarval shell covered by fine pits.

*Discussion. –* The genus is closely similar to *Paldiskia* Gorjansky, 1969, in the general shape and outline of the shell, as well as in the characters of the ventral interior, and in the possession of a finely pitted external surface. *Thysanotus* differs in having marginal spines and in lacking radial ornamentation. The genus *Thysanobolus* Havlíček, was described from the Třenice and Mílina formations in Bohemia. According to Havlíček (1982, p. 21), *Thysanotos* differs from *Thysanobolus* 'in having a subcircular to transversely elliptical shell and a visceral field supported by a low callosity in the pedicle valve'. However, the morphology of the ventral visceral field is highly variable in *Thysanotos*, and the outline of the shell also varies; moreover, the examination of topotype specimens (kindly supplied by Michal Mergl, Plzen) of the type species, *Thysanobolus lingulides* suggests that the elongate shell may be the result of *post mortem* deformation.

Two other species, *Thysanobolus giganteus* (Koliha, 1937) and *T. pirolus* Havlíček, 1982, are based on poorly preserved material and can be distinguished from the type species by minor differences in size and details of ornamentation. *Thysanotos primus* (Koliha, 1924), from the Třenice Formation, also differs from the type species only in relative size and ornamentation (Havlíček 1982). We suggest here that all these differences can be accomodated within the range of variation of *T. siluricus,* and all the species discussed above are considered to be junior synonyms of the type species.

As noted above (p. 32), the Tremadoc age of the records from Poland and Bohemia is here questioned. As far as can be determined, there does not seem to be any evidence to support the pre-Arenig records.

*Occurrence.* – Ordovician (Arenig); Germany, Bohemia, Serbia, ?Poland, Estonia, Ingria, South Urals.

## *Thysanotos siluricus* (Eichwald, 1840)
Figs. 43–44

*Synonymy.* – □1840 *Obolus siluricus* sp. nov. – Eichwald, p. 195. □1843 *Obolus siluricus* – Eichwald, p. 7, Pl. 1:15a–c. □1845 *Obolus siluricus* Eichwald – de Verneuil, p. 291 [in part]. □1848 *Aulonotreta polita* sp. nov. – Kutorga, p. 279 [in part]. □1859 *Obolus siluricus* Eichwald – Eichwald, Pl. 37:6, 7a–b. □1860 *Obolus siluricus* Eichwald – Eichwald, p. 297. □1896 *Obolus* (*Thysanotos*) *siluricus* Eichwald – Mickwitz, p. 195, Pl. 3:1–9. □1912 *Obolus* (*Mickwitziella*) *siluricus* Eichwald – Walcott, p. 434, Pl. 30:1–4 [synonymy]. □1924 *Obolus* (*Lingulobolus*) *Feistmanteli* (Barrande) var. *Barrandei prima* var. nov. – Koliha, p. 19, Pl. 1:6. □1937 *Obolus giganteus* sp. nov. – Koliha, p. 481. □1955 *Obolus* (*Thysanotos*) *siluricus* Eichwald – Sdzuy, p. 7, Pl. 1:1–4 [synonymy]. □?1962 *Obolus* (*Thysanotos*) *siluricus* Eichwald – Bednarczyk, p. 157, Pl. 30:3, 4. □?1964 *Obolus* (*Thysanotos*) *siluricus* Eichwald – Bednarczyk, p. 34, Pls. 1:1–14; 9:8–12. □1969 *Thysanotos siluricus* (Eichwald) – Gorjansky, p. 32, Pl. 4:1–10 [synonymy]. □1982 *Thysanotos siluricus* (Eichwald) – Havlíček, p. 24, Pl. 2:1–6 [synonymy of Bohemian forms]. □1982 *Thysanobolus linguides* sp. nov. Havlíček, p. 21, Pl. 2:10–13. □1982 *Thysanobolus giganteus* (Koliha) – Havlíček, p. 22, Pls. 2:9; 4:13. □1982 *Thysanobolus pirolus* sp. nov. Havlíček, p. 23, Pl. 10:10–11. □1982 *Thysanotos primus* (Koliha) – Havlíček, p. 24, Pl. 2:7–8.

*Lectotype.* – Selected by Gorjansky (1969, p. 33); ventral valve illustrated by Eichwald (1843, Pl. 1:15a, b); specimen in the University of St. Petersburg, Department of Geology; lower Leetse beds; Pakerort near Paldiski, northern Estonia.

*Material.* – Figured from South Urals. Complete shells: RM Br 135939 (LI 3.2, WI 19.2, ML 2.6, WG 2); RM Br 135944 (L 39.7, T 6.2). Ventral valves: RM Br 135940, RM Br 135941, RM Br 135942, RM Br 135943. Dorsal valves: RM Br 135945, RM Br 135947, RM Br 135948. Fragment: RM Br 135946. Total of 9 complete shells, 22 ventral valves, and 17 dorsal valves.

Figured from northern Estonia. Complete shell: GT Br3501 (L 20, W 18.3, T 6.2). Ventral valves: GT Br3503 (L 21.7, W 22.9, LI 5, ML 3, WI 14.1, WG 2.5, LV 10); GT Br3504 (L 24.5, W 26.7, LV 12.3). Dorsal valves: GT Br3502 (L 20, W 21.3, LI 4.2, ML 1.5, WI 12.4, WG 6, LV 12.5), LO 6560t (L 19.4, W 28.5).

*Diagnosis.* – As for genus.

*Description of South Ural material.* – The shell is dorsibiconvex, about 16% as thick as long in one specimen; it is transversly suboval to subcircular in outline, about 75–101% as long as wide, with the maximum width near mid-length (Fig. 43F, G). The ventral valve is weakly convex, with the maximum height at the posterior third. The ventral pseudointerarea is widely triangular, orthocline, about 17% as long as wide. The pedicle groove is deep, with steep lateral margins, narrowly triangular, occupying about 10% of the width of the pseudointerarea. The propareas are raised above the valve floor, with well-defined flexure lines (Fig. 43A, B, J). The ventral visceral field is subtriangular and does not extend to the centre of the valve; it is slightly raised anteriorly (Fig. 43B). The *vascula lateralia* are arcuate and submarginal.

The dorsal valve is gently convex in lateral profile, with maximum height somewhat anterior to the umbo. The dorsal pseudointerarea is anacline, with a wide, concave median grove and raised propareas with flexure lines (Fig. 43K). The dorsal visceral field has a median tongue, usually extending somewhat anterior to the centre of the valve; this tongue is bounded laterally by two arcuate ridges. The *vascula lateralia* are submarginal, arcuate, and widely divergent in their proximal parts. The *vascula media* are long and divergent.

The larval shells are not preserved on adults, but on some juvenile valves it is close to circular, around 0.4 mm across (Fig. 43C, D, I). The larval and postlarval shell is finely pitted, with large, rounded pits of subequal size, up to about 14 µm across; the larger pits are surrounded by raised ridges, sometimes bearing clusters of minute pits, around 800 nm across (Fig. 43L–N). The shell surface is covered by evenly spaced concentric rugellae, about 4–6 per millimetre in adults, superposed on evenly spaced growth lamellae, up to about 6 mm apart, with rows of marginal spines, up to about 1 mm long; at the distance of 25 mm from the posterior margin, there are about 4–7 spines per 3 mm (Fig. 43A, F–H).

*Discussion.* – The material of *T. siluricus* from the South Urals is identical in its main characters to topotypes from northern Estonia (Figs. 43O, P, 44). The only differences are related to the maximum shell size: the specimens from the South Urals

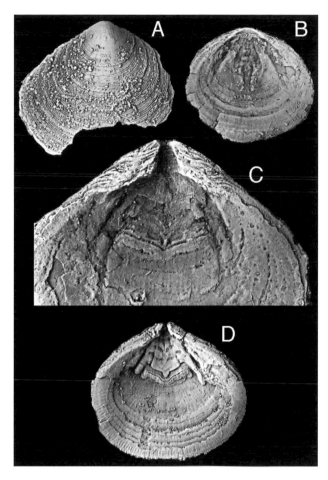

*Fig. 44. Thysanotos siluricus* (Eichwald); Leetse beds; northern Estonia (coll. A. Öpik). □A. Ventral view of complete shell; Hundikuristik; GT Br 3501; ×2. □B. Dorsal interior; Paldiski; GT Br 3502; ×1.5. □C. Detail of ventral valve, showing pseudointerarea and visceral area; Mäeküla; GT Br 3503; ×3.5. □D. Ventral interior; Pakri island; GT Br 3504; ×1.5.

*Fig. 43. Thysanotos siluricus* (Eichwald). □A. Dorsal view of complete shell; Medes River (sample G-237-1); RM Br 135939; ×2. □B. Lateral view of ventral interior of juvenile; Medes River (sample G-212); RM Br 135940; ×110. □C. Ventral exterior of juvenile; Medes River (sample G-212); RM Br 135943; ×25. □D. Ventral exterior of juvenile; Akbulaksai Formation; Alimbet River (sample B-780-1); RM Br 135942; ×23. □E. Dorsal internal mould; Kidryas Formation; Tyrmantau Ridge (sample G-135-1); RM Br 135948; ×2. □F. Ventral view of complete shell; Alimbet Formation; Alimbet Farm (sample B-578-2); RM Br 135944; ×2. □G. Dorsal view of F; ×2. □H. Detail of F, showing marginal spines; ×8. □I. Dorsal exterior of juvenile; Akbulaksai Formation; Alimbet River (sample B-780-1); RM Br 135945; ×23. □J. Ventral interior; Medes River (sample G-212); RM Br 135941; ×10. □K. Dorsal pseudointerarea with median groove; Akbulaksai Formation; Alimbet River (sample B-780-1) RM Br 135947; ×15. □L. Oblique lateral view of broken valve, showing ornamentation with marginal spines; Akbulaksai Formation; Alimbet River (sample B-780-1); RM Br 135946; ×38. □M. Detail of L, showing pitted micro-ornament; ×135. □N. Pitted micro-ornament of L; ×1200. □O. Detail of dorsal valve, showing ornamentation with marginal spines; Leetse beds, probably from Leetse (coll. Törnquist), northern Estonia; LO 6568t; ×32. □P. Detail of surface, showing rugellae with pitted micro-ornament; ×50.

are up to about 40 mm long, while the maximum length recorded from Estonia is somewhat more than 25 mm; the Estonian shells also appear to be somewhat more biconvex and 31% as long as high in one specimen, while the corresponding ratio in the only complete shell from the South Urals is 16%. Moreover, the growth lamellae appear to be more numerous in the Estonian material, where a fully grown shell has up to about 7–8 lamellae (Fig. 44A); the lamellae appear to be fewer and farther apart in the specimens from the South Urals (Fig. 43F–H). The growth lamellae represent temporary interruptions in growth that perhaps are related to spawning, and similar types of lamellae are known from many other Lower Palaeozoic lingulids (Holmer & Popov, unpublished observations).

*Occurrence.* – The distribution of the *Thysanotus–Leptembolon* assemblage is discussed above (p. 32). *T. siluricus* is common and widespread in the lower Leetse beds throughout northern Estonia. In the South Urals, it is present in the Akbulaksai and Alimbet formations at the following localities: Alimbet Farm (Fig. 13; Appendix 1B); Alimbet River (Fig. 16; Appendix 1C); Akbulaksai River (Fig. 18; Appendix 1D); Karabutak and Kosistek rivers (Appendix 1E); Koagash River (Appendix 1F). It is also present in the upper part of the Kidryas Formation at the Tyrmantau Ridge (Fig. 11; Appendix 1A), and at the Medes River.

## Subfamily Lingulellinae Schuchert, 1893

*Diagnosis.* – Shell elongate oval to subtriangular. Pseudointerareas usually not raised above valve floor. Dorsal pseudointerarea forming concave undivided plate.

*Remarks.* – Havlíček (1982, p. 13) provided a recent summary of the main morphological differences between the obolid subfamilies Lingulellinae, Obolidae, and Glossellinae.

## Genus *Lingulella* Salter, 1866

*Type species.* – Subsequent designation by Dall (1870, p. 159); *Lingula davisii* M'Coy, 1851; Upper Cambrian (Merioneth Series) Festiniog Beds; south of Penmorfa, Gwynedd, Wales.

*Diagnosis.* – Shell elongate oval to subtriangular, thin-shelled, subacuminate. Ventral pseudointerarea with narrow, triangular pedicle groove and well-defined flexure lines. Dorsal visceral field with low median ridge; dorsal scars of anterior lateral and central muscles closely spaced. *Vascula lateralia* of both valves peripherally placed.

*Discussion.* – As noted by Krause & Rowell (1975, p. 14), the plethora of Lower Palaeozoic species that have been placed within the genus usually includes any smooth-shelled, elongate obolids. In the absence of any information on the interior characters of most of these taxa, it is impossible to evaluate their taxonomic relationships.

It is important to note that the outline and proportions of valves of both recent and extinct lingulids vary significantly during ontogeny and also from one population to another; these characters may by used for definition of lingulid species only together with distinctive characters of the internal morphology and pseudointerareas.

## *Lingulella antiquissima* (Jeremejew, 1856)

Figs. 45A–M, 46, 47A–M

*Synonymy.* – □1856 *Lingula antiquissima* sp. nov. – Jeremejew, p. 80, Fig. 6. □?1866 *Lingulella lepis* sp. nov. – Salter *in* Ramsay, p. 334, Fig. 11. □1877 *Lingulella nicholsoni* sp. nov. – Callaway, p. 668, Pl. 24:11, 11a, b. □1882 *Lingulella? lepis* Salter – Brøgger, p. 44, Pl. 10:5a–b. □1906 *Lingulella lepis?* Salter – Moberg & Segerberg, p. 62, Pl. 1:20. □1906 *Lingula? corrugata* sp. nov. – Moberg & Segerberg, p. 63, Pl. 1:21. □1906 *Lingula? producta* sp. nov. – Moberg & Segerberg, p. 63, Pl. 1:23. □1906 *Lingula? ordovicensis* sp. nov. – Moberg & Segerberg, p. 63, Pl. 1:24. □1909 *Lingulella lepis* Salter – Westergård, p. 57, Pl. 2:20–22. □1909 *Lingula? corrugata* Moberg & Segerberg – Westergård, p. 57, Pl. 2:23. □1912 *Lingulella lepis* Salter – Walcott, p. 514, Pl. 31:4, 4a–f [synonymy]. □1912 *Lingulella nicholsoni* Callaway – Walcott, p. 522, Pl. 30:3, 3a–f [synonymy]. □1922 *Lingulella lepis* Salter – Poulsen, p. 16, Fig. 14. □?1960 *Lingulella* cf. *fuchsi* Redlich – Poulsen, p. 6, Pl. 1:5–9. □not 1973 *Lingulella nicholsoni* Callaway – Rushton *in* Bulman & Rushton, Pl. 5:1–4. □?1978 *Lingulella lepis* Salter – Cocks, p. 15 [synonymy]. □1978 *Lingulella nicholsoni* Callaway – Cocks, p. 15 [synonymy]. □?1982 *Lingulella lepis* Salter – Rushton & Bassett *in* Owens *et al.*, p. 23, Pl. 5p–u. □1982 *Lingulella nicholsoni* Callaway – Rushton & Bassett *in* Owens *et al.*, p. 21, Pl. 5j–o. □1989 *Lingulella antiquissima* (Jeremejew) – Popov & Khazanovitch *in* Popov *et al.*, p. 124, Pls. 4:1, 2, 4; 8:6–10.

*Neotype.* – Selected by Popov & Khazanovich (*in* Popov *et al.* 1989); CNIGR 180/12348; ventral valve (L 11.2, W 8.6, LI 2.8, WI 4.8, ML 1.7) from the Tosna Formation (*Cordylodus proavus* Biozone); Syas' River near Rebrovo village, Ingria.

*Material.* – Figured from southern Kendyktas Range. Ventral valves: RM Br 135960 (L 13.7, W 10.4, LI 1.5, WI 3.4); RM Br 135962; RM Br 135963; RM Br 135964 (L 11.5, W 6.56, LI 1.6, WI 3.36, LV 5.6, WV 4.88); RM Br 135966; RM 135967; RM Br 135968. Dorsal valve: RM Br 135965 (L 12.8, W 6.56, LI 1.96, WI 3.6, LV 6.8). Total of 39 ventral and 49 dorsal valves.

Figured from Scandinavia. Ventral valves: RM Br 1804a (L 12.5, W 10.3); RM Br 133943 (L 5.4, W 3.3); PMO 19108; LO 2238t; LO 1770T; ?LO 1772T; LO 6506t; LO 6507t, LO 6508t. Dorsal valves: RM Br 133944 (L 4.8, W 3.8); RM Br 133945; RM Br 133946; LO 6504t; LO 6505t; SGU 8496 (L 6.8, W 5.7, LI 1.1, WI 4, ML 0.6, WG 1.5, LV 4.6). Total of 3 complete shells, 27 dorsal valves, and 47 ventral valves.

Figured from Britain. Ventral valves: BM B47864 (L 3.5, W 2.7); BM B14360 (L 2.3, W 1.7); NMW 77.1G.48 (L 5.4, W 4, LI 1.1, WI 2.8); USNM 51748a. Dorsal valves: BM B47860 (W 4.7); BM B47862; NMW 77.1G.49 (L 3.8, W 3.6, LI 0.9, WI 2.5, LV 3.3).

*Measurements.* – See Tables 1–4.

*Diagnosis.* – Shell elongate subtriangular, with maximum width somewhat anterior to mid-length. Ventral pseudointerarea subtriangular in outline; pedicle groove shallow with subparallel or slightly divergent lateral margins. Pseudo-interareas of both valves generally occupying more than half of valve width. Ventral visceral field extending anterior to mid-length. Dorsal visceral field with narrow, long median tongue, extending to over 60% of valve length.

*Description of Scandinavian and South Kazakhstan material.* – The shell is flattened, on average 20% as thick as long in three complete shells (sample Öl-1). The outline is elongate suboval to subtriangular. The ventral valve is about 1.5 times as long as wide (Tables 1, 3). The ventral pseudointerarea is narrow, triangular, somwhat less than half as long as wide (Tables 1, 3), occupying on average 40% and 71% of the valve width, and 11% and 25% of the valve length, in two samples (Tables 1, 3). The pedicle groove is narrow and shallow, with subparallel to slightly divergent lateral margins; the flexure lines are well developed (Fig. 46E, F). The ventral internal characters are usually weakly impressed. The ventral visceral field is subtriangular in outline, extending to about half of the valve length and occupying about three quarters of the width (Tables 1, 3). In adults, the ventral visceral field is bisected by

*Fig. 45.* □A–M. *Lingulella antiquissima* (Jeremejew). □A. Ventral exterior; Bjørkåsholmen Limestone; Ottenby, Öland; RM Br 133943; ×12. □B. Dorsal exterior; Bjørkåsholmen Limestone; Ottenby, Öland; RM Br 133944; ×12. □C. Dorsal exterior; Bjørkåsholmen Limestone; Ottenby, Öland; RM Br 133945; ×12. □D. Partly exfoliated ventral exterior; figured by Brøgger (1882, Pl. 10:5a); Bjørkåsholmen Limestone; Vestfossen, Oslo region; PMO 19108; ×7. □E. Ventral internal mould; figured by Westergård (1909, Pl. 2:21); Dictyonema Shale; Fågelsång, Scania; LO 2238t; ×11. □F. Ventral internal mould; Kendyktas Formation (sample 554); RM Br 135966; ×10. □G. Ventral exterior; Kendyktas Formation (sample 554); RM Br 135967; ×10. □H. Ventral external mould; Kendyktas Formation (sample 554); RM Br 135968; ×10. □I. Ventral exterior; Dictyonema Shale; Fågelsång, Scania; RM Br 1804a; ×3.5. □J. Dorsal internal mould; Bjørkåsholmen Limestone; Bjørkåsholmen, Oslo region; RM Br 133946; ×10. □K. Ventral exterior; holotype by monotypy of *Lingula? corrugata* Moberg & Segerberg (1906, Pl. 1:21); Dictyonema Shale; Flagabro, Scania; LO 1770T; ×8. □L. Ventral exterior; holotype by monotypy of *Lingula? producta* Moberg & Segerberg (1906, Pl. 1:23); Bjørkåsholmen Limestone; Fågelsång, Scania; LO 1772T; ×12. □M. Anterior view of L; ×12. □N. *Broeggeria salteri* (Holl); dorsal exterior; holotype by monotypy of *Obolus? inflatus* Westergård (1909, Pl. 5:25); Bjørkåsholmen Limestone; Jerrestad, Scania; LO 2302T; ×5. □O. Lateral view of N; ×5. □P–S. *Ralfia? bryograptorum*; Dictyonema Shale; Fågelsång, Scania. □P. Dorsal internal mould; holotype by monotypy; figured by Moberg & Segerberg (1906, Pl. 1:22); LO 1771T; ×5. □Q. Ventral internal mould; RM Br 1804b; ×6. □R. Ventral? internal mould; RM Br1804c; ×4.5. □S. Ventral? external mould; RM Br 1804f; ×7.

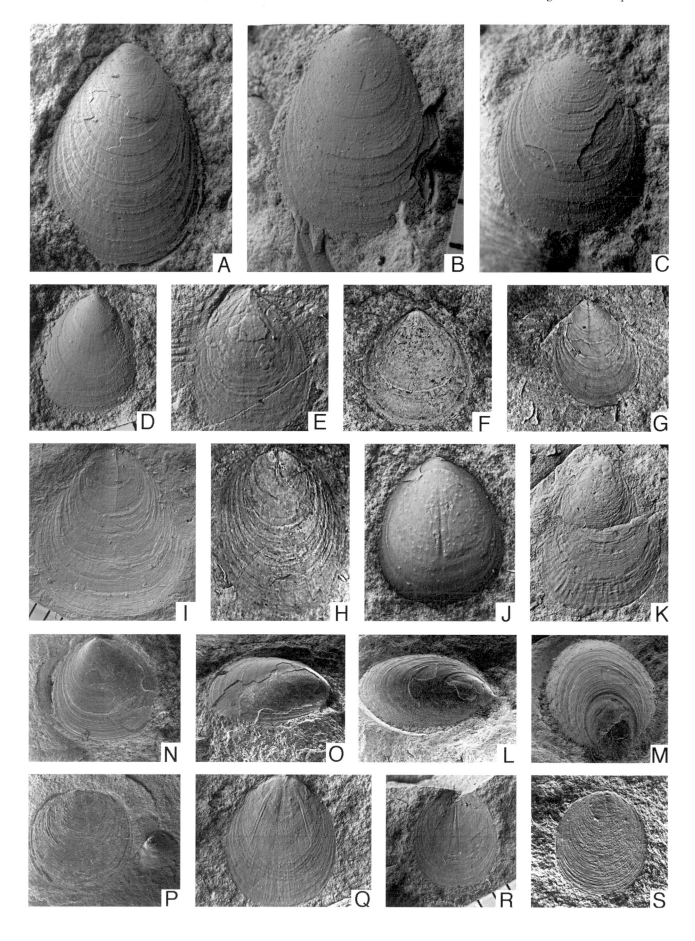

two slightly divergent furrows extending to the anterior margin of the field, at a point about one-third of the valve length from the posterior margin. The *vascula lateralia* are submarginal, straight, diverging proximally, and becoming arcuate distally (Fig. 45E, F).

The dorsal valve is on average 1.2–1.3 times as long as wide in two samples (Tables 2, 4). The dorsal pseudointerarea forms a narrow, crescent-shaped strip, in two samples occupying on average 42% and 64%, respectively, of the valve width and 7% and 21%, respectively, of the valve length (Tables 2, 4). The dorsal median groove is wide, but poorly defined from the reduced propareas (Fig. 46B, C, J). The dorsal visceral field has a narrow, long median tongue, on average extending to 60% and 74% of the valve length in two samples (Tables 1, 3), and bisected by a long median ridge (Figs. 45J, 46B–D). The dorsal *vascula lateralia* are weakly impressed, arcuate, and marginal; the *vascula media* are short and divergent. The entire visceral area of both valves have distinctive rounded to hexagonal, closely packed depressions, up to about 35 μm across (Fig. 46D).

*Discussion.* – This previously poorly known species was redescribed recently by Popov & Khazanovich (*in* Popov *et al.* 1989). It occurs in the Tosna Formation, which probably represents near-shore deposits, dominated by cross-bedded sands with numerous *Skolithos* burrows (Popov *et al.* 1989).

Table 1. *Lingulella antiquissima*, average dimensions and ratios of ventral valves.

|  | L | W | LI | WI | LV | $L/W$ | $LI/WI$ | $LI/L$ | $WI/W$ | $LV/L$ |
|---|---|---|---|---|---|---|---|---|---|---|
| Sample Öl-1 | | | | | | | | | | |
| N | 4 | 4 | 3 | 3 | 1 | 4 | 3 | 3 | 3 | 1 |
| X | 3.94 | 2.65 | 0.88 | 1.73 | 1.88 | 146% | 49% | 25% | 71% | 53% |
| S | 1.30 | 0.57 | 0.38 | 0.46 | | 0.20 | 0.10 | 0.04 | 0.06 | |
| MIN | 2.36 | 2.00 | 0.48 | 1.28 | 1.88 | 118% | 38% | 20% | 64% | 53% |
| MAX | 5.40 | 3.30 | 1.24 | 2.20 | 1.88 | 164% | 56% | 28% | 76% | 53% |

Table 2. *Lingulella antiquissima*, average dimensions and ratios of dorsal valves.

|  | L | W | LI | WI | LV | $L/W$ | $LI/WI$ | $LI/L$ | $WI/W$ | $LV/L$ |
|---|---|---|---|---|---|---|---|---|---|---|
| Sample Öl-1 | | | | | | | | | | |
| N | 6 | 6 | 4 | 4 | 5 | 6 | 4 | 4 | 4 | 5 |
| X | 3.60 | 2.93 | 0.80 | 1.93 | 2.81 | 123% | 42% | 21% | 64% | 74% |
| S | 0.60 | 0.52 | 0.07 | 0.19 | 0.28 | 0.08 | 0.02 | 0.01 | 0.04 | 0.03 |
| MIN | 2.60 | 2.12 | 0.72 | 1.80 | 2.48 | 114% | 40% | 20% | 61% | 71% |
| MAX | 4.28 | 3.60 | 0.88 | 2.20 | 3.16 | 138% | 44% | 23% | 70% | 78% |

Table 3. *Lingulella antiquissima*, average dimensions and ratios of ventral valves.

|  | L | W | LI | WI | LV | WV | $L/W$ | $LI/WI$ | $LI/L$ | $WI/W$ | $LV/L$ | WV/W |
|---|---|---|---|---|---|---|---|---|---|---|---|---|
| Sample 560 | | | | | | | | | | | | |
| N | 5 | 5 | 5 | 5 | 5 | 1 | 5 | 5 | 5 | 5 | 5 | 1 |
| X | 11.54 | 7.84 | 1.31 | 3.09 | 5.63 | 5.40 | 148% | 42% | 11% | 40% | 49% | 76% |
| S | 1.66 | 1.18 | 0.38 | 0.45 | 0.57 | | 0.21 | 0.09 | 0.03 | 0.04 | 0.05 | |
| MIN | 8.80 | 7.12 | 0.80 | 2.48 | 4.88 | 5.40 | 122% | 29% | 8% | 35% | 43% | 76% |
| MAX | 13.20 | 9.92 | 1.68 | 3.60 | 6.40 | 5.40 | 174% | 53% | 14% | 46% | 55% | 76% |

Table 4. *Lingulella antiquissima*, average dimensions and ratios of dorsal valves.

|  | L | W | LI | WI | LV | $L/W$ | $LI/L$ | $WI/W$ | $LV/L$ |
|---|---|---|---|---|---|---|---|---|---|
| Sample 560 | | | | | | | | | |
| N | 5 | 5 | 3 | 2 | 3 | 5 | 3 | 2 | 3 |
| X | 10.63 | 8.08 | 0.72 | 3.04 | 6.53 | 131% | 7% | 42% | 60% |
| S | 2.19 | 1.29 | 0.29 | 0.79 | 0.83 | 0.16 | 0.01 | 0.06 | 0.07 |
| MIN | 7.28 | 6.56 | 0.40 | 2.48 | 5.60 | 111% | 5% | 38% | 55% |
| MAX | 13.10 | 10.00 | 0.96 | 3.60 | 7.20 | 155% | 8% | 47% | 67% |

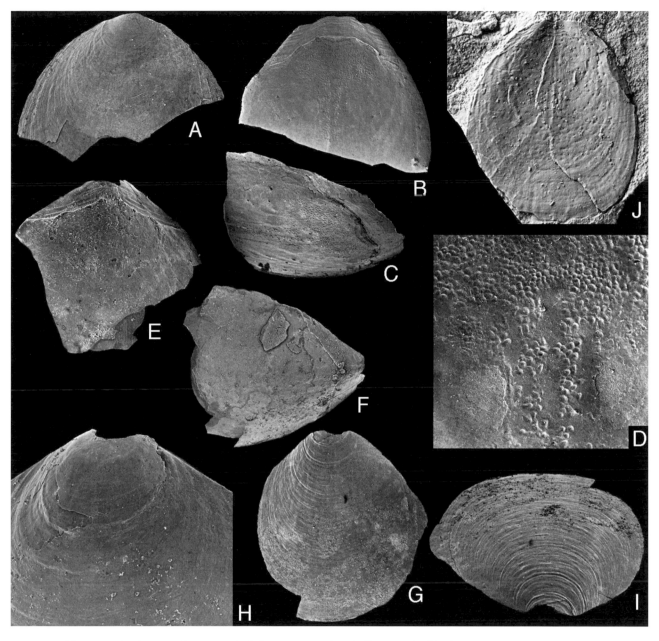

*Fig. 46. Lingulella antiquissima* (Jeremejew). □A–J. Bjørkåsholmen Limestone (sample Öl-1); Ottenby, Öland. □A. Dorsal exterior; LO 6504t; ×27. □B. Dorsal interior; LO 6505t; ×16. □C. Lateral view of B; ×22. □D. Detail of B, showing pitted visceral field with central muscle scars and median tongue; ×56. □E. Ventral interior; LO 6506t; ×32. □F. Lateral view of ventral interior; LO 6507t; ×18. □G. Ventral exterior; LO 6508t; ×16. □H. Detail of G, showing apex with larval shell; ×75. □I. Posterior view of G; ×22. □J. Dorsal interior; Dictyonema Shale, Storby, Östergötland; SGU 8496; ×7.

In these deposits, both juveniles and adult specimens (maximum L 11) of *L. antiquissima* are quite abundant. Juveniles, about 3–6 mm long, are closely comparable with lingulids, referred usually to *L. lepis* in the Dictyonema and Ceratopyge shales and the Bjørkåsholmen Limestone in Scandinavia (Figs. 45A–D, J–M; 46), and to *L. nicholsoni* and *L. lepis* in the Tremadoc in Britain and Canada (Fig. 47A–H). Only a few larger specimens (maximum L 12) of *Lingulella* have been found in the Dictyonema Shale in Scania (Fig. 45I), and the only difference from *L. antiquissima* lies in the elongate oval, rather than subtriangular outline of the shells.

In the Kendyktas Formation, juvenile lingulid shells (Fig. 45F–H) are usually associated with the *Broeggeria* assemblage and the *Hysterolenus* assemblage of trilobites. These lingulids are also more or less identical in size (3–6 mm long) and morphology to those usually described as *L. lepis* and *L. nicholsoni* from the Tremadoc of Britain and Scandinavia.

In the overlying Agalatas Formation, cross-bedded sandstones contain abundant valves of larger (maximum L 13.7) *Lingulella* (Fig. 47I–M). In all important characters of morphology, these are also identical to *L. antiquissima* from the East Baltic. We suggest here that the larger shells of *Lingulella*

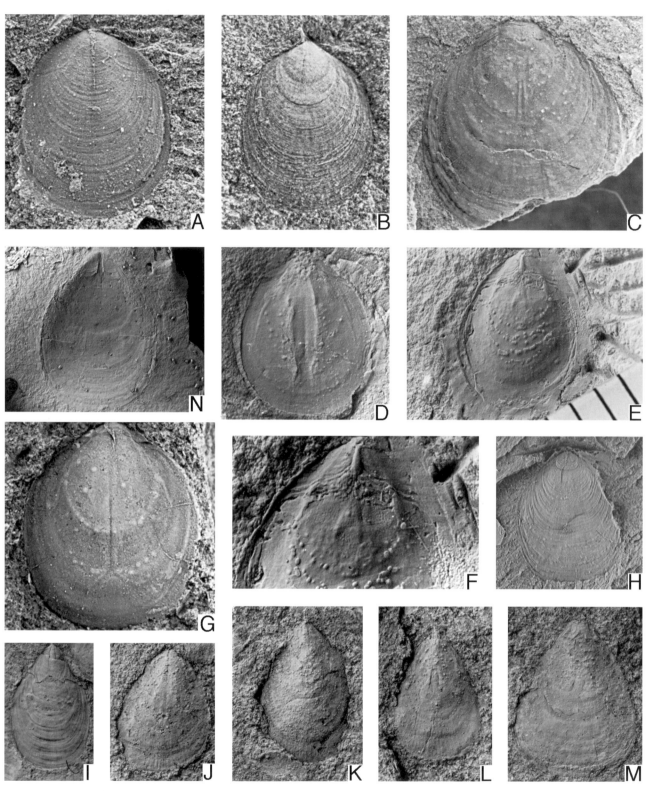

*Fig. 47.* □A–H. *Lingulella antiquissima* (Jeremejew); Tremadoc, Britain. □A. Ventral exterior; Shineton Shale; Salop, Shropshire; BM B47864; ×15. □B. Ventral exterior; Shineton Shale; Salop, Shropshire; BM B14360; ×20. □C. Dorsal internal mould; Shineton Shale; Salop, Shropshire; BM B47860; ×11. □D. Dorsal internal mould; figured by Owens *et al.* (1982, Pl. 5m); *Clonograptus tenellus* Biozone; Dyfed, Carmarthen district, South Wales; NMW 77.1G.49; ×10. □E. Ventral internal mould; figured by Owens (*et al.*1982, Pl. 5o); *Clonograptus tenellus* Biozone; Dyfed, Carmarthen district, South Wales; NMW 77.1G.48; ×8. □F. Detail of E, showing pseudointerarea; ×15. □G. Dorsal internal mould; Shineton Shale; Shineton, Salop, Shropshire; BM B47862; ×11. □H. Ventral exterior; figured by Walcott (1912, Pl. 30:3b); Shineton Shale; Shineton, Salop, Shropshire; USNM 51748a; ×10. □I–M. *Lingulella antiquissima* (Jeremejew); Agalatas Formation (sample 560); Kendyktas Range, Kazakhstan. □I. Partial ventral internal mould; RM Br 135963; ×3. □J. Dorsal internal mould; RM Br 135965; ×3. □K. Ventral internal mould; RM Br 135964; ×3. □L. Ventral internal mould; RM Br 135962; ×3. □M. Ventral internal mould; RM Br 135960; ×3. □N. *Palaeobolus quadratus* (Bulman, 1927); latex cast of ventral internal mould; figured by Owens (*et al.*1982, Pl. 6f); *Clonograptus tenellus* Biozone; Dyfed, Carmarthen district, South Wales; NMW 77.1G.26a; ×3.

recorded from coeval beds in the studied areas represent mature individuals (probably living in relatively more near-shore environments) of *L. antiquissima*, and that the smaller specimens, usually referred to either *L. lepis* or *L. nicholsoni* in Scandinavia and Britain are juveniles (or stunted adults) of *L. antiquissima*.

*L. nicholsoni* was revised most recently by Rushton & Bassett (*in* Owens *et al.* 1982), mainly based on new, well-preserved material from the Tremadoc of the Carmarthen district, South Wales. In all important characters, such as, the shape and outline of the valves and the shape and size of pseudointerareas and visceral areas, the British species is closely similar to the type material of *L. antiquissima* from the Tosna Formation.

The relationship between *L. antiquissima* and *L. lepis* Salter, 1866, from the lower Tremadoc of North Wales is more difficult to ascertain in view of the extremely poor preservation of the type material; the latter species was also revised by Rushton & Bassett (*in* Owens *et al.* 1982). It is here tentatively placed in synonymy with *L. antiquissima*, but it might be better considered as a *nomen dubium*. Outside of Britain, *L. lepis* seems to have been used more commonly than *L. nicholsoni* for the forms referred here to *L. antiquissima* (see, e.g., Brøgger 1882; Moberg & Segerberg 1906; Walcott 1912).

Moberg & Segerberg (1906) described the new species *Lingula? producta* (Fig. 45L–M), *Lingula? ordovicensis* and *Lingula? corrugata* (Fig. 45K) from Scania; they were all considered to be junior synonyms of *Lingulella lepis* by Walcott (1912), and they are here tentatively referred to *L. antiquissima*.

*L. lepis* and a plethora of other species of *Lingulella* have been recorded from the supposed Tremadoc (here referred to the Arenig) of the Holy Cross Mountains, but this material is so poorly illustrated (e.g., Bednarczyk 1964, p. 44, Pls. 4:7–8; 8:14–15), that it cannot be compared in any meaningful way with *L. antiquissima*.

Poulsen (1960) described *Lingulella* cf. *fuchsi* Redlich from rocks of supposed early Cambrian age in Mauritania, but it is probable that the fauna is Lower Ordovician in age, and these specimens can be referred questionably to *L. antiquissima*.

It appears that *L. antiquissima* is also related closely to the type species *L. davisii*. However, this species has never been revised subsequent to Walcott (1912) and remains poorly known. The shape and outline of the valves of *L. davisii*, occurring in Upper Cambrian shales, are invariably more or less deformed. The two species are similar in having (1) a ventral pseudointerarea with a narrow, triangular pedicle groove, and well-defined flexure lines; (2) a dorsal visceral field with a low median ridge, and anterior lateral and central muscles scars that are closely spaced; and (3) *vascula lateralia* that are peripherally placed (see, e.g., Walcott 1912, Pl. 31:6).

*L. davisii* differs mainly *L. antiquissima* in having pseudointerareas that are longer and wider and occupy a greater proportion of the shell width and length.

Some specimens of *Palaeobolus quadratus* (Bulman) seem very close to *L. antiquissima* in morphology of the pseudo-interarea and other characters. It seems to differ mainly in the general outline of the shell. *P. quadratus* was redescribed by Rushton & Bassett (*in* Owens *et al.* 1982) but was not investigated during this study; one ventral interior (Fig. 47N) is illustrated for comparison with *L. antiquissima*.

The rounded to hexagonal depressions described here from the visceral area of *L. antiquissima* appear to be identical to those described from *Lingulella* sp. by Curry & Williams (1983; see also McClean 1988) and may represent epithelial moulds.

*Occurrence.* – The species occurs in the Tosna Formation in Ingria. It is relatively abundant in the Kendyktas and Agalatas formations (Figs. 31, 35; Appendix 1J). In Scandinavia, it is recorded from almost all studied localities in the Dictyonema and Ceratopyge shales and Bjørkåsholmen Limestone (Figs. 5–8; Appendix 1K). The species also occurs in Great Britain (Wales, Shropshire), Canada (Nova Scotia, Newfoundland), and possibly in Mauritania.

# Genus *Leptembolon* Mickwitz, 1896

*Synonymy.* – □1896 *Obolus* (*Leptembolon*) subgen. nov. – Mickwitz, p. 130. □1965 *Lingulella* Salter – Rowell, p. H266 [in part]. □1969 *Lingulella* (*Leptembolon*) (Mickwitz) – Gorjansky, p. 37 [synonymy]. □1982 *Leptembolon* Mickwitz – Havlíček, p. 38.

*Type species.* – Original designation by Mickwitz (1896, p. 130); *Obolus* (*Leptembolon*) *lingulaeformis*; Leetse beds; northern Estonia.

*Diagnosis.* – Shell elongate oval or subtriangular; subacuminate. Ventral pseudointerarea with triangular, narrow pedicle groove and well-defined flexure lines. Ventral visceral field rhomboidal in outline, extending to mid-length, thickened anteriorly. Dorsal visceral field slightly thickened anteriorly, with median tongue, extending close to anterior margin, bisected by low median ridge. *Vascula lateralia* of both valves submarginal and slightly arcuate.

*Species included.* – *Obolus* (*Leptembolon*) *lingulaeformis* Mickwitz, 1896; *Lingula insons* Barrande, 1879; *Lingula testis* Barrande, 1879.

*Discussion.* – *Leptembolon* has almost invariably been considered to be a subgenus of *Lingulella* or *Obolus* (e.g., Mickwitz 1896; Walcott 1912). Rowell (1965, p. H266) placed it as a junior synonym of *Lingulella*, while Gorjansky (1969, p. 37) again used *Leptembolon* as a subgenus of that genus. Gorjansky (1969) also gave the only recent review and discussion of previous records of the genus. Havlíček (1982, p. 38) elevated *Leptembolon* in rank to genus, and this practice is followed here.

As noted above (p. 32), the age of the supposed Tremadoc records of the genus from Poland and Bohemia is here questioned. As far as can be determined, there does not seem to be any evidence to support the pre-Arenig records.

*Occurrence.* – Ordovician (Arenig); Bohemia, ?Serbia, ?Poland, Estonia, South Urals.

## *Leptembolon lingulaeformis* (Mickwitz, 1896)

Figs. 48, 49

*Synonymy.* – □1896 *Obolus* (*Leptembolon*) *lingulaeformis* sp. nov. – Mickwitz, p. 200, Pl. 3:10–17. □1896 *Obolus* (*Leptembolon*) *lingulaeformis* var. *solidus* var. nov. – Mickwitz, p. 204, Pl. 3:17, 18. □1912 *Lingulella* (*Leptembolon*) *lingulaeformis* (Mickwitz, 1896) – Walcott, 1912, p. 542, Pl. 14:5, 5a [synonymy]. □?1964 *Lingulella sancta-crucensis* sp. nov. – Bednarczyk, p. 44, Pl. 6:1–5. □?1964 *Lingulella zejszneri* sp. nov. – Bednarczyk, p. 46, Pls. 7:1–14; 8:1–4. □1969 *Lingulella* (*Leptembolon*) *lingulaeformis* (Mickwitz, 1896) – Gorjansky, p. 38, Pl. 5:5–10 [synonymy]. □1969 *Lingulella* (*Leptembolon*) *recta*, sp. nov. – Gorjansky, p. 39, Pl. 5:11–17. □1977 *Obolus razumovskii* Lermontova – Nassedkina *in* Ancygin *et al.*, 1977, p. 191, Pl. 26:10.

*Lectotype.* – Selected by Gorjansky (1969, p. 38); CNIGR 105/10892 (L 18.6, W 14.1, H 2.9); ventral valve (Mickwitz 1896, Pl. 3:10); Leetse beds; coastal section at Leetse, close to Paldiski, northern Estonia.

*Material.* – Figured from South Urals. Complete shells: RM Br 135951 (L 22.7, W 22.4, T 8.3); RM Br 135952 (L 22.7, W 22.9, T 8.2). Ventral valves: RM Br 135955; RM Br 135958; RM Br 135959 (L 17.5, W 13.95, WI 8.1, LV 9.05, WV 7.8). Dorsal valves: RM Br 135953; RM Br 135954; RM Br 135956; RM Br 135957 (L 17.65, W 17.3, WI 11, LV 13.6, WV 10.6). Total 2 complete shells, 191 ventral valves, and 143 dorsal valves.

Figured from northern Estonia. Ventral valve: GT Br3505 (L 17.2, W 13.2, LI 3.7, WI 7, WG 1.5, ML 2.3, LV 10). Dorsal valves: RM Br 136258 (L 18.3, W 15); GT Br3506 (L 17.4, W 16, LI 3.2, WG 4.8, ML 0.8, LV 15.6).

*Measurements.* – See Tables 5–8.

*Diagnosis.* – Shell subequally biconvex, about one-third as thick as long; transversely oval to elongate subtriangular, with maximum width anterior to mid-length. Ventral pseudointerarea triangular; pseudointerareas of both valves occupying somewhat more than half of valve width. Ventral visceral area forming subtriangular platform extending to mid-length. Dorsal visceral area thickened, long, extending for about ⅔–⅘ of valve length; dorsal central and anterior lateral muscle scars placed far apart.

*Description of South Ural material.* – The shell is evenly and subequally biconvex, slightly inequivalved, 36% as thick as long (*N* 2) (Fig. 48B, E). The outline of the shell is strongly variable, with adults being elongate, subtriangular, with maximum width at about one-third of the maximum shell length from the anterior margin. The ventral valve is slightly longer than the dorsal, on average 1.4 times as long as wide (*N* 58; Table 5). The ventral pseudointerarea is not elevated above the valve floor; it occupies about 58% of the maximum valve width in one valve (Fig. 48M). The pedicle groove is deep and narrowly triangular. The propareas are slightly elevated and bisected by widely divergent flexure lines (Fig. 48L, M). The ventral visceral platform is subtriangular, slightly raised anteriorly, occupying about 52% of the valve length in one specimen (Fig. 48M). The *vascula lateralia* are marginal, arcuate, with the proximal parts slightly divergent anteriorly.

The dorsal valve is on average 1.4 times as long as wide (*N* 62; Table 6). The dorsal pseudointerarea is crescent-shaped, forming a strip about 64% of the total valve width in one specimen (Fig. 48K); it is not raised above the inner surface of the valve. The median groove is wide and poorly defined, and the propareas are reduced. The dorsal visceral area is slightly thickened, with elongate median tongue extending for about 62% of the valve length in one specimen (Fig. 48K). The dorsal central muscle scars are thickened, elongate oval in outline, slightly divergent, and extending anteriorly to the centre of the valve. The anterior lateral muscle scars are small, lanceolate, and bisected by the median ridge. The posterolateral muscle fields are elongated and thickened, forming widely divergent ridges along the posterolateral margins of the valve interior. The dorsal *vascula lateralia* are submarginal and subparallel, slightly arcuate; the *vascula media* are short and divergent.

*Discussion.* – The new material of *L. lingulaeformis*, and numerous topotype specimens from northern Estonia (Fig. 49; Tables 7–8), show a wide range of morphological variation; among other features, the shell outline in both regions varies from transversely oval to subtriangular; *L. recta* was

*Fig. 48.* *Leptembolon lingulaeformis* (Mickwitz); South Urals. □A. Ventral view of complete shell; Akbulaksai Formation; Alimbet Farm (sample B-578-4); RM Br 135951; ×3. □B. Lateral view of A; ×3. □C. Dorsal view of A; ×3. □D. Dorsal view of complete shell; Akbulaksai Formation; Alimbet Farm (sample B-578-4); RM Br 135952; ×3. □E. Lateral view of D; ×3. □F. ventral view of D; ×3. □G. Dorsal exterior; Akbulaksai Formation; Akbulaksai River (sample B-607-1); RM Br 135953; ×2. □H. Dorsal exterior; Akbulaksai Formation; Akbulaksai River (sample B-607-1); RM Br 135954; ×2. □I. Ventral exterior; Akbulaksai Formation; Akbulaksai River (sample B-607-1); RM Br 135955; ×5. □J. Dorsal interior; Akbulaksai Formation; Akbulaksai River (sample B-607-1); RM Br 135956; ×2. □K. Dorsal internal mould; Kidryas Formation; Tyrmantau Ridge (sample G-141-1); RM Br 135957; ×3. □L. Ventral exterior; Akbulaksai Formation; Akbulaksai River (sample B-607-1); RM Br 135958; ×3. □M. Ventral internal mould; Kidryas Formation; Tyrmantau Ridge (sample G-141-1); RM Br 135959; ×3.

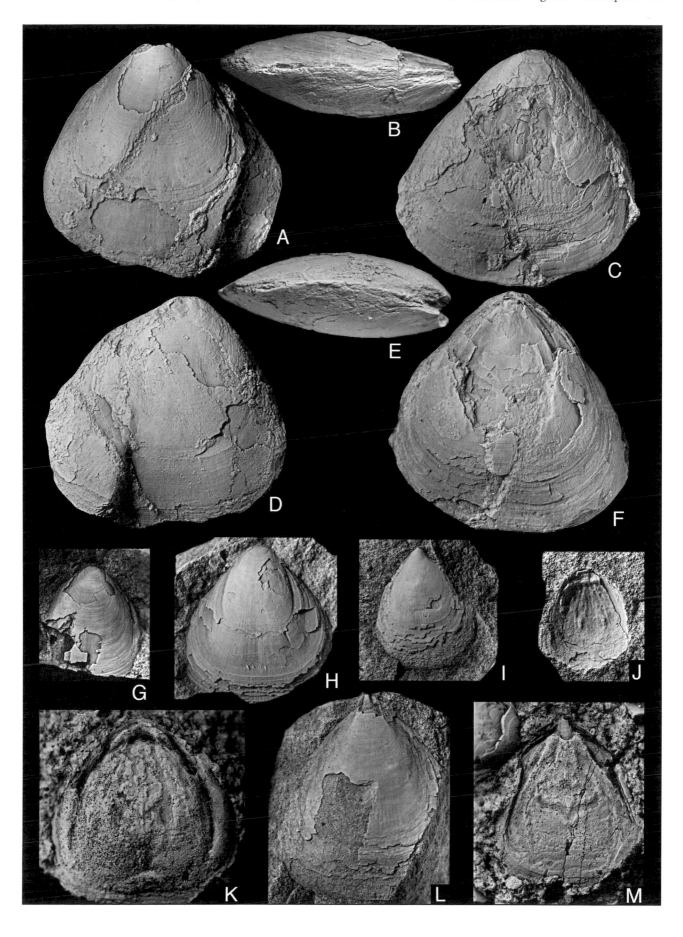

established by Gorjansky (1969, p. 39), as a separate species, based almost entirely on the 'more elongate shell', and it is here considered to be a junior synonym of *L. lingulaeformis*.

It is difficult to compare *L. lingulaeformis* with *L.? sanctacrucensis* (Bednarczyk, 1964) and *L.? zejszneri* (Bednarczyk, 1964) in view of the poor illustrations of the latter two species; they would appear to differ mainly in having more strongly thickened visceral fields, but their exact relationship to the type species is unclear. *L. insons insons* (Barrande, 1879) and *L. insons testis* (Barrande, 1879) from the Bohemian Lower Ordovician Mílina and Klabava formations, re-

spectively (Havlíček 1982), are closely similar to *L. lingulaeformis* when comparing specimens of equal size; however, gerontic specimens of the latter species are significantly larger and generally have a subtriangular outline. Other distinctive features of *L. lingulaeformis* include the strongly thickened dorsal central muscle scars that sometimes form wide, narrowly divergent ridges on the visceral field.

The specimens referred to *Obolus razumovskii* Lermontova by Nassedkina (*in* Ancygin *et al.* 1977) are more or less identical with *L. lingulaeformis*.

*Occurrence.* – The distribution of the *Thysanotus–Leptembolon* assemblage is discussed above (p. 32). The species is common in the lower Leetse beds throughout northern Estonia. It is also one of the most abundant lingulate species in the Akbulaksai Formation, where it occurs at the following localities: Alimbet Farm (Fig. 13; Appendix 1B), Alimbet (Fig. 16; Appendix 1C), Akbulaksai (Fig. 18; Appendix 1D), Karabutak (Appendix 1E), and Koagash rivers (Appendix 1F); *L. lingulaeformis* also occurs in the Kidryas Formation at the Tyrmantau Ridge (Fig. 11; Appendix 1A).

*Table 5. Leptembolon lingulaeformis*, average dimensions and ratios of ventral valves.

| | L | W | L/W |
|---|---|---|---|
| Samples B-578-4, B-603-1, B605, B607-1, B607-8 | | | |
| N | 58 | 58 | 58 |
| X | 10.31 | 7.63 | 141% |
| S | 3.97 | 3.87 | 0.19 |
| MIN | 3.40 | 2.40 | 104% |
| MAX | 22.60 | 20.30 | 181% |

*Table 6. Leptembolon lingulaeformis*, average dimensions and ratios of dorsal valves.

| | L | W | L/W |
|---|---|---|---|
| Samples B-578-4, B-603-1, B605, B607-1, B607-8 | | | |
| N | 62 | 62 | 62 |
| X | 8.71 | 6.42 | 141% |
| S | 2.98 | 2.91 | 0.17 |
| MIN | 4.20 | 2.80 | 104% |
| MAX | 18.80 | 16.90 | 195% |

*Table 7. Leptembolon lingulaeformis*, average dimensions and ratios of ventral valves.

| | L | W | H | L/W | H/L |
|---|---|---|---|---|---|
| northern estonia | | | | | |
| N | 10 | 10 | 10 | 10 | 10 |
| X | 17.19 | 13.28 | 2.21 | 133% | 13% |
| S | 2.50 | 2.99 | 0.75 | 0.16 | 0.05 |
| MIN | 11.00 | 7.00 | 1.20 | 113% | 6% |
| MAX | 20.00 | 16.00 | 3.60 | 160% | 20% |

*Table 8. Leptembolon lingulaeformis*, average dimensions and ratios of dorsal valves.

| | L | W | H | L/W | H/L |
|---|---|---|---|---|---|
| Northern Estonia | | | | | |
| N | 7 | 7 | 7 | 7 | 7 |
| X | 16.94 | 13.26 | 2.44 | 128% | 14% |
| S | 0.69 | 1.19 | 0.92 | 0.09 | 0.05 |
| MIN | 16.00 | 11.00 | 1.00 | 118% | 6% |
| MAX | 17.70 | 14.70 | 3.90 | 145% | 22% |

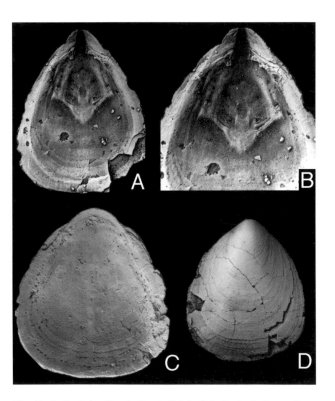

*Fig. 49. Leptembolon lingulaeformis* (Mickwitz); Leetse beds; northern Estonia. □A. Ventral interior; Iru; GT Br 3505; ×2.5. □B. Detail of A, showing psedointerarea and visceral field; ×3.5. □C. Dorsal interior; Leetse; GT Br 3506; ×2.5. □D. Dorsal exterior; Maardu; RM Br 136258; ×2.

## Genus *Agalatassia* gen. nov.

*Name.* – After the Agalatas River, close to the type locality.

*Type and only species.* – *Agalatassia triangularis* gen. et sp. nov.; Agalatas Formation; southern Kendyktas Range.

*Diagnosis.* – Shell elongate triangular, ornamented by fine rugellae, forming zig-zag pattern. Ventral pseudointerarea small, triangular, with shallow, broadly triangular pedicle groove, and reduced propareas. Ventral visceral area subtriangular, slightly thickened anteriorly, extending to mid-length; ventral *vascula lateralia* submarginal. Dorsal pseudointerarea poorly developed, not raised above valve floor. Dorsal umbonal scar undivided, placed on platform; dorsal visceral area with narrow median tongue, extending some-what anterior to mid-length; dorsal *vascula lateralia* submarginal and long; *vascula media* long.

*Occurrence.* – At the type locality only.

## *Agalatassia triangularis* sp. nov.
Figs. 50A–J

*Name.* – Latin *triangularis*, alluding to the triangular outline of the shell.

*Holotype.* – Fig. 50I; RM Br 135982; dorsal valve (L 7.5, W 7, LV 5, WV 4.2); Agalatas Formation (sample 564); Keskentas Ridge.

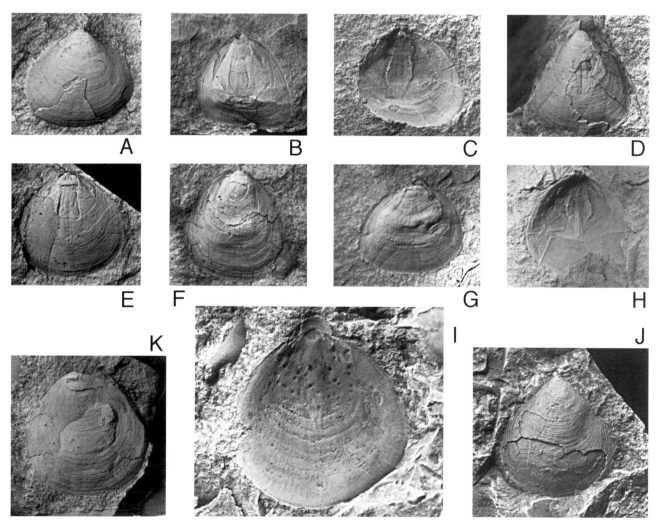

*Fig. 50.* □A–J. *Agalatassia triangularis* sp. nov.; Kendyktas Range, Kazakhstan. □A. Ventral exterior; Kurdai Formation (sample 559); RM Br 135978; ×3. □B. Dorsal internal mould; Kurdai Formation (sample 559); RM Br 135975; ×3. □C. Dorsal interior; Kurdai Formation (sample 559); RM Br 135976; ×3. □D. dorsal exterior; Kurdai Formation (sample 559); RM Br 135977; ×3. □E. Dorsal internal mould; Kurdai Formation (sample 559); RM Br 135979; ×3. □F. Ventral exterior; Kurdai Formation (sample 559); RM Br 135980; ×3. □G. Ventral internal mould; Kurdai Formation (sample 559); RM Br 135974; ×3. □H. Dorsal interior; Kurdai Formation (sample 559); RM Br 135973; ×3. □I. Holotype; dorsal interior; Agalatas Formation (sample 564); RM Br 135982; ×5. □J. Ventral exterior, showing ornamentation; Agalatas Formation (sample 564); RM Br 135981; ×5. □K. *Westonia aurora* (Hall); latex cast of dorsal exterior; Upper Cambrian Lodi Shale Member, Trempealeau Formation; Praire du Sac, Wisconsin; USNM; ×2.

*Paratypes.* – Figured. Ventral valves: RM Br 135974 (L 8.5, W 8.35, LV 3.25, WV 4.6); RM Br 135978; RM Br 135980; RM Br 135981 (L 4.9, W 5.25). Dorsal valves: RM Br 135973 (L 8.3, W 7.8, LV 5.2, WV 3.7, LM1 4.6, WM1 1.8); RM Br 135975 (L 8.65, W 8.65, LV 5.6, WV 4.1, LM1 4.6, WM1 2.1); RM Br 135976; RM Br 135977 (L 9.1, W 8.3, LV 5.7, LM1 4.5), RM Br 135979. Total of 34 ventral and 24 dorsal valves.

*Measurements.* – See Tables 9–10.

*Diagnosis.* – As for genus.

*Description.* – The shell is elongate triangular and flattened. The ventral valve is about as long as wide (*N* 10; Table 9). The ventral pseudointerarea is low and triangular, with an extremely wide, shallow pedicle groove. The ventral visceral field is subtriangular and distinctly raised anteriorly (Fig. 50G), extending on average 42% of the valve length (*N* 9) and occupying 54% of the valve width (Table 9). The ventral *vascula lateralia* are submarginal.

The dorsal valve is almost as long as wide (*N* 9; Table 10) and gently convex, with the maximum height somewhat anterior to the beak. The dorsal pseudointerarea is reduced to form a low undifferentiated strip (Fig. 50I). The dorsal visceral area extends on average 63% of the valve length (*N* 9; Table 10). The dorsal umbonal muscle scar is undivided and placed on a thickened platform (Fig. 50I). The dorsal central muscle scars are elongate oval, bounded laterally by two low, slightly divergent ridges, originating anterior to the umbonal muscle scar. The dorsal posterolateral muscle fields form slightly thickened, lanceolate, divergent ridges (Fig. 50B–E, H). The dorsal *vascula lateralia* are long and submarginal, and the *vascula media* are long and divergent. The postlarval shell is ornamented by fine, closely spaced concentric rugellae, forming a zig-zag pattern (Fig. 50J).

*Table 9. Agalatassia triangularis, average dimensions and ratios of ventral valves.*

|  | L | W | LV | WV | L/W | VI/L | WV/W |
|---|---|---|---|---|---|---|---|
| **Sample 564** | | | | | | | |
| N | 10 | 10 | 9 | 5 | 10 | 9 | 5 |
| X | 8.45 | 8.52 | 3.48 | 4.61 | 100% | 42% | 54% |
| S | 0.43 | 0.70 | 0.40 | 0.49 | 0.09 | 0.05 | 0.02 |
| MIN | 7.84 | 7.28 | 2.88 | 4.08 | 85% | 34% | 53% |
| MAX | 9.12 | 9.44 | 4.32 | 5.04 | 115% | 53% | 56% |

*Table 10. Agalatassia triangularis, average dimensions and ratios of dorsal valves.*

|  | L | W | LV | WV | L/W | LV/L | WV/W |
|---|---|---|---|---|---|---|---|
| **Sample 559** | | | | | | | |
| N | 9 | 9 | 9 | 7 | 9 | 9 | 7 |
| X | 8.22 | 8.06 | 5.20 | 3.76 | 102% | 63% | 48% |
| S | 0.63 | 0.52 | 0.92 | 0.31 | 0.06 | 0.11 | 0.03 |
| MIN | 7.20 | 7.52 | 3.84 | 3.20 | 93% | 53% | 43% |
| MAX | 9.12 | 8.96 | 6.76 | 4.16 | 110% | 87% | 51% |

*Discussion.* – *A. triangularis* is similar to species of *Westonisca* Havlíček and *Westonia* Walcott in the shape of the shell and in having a zig-zag ornamentation like that of, e.g., *Westonia aurora* (see, e.g., Walcott 1912, Pl. 46:1g–h; Fig. 50K herein). This type of ornamentation was interpreted as burrowing sculptures by Savazzi (1986). *A. triangularis* differs from species of *Westonia* mainly in having reduced pseudointerareas lacking flexure lines; it differs from *Westonisca* in having (1) a raised ventral visceral area, (2) submarginal ventral *vascula lateralia*, (3) a dorsal visceral area extending somewhat anterior to mid-length, (4) dorsal anterior lateral muscle scars situated close to those of the central muscles, (5) an undivided dorsal umbonal muscle scar situated on an elevated platform, and (6) long divergent *vascula media*.

*Occurrence.* – At the type locality only (Fig. 35; Appendix 1J).

## Genus *Palaeoglossa* Cockerell, 1911

*Type species.* – Original designation by Cockerell (1911, p. 96); *Lingula attenuata* Sowerby, 1839; Rorrington beds (Llandeilo–Costonian); Rorrington, Shelve, Shropshire.

*Diagnosis.* – See Williams (1974, p. 31).

*Discussion.* – This poorly known genus was reviewed most recently by Williams (1974). It is possibly a junior synonym of *Lingulella*, but this question needs further investigations.

## *Palaeoglossa? razumovskii* (Lermontova, 1933)

Fig. 51[2]

*Synonymy.* – □1933 *Obolus razumovskii* sp. nov. – Lermontova *in* Lermontova & Razumovskij, p. 205, Pl. 2:14–18.

*Lectotype.* – Selected here: CNIGR 17/6565; ventral internal mould (L 4.75, W 3.75, LI 1.25, WI 2.63, LV 2.38, WV 2.75); illustrated by Lermontova (1933, Pl. 2:18); Kidryas Formation (sample 2594 *in* Lermontova & Razumovskij 1933); Tyrmantau Ridge.

*Material.* – Figured. Ventral valves: CNIGR 50/6565 (L 5, W 3.75); RM Br 135970. Dorsal valves: CNIGR 56/6565 (L 5, LV 3.38); CNIGR 57/6565 (L 5.25, W 4); RM Br 135969; RM Br 135971. Total of 7 ventral and 4 dorsal valves.

*Diagnosis.* – Valves elongate suboval, about 1.5 times as long as wide. Ventral pseudointerarea occupying more than half of valve width. Ventral visceral area thickened, extending anterior to mid-length. Dorsal visceral area extending more than half of valve length, bisected by fine median ridge.

*Description.* – The shell is subequally biconvex, slightly inequivalved, and elongate suboval with subparallel lateral margins. The ventral valve is gently convex, about 1.3–1.4 as

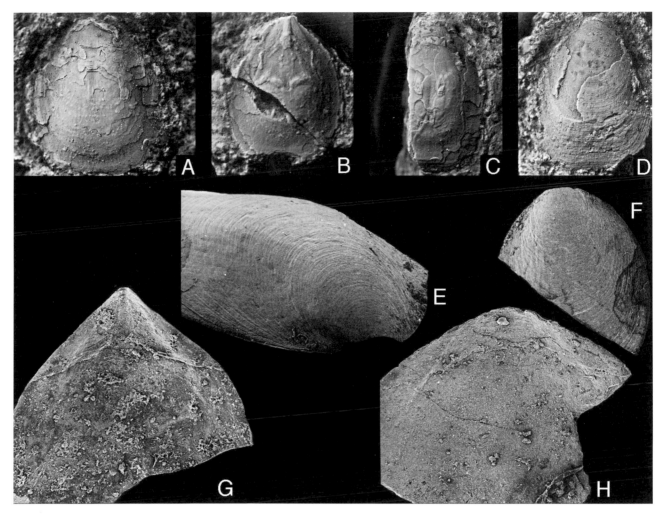

*Fig. 51. Paleoglossa? razumovskii* (Lermontova); Kidryas Formation; Tyrmantau Ridge, South Urals. □A. Dorsal exterior (sample 2594); CNIGR 57/6565; ×8. □B. Lectotype; ventral internal mould (sample 2594); CNIGR 17/6565; ×8. □C. Dorsal internal mould (sample 2594); CNIGR 56/6565; ×8. □D. Ventral exterior (sample 2594); CNIGR 50/6565 ; ×8. □E. Posterior view of dorsal exterior (sample B-768-1); RM Br 135969; ×21. □F. Dorsal exterior of E; ×10. □G.Ventral interior (sample B-768-1); RM Br 135970; ×32. □H. Dorsal interior (sample B-768-1); RM Br 135971; ×27.

long as wide. The ventral pseudointerarea occupies about 25% of the total valve length, and somewhat more than 60% of the maximum valve width; it is not elevated above the valve floor. The pedicle groove is shallow and narrow, with subparallel lateral margins; the propareas have well-defined flexure lines (Fig. 51G). The ventral visceral field is subtriangular, extending to centre of the valve, and bisected by two fine, subparallel furrows. The mantle canals are baculate, with arcuate, submarginal *vascula lateralia* (Fig. 51B).

The dorsal valve is gently convex, about 1.3–1.3 times as long as wide. The dorsal pseudointerarea forms a wide, concave, undivided rim that is not elevated above the valve floor (Fig. 51H). The dorsal visceral field is weakly impressed, but has a median tongue, extending about three-fifths of the valve length. The dorsal central muscle scars are small, elongate oval in outline, situated at about mid-length.

A fine median ridge bisects the anterior part of the dorsal visceral area. The dorsal anterior lateral muscle scars are very small, elongate oval (Fig. 51C). The dorsal *vascula lateralia* are submarginal and subparallel; the *vascula media* are widely divergent.

*Discussion.* – This species is poorly known; more material is needed in order to understand its systematic position. In the size and shape of the shell, *P.? razumovskii* is closely comparable with *P. myttonensis* (Williams, 1974, p. 33, Pl. 4:1–5). The species from the South Urals differs in the proportionally longer ventral pseudointerarea and more strongly impressed visceral fields, and in having a dorsal median ridge that is well defined only in the posterior part of the visceral area.

*Occurrence.* – At the type locality only (Fig. 11; Appendix 1A).

# Genus *Ralfia* Popov & Khazanovitch, 1989

*Type species.* – Original designation by Popov & Khazanovitch 1989 (*in* Popov et al.), p. 126; *Ungula ovata* Pander, 1830; Upper Cambrian Ladoga Formation; Izhora River, Ingria.

*Diagnosis.* – Shell triangular, rounded. Ventral pseudointerarea small, rhomboidal, with narrow, deep pedicle groove and subparallel flexure lines. Ventral visceral area slightly thickened, extending to mid-length, bisected by two submedian furrows. Ventral *vascula lateralia* subparallel, submarginal. Dorsal pseudointerarea reduced, rhomboidal. Dorsal visceral area with long median tongue; *vascula media* short, divergent.

*Species included.* – *Ungula ovata* Pander, 1830; ?*Lingula*? *bryograptorum* Moberg & Segerberg, 1906.

*Occurrence.* – Upper Cambrian –?Lower Ordovician (Tremadoc); Ingria, Sweden.

## *Ralfia*? *bryograptorum* (Moberg & Segerberg, 1906)
Fig. 45P–S

*Synonymy.* – ☐1906 *Lingula*? *bryograptorum* sp. nov. – Moberg & Segerberg, p. 63, Pl. 1:22.

*Holotype.* – By monotypy; LO 1771T; partial internal mould of dorsal valve; the original of Moberg & Segerberg (1906, Pl. 1:22); Dictyonema Shale; Fågelsång, Scania.

*Paratypes.* – Figured. Ventral valves: RM Br 1804b; RM Br 1804f; RM 1804c. Total of 3 ventral and 1 dorsal valve.

*Description.* – The shell is close to equivalved and rounded, strongly flattened and equi-biconvex. The ventral pseudointerarea is extremely small and narrow, with a narrow and deep pedicle groove (Fig. 45Q). The ventral visceral field is short, extending only to about 30% of the total valve length in one specimen (Fig. 45Q), and bisected by two strongly developed and slightly divergent median ridges. The ventral *vascula lateralia* are divergent and straight (Fig. 45Q, R). The dorsal pseudointerarea is poorly visible in the holotype, but apparently reduced (Fig. 45P). The dorsal interior is not known.

*Discussion.* – This species is comparable with *R. ovata* in the shape of the shell and the development of the ventral pseudointerarea as well as interior characters. The poorly preserved, rare material does not permit a closer comparison. Walcott (1912, p. 514) considered the species to be a junior synonym of *Lingulella lepis*, but it is clearly different from *Lingulella* in having reduced pseudointerareas.

*Occurrence.* – At the type locality only.

# Subfamily Elliptoglossinae subfam. nov.

*Diagnosis.* – Shell subequally biconvex, elongate oval. Pseudointerareas absent in both valves, or reduced, forming short, undivided, crescent-shaped strip, bounding posterior margin. Visceral area of both valves large, extending anterior to mid-length; distinct limbus in both valves.

*Discussion.* – *Elliptoglossa* and *Lingulops* have usually been included in the Family Paterulidae; however, they differ from paterulids in having (1) a smooth, non-pitted larval and postlarval shell, (2) rudimentary pseudointerareas in one or both valves, and (3) no pedicle notch. These distinctive features suggest that the elliptoglossines are more closely related to the Family Obolidae than to the Paterulidae. At the same time, they differ from the obolids in having an equivalved shell with extremely reduced pseudointerareas, and a large ventral visceral field that extends anterior to the mid-length. These features, together with the extremely small size and the poorly developed or absent pedicle opening (which may be due to the complete reduction of the pedicle), are characteristics of obolid larvae and may suggest a paedomorphic origin of the Elliptoglossinae.

*Genera assigned.* – *Elliptoglossa* Cooper, 1956; *Lingulops* Hall, 1871.

*Occurrence.* – Lower Ordovician – Silurian.

## Genus *Elliptoglossa* Cooper, 1956

*Type species.* – Original designation by Cooper (1956, p. 241); *Leptobolus*? *ovalis* Bassler, 1919; Martinsburg Shale; Pennsylvania.

*Diagnosis.* – Visceral platform lacking.

*Discussion.* – *Elliptoglossa* is very similar to *Lingulops*, and the main difference seems to be that the latter has a well-developed, thickened visceral area ('platform') in both valves. Krause & Rowell (1975) and Percival (1978) provided the most recent information on the genus. As noted by Krause & Rowell (1975) it is sometimes difficult to tell the dorsal and ventral valves apart, especially when the muscle scars are poorly developed. In general, the ventral valve seems to be less convex and more pointed as compared with the dorsal.

*Occurrence.* – Ordovician; North America, Baltoscandia, Bohemia, South Urals, Kazakhstan, Australia.

## *Elliptoglossa linguae* (Westergård, 1909)
Figs. 52–53

*Synonymy.* – ☐1909 *Lingula*? *linguae* sp. nov. – Westergård, p. 77, Pl. 5:26. ☐1982 '*Lingula*' cf. *linguae* Westergård – Rushton & Bassett *in* Owens et al., p. 19, Pl. 5a–g. ☐?1986 *Paterula*? *delicata* – Bednarczyk, p. 412, Pl. 6:1–3.

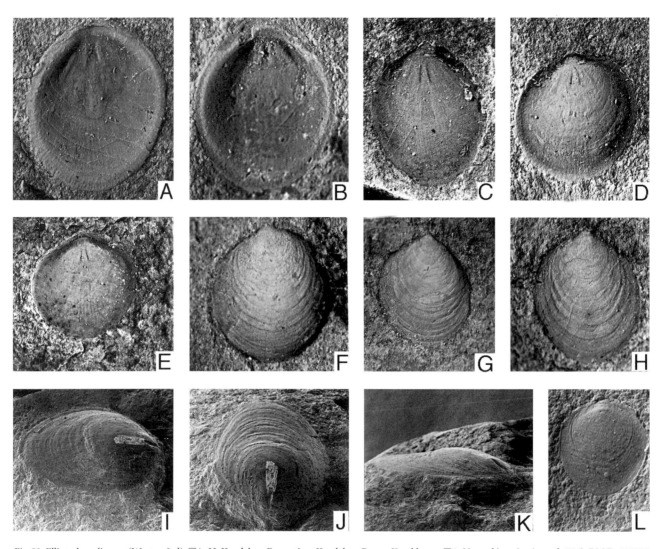

*Fig. 52. Elliptoglossa linguae* (Westergård). □A–H. Kendyktas Formation; Kendyktas Range, Kazakhstan. □A. Ventral interior (sample 554); RM Br 135983; ×5. □B. Ventral interior (sample 554); RM Br 135984; ×15. □C. Ventral internal mould (sample 553); RM Br 135985; ×10. □D. Ventral internal mould (sample 554); RM Br 135986; ×10. □E. Ventral internal mould (sample 554); RM Br 135987; ×10. □F. Ventral exterior (sample 554); RM Br 135988; ×10. □G. Ventral exterior (sample 554); RM Br 135989; ×10. □H. Ventral exterior (sample 554); RM Br 135990, ×10. □I. Ventral partial internal mould; holotype by monotypy; figured by Westergård (1909, Pl. 5:26); Bjørkåsholmen Limestone; Jerrestad, Scania; LO 2303T; ×11. □J. Posterior view of I; ×11. □K. Lateral view of I; ×11. □L. Dorsal internal mould; figured by Owens *et al.* (1982, Pl. 5e); *Clonograptus tenellus* Biozone; Dyfed, Carmarthen district, South Wales; NMW 77.1G.53; ×13.

*Holotype.* – By monotypy; LO 2303T; ventral valve illustrated by Westergård (1909, Pl. 5:26; Fig. 52I–K herein); Bjørkåsholmen Limestone; Jerrestad, Scania.

*Material.* – Figured from southern Kendyktas Range. Ventral valves: RM Br 135983 (L 2.8, W 2.4, LI 0.12, WI 1.4, LV 1.68, WV 1.24); RM Br 135984; RM Br 135985 (L 2.56, W 2.04, LI 0.56, WI 1.06, WG 0.24); RM Br 135986 (L 2.24, W 2.16); RM Br 135987 (L 2.88, W 2.72); RM Br 135988; RM Br 139989 (L 2.88, W 2.4); RM Br 135990 (L 2.26, W 1.88). Total of 3 complete shells and about 28 valves, most of which seem to be ventral.

　　Figured from Scandinavia. Dorsal valves: LO 6509t; LO 6510t; LO 6511t; LO 6512t; LO 6513t; LO 6514t. Total of 6 dorsal valves.

Figured from Britain. Dorsal valve: NMW 77.1G.53.

*Measurements.* – See Tables 11–12.

*Diagnosis.* – Shell elongate oval to subcircular, about 1.2 times as long as wide. Pseudointerareas of both valves vestigial. Ventral interior with two subparallel ridges bisecting visceral area.

*Description.* – The shell is elongate oval and close to equibiconvex. The ventral valve is on average 1.2 times as long as wide (*N* 11; Table 11), gently and evenly convex in lateral profile. The ventral pseudointerarea forms a very short crescent-shaped strip bounding the posterior margin; the pedicle groove is poorly defined (Fig. 52A). The ventral posterior margin is somewhat acute at the umbo. The ventral

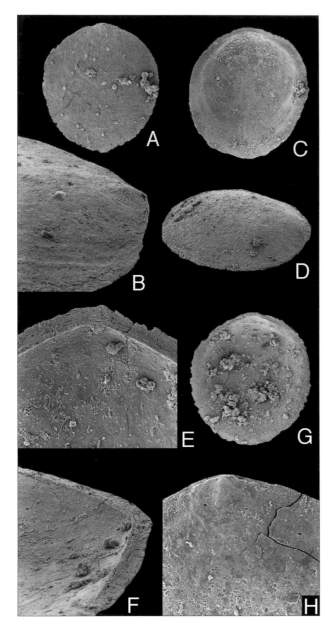

*Fig. 53.* *Elliptoglossa linguae* (Westergård); Bjørkåsholmen Limestone (sample Sk-3); Fågelsång, Scania. □A. Dorsal exterior; LO 6509t; ×28. □B. Lateral view of apex of A; ×100. □C. Dorsal interior; LO 6510t; ×28. □D. Lateral view of dorsal exterior; LO 6511t; ×28. □E. Detail of dorsal pseudointerarea; LO 6512t; ×90. □F. Lateral view of E; ×90. □G. Dorsal interior; LO 6513t; ×28. □H. Detail of dorsal apex; LO 6514t; ×100.

visceral field is large, but poorly defined, extending somewhat anterior to the centre of the valve. The ventral umbonal muscle scars are situated on low, short, slightly divergent ridges, and two fine ridges, bisect the full length of the visceral field (Fig. 52A–E, I, J).

The dorsal valve is on average 1.1 times as long as wide (*N* 8; Table 12), with an evenly curved posterior margin. The dorsal pseudointerarea is vestigial, undivided, and not raised above the valve floor (Fig. 53E, F). A weakly defined median ridge bisects the dorsal visceral field.

*Discussion.* – There are only a few characters that can be used for definition of species of *Elliptoglossa*. The most distinctive character of *E. linguae* is the presence of reduced, but well-defined, pseudointerareas in both valves. In *E. ovalis* (Bassler) and *E. sylvanica* Cooper, the ventral pseudointerarea is reduced to a narrow, undivided plate near the umbo, and dorsal pseudointerareas are possibly absent (Cooper 1956).

There are no important differences in morphology between the examined material from Scania (Figs. 52I–K, 53), Kazakhstan (Fig. 52A–H), and Britain (Fig. 52L).

*Occurrence.* – *E. linguae* is one of the characteristic species in the *Broeggeria* assemblage (see above, p. 31). It is relatively common in the Kendyktas Formation (Fig. 35; Appendix 1J). In Scandinavia, it is a rare species that is restricted to the Bjørkåsholmen Limestone in Scania (Jerrestad and Fågelsång; Appendix 1K). In Great Britain, it occurs in the Tremadoc of Carmarthen, South Wales (Rushton & Bassett *in* Owens *et al.* 1982).

# Family Elkaniidae Walcott & Schuchert, 1908

*Diagnosis.* – Shell dorsi-biconvex or subequally biconvex, thickened posteriorly. Post-larval shell with finely pitted micro-ornament. Ventral proction as wide, with well-defined flexure lines. Dorsal pseudointerarea forming wide, crescent-shaped strip along posterior margin; dorsal median groove wide, slightly concave. Visceral fields of both valves strongly thickened, forming elevated platforms; median

*Table 11. Elliptoglossa linguae*, average dimensions and ratios of ventral valves.

|  | L | W | LI | WI | LV | WV | $L/W$ | $LI/WI$ | $WI/W$ | $LV/L$ | WV/W |
|---|---|---|---|---|---|---|---|---|---|---|---|
| Samples 551, 553, 554 | | | | | | | | | | | |
| N | 11 | 11 | 2 | 2 | 2 | 2 | 11 | 2 | 2 | 2 | 2 |
| X | 2.64 | 2.22 | 0.14 | 1.41 | 1.60 | 1.14 | 119% | 10% | 65% | 59% | 52% |
| S | 0.26 | 0.21 | 0.03 | 0.07 | 0.11 | 0.14 | 0.11 | 0.03 | 0.06 | 0.01 | 0.01 |
| MIN | 2.24 | 1.92 | 0.12 | 1.36 | 1.52 | 1.04 | 97% | 8% | 61% | 58% | 52% |
| MAX | 3.04 | 2.64 | 0.16 | 1.46 | 1.68 | 1.24 | 132% | 12% | 69% | 60% | 53% |

*Table 12. Elliptoglossa linguae*, average dimensions and ratios of dorsal valves.

|  | L | W | $L/W$ |
|---|---|---|---|
| Samples 553, 554 | | | |
| N | 8 | 8 | 8 |
| X | 2.55 | 2.32 | 111% |
| S | 0.74 | 0.73 | 0.06 |
| MIN | 1.46 | 1.36 | 103% |
| MAX | 3.84 | 3.56 | 122% |

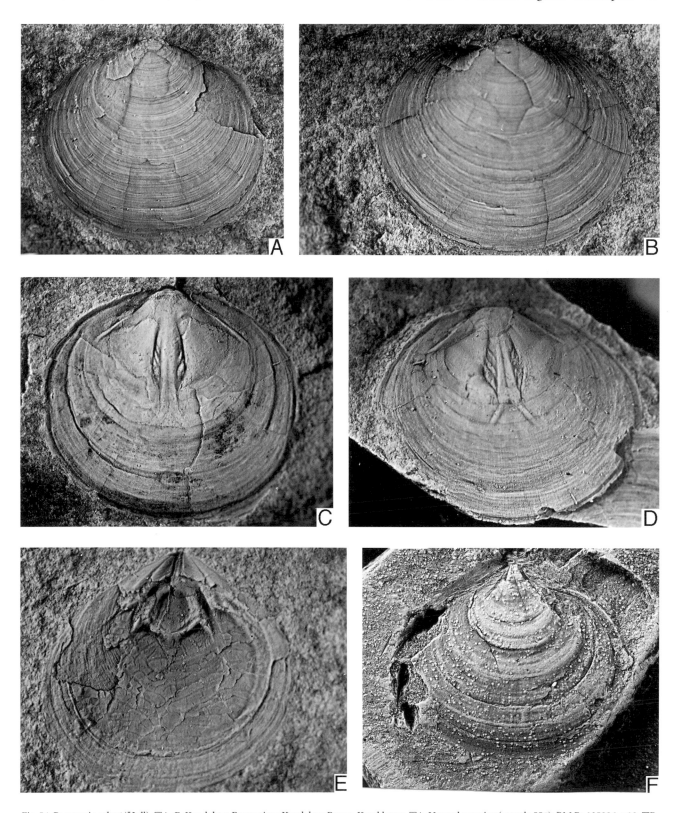

*Fig. 54. Broeggeria salteri* (Holl). □A–E. Kendyktas Formation; Kendyktas Range, Kazakhstan. □A. Ventral exterior (sample 556); RM Br 135996; ×10. □B. Dorsal exterior (sample 556); RM Br 135997; ×10. □C. Dorsal internal mould (sample 556); RM Br 135998; ×10. □D. Dorsal internal mould (sample 556); RM Br 135999; ×10. □E. Ventral interior (sample 551); RM Br 136000; ×10. □F. Ventral internal mould; figured by Owens *et al.* (1982, Pl. 7i); Merioneth Series, White-Leaved-Oak Shales; Malvern Hills, South Wales; IGS GSM85075; ×8.

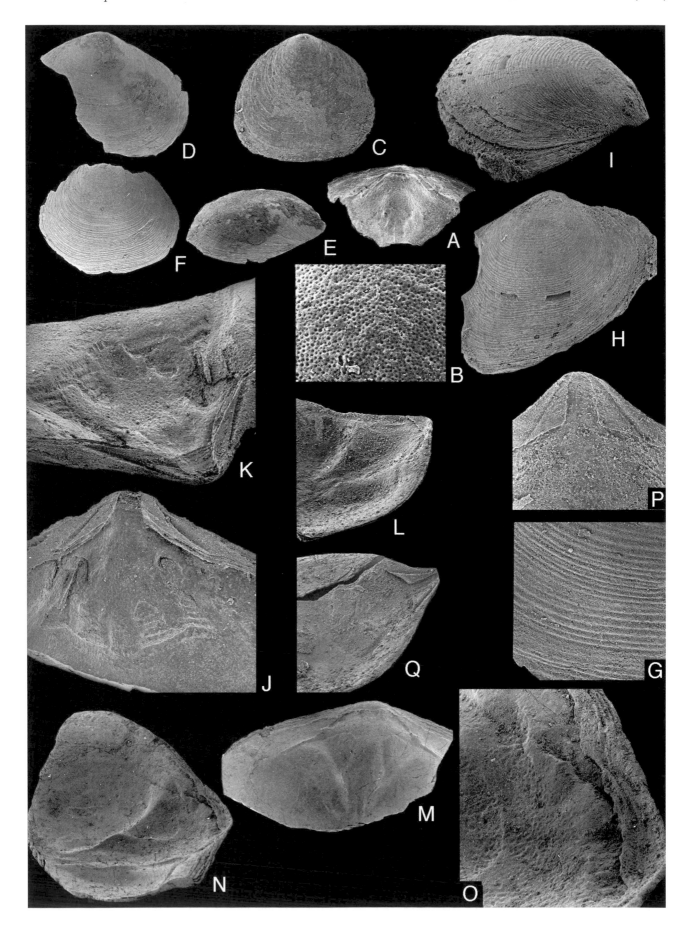

tongue of dorsal visceral field not extending anterior to mid-length; *vascula lateralia* of both valves widely divergent in posterior half, arcuate; dorsal *vascula media* long, widely divergent.

*Genera included.* – See lists of Holmer (1991, 1993) in addition to *Keskentassia* gen. nov.

*Discussion.* – See Holmer (1991, 1993) for brief reviews and discussions of the main characters of the family.

*Occurrence.* – Upper Cambrian – Upper Ordovician (Ashgill).

## Genus *Broeggeria* Walcott, 1902

*Type and only species.* – Original designation by Walcott (1902, p. 605); *Obolella salteri* Holl, 1885; Upper Cambrian (Merioneth Series); Malvern Hills, Herefordshire.

*Diagnosis.* – Shell rounded to subtriangular, moderately biconvex. Post-larval micro-ornament of irregularly distributed circular pits. Ventral pseudointerarea broadly triangular, with widely triangular pedicle groove. Dorsal pseudointerarea, crescent-shaped, with wide median groove. Visceral areas of both valves moderately thickened.

*Discussion.* – As noted by Rushton & Bassett (*in* Owens *et al.* 1982, p. 23), *Broeggeria* resembles *Elkania* in outline and ornamentation, but differs mainly in being less biconvex; one complete shell of *Elkania hamburgensis* from Nevada (USNM 17286b; L 6.1, W 6.1, T 4.2; Rowell 1965, Fig. 164:1d–f) is about 69% as thick as wide, while the maximum convexity of *Broeggeria* is about 20%. As noted by Holmer (1991, 1993), there seems to be a general tendency to form strongly biconvex shells within the family. This tendency is particularly well illustrated by *Lamanskya* Moberg & Segerberg, and *Volborthia* von Möller that was referred to the Elkaniidae by Holmer (1991, 1993).

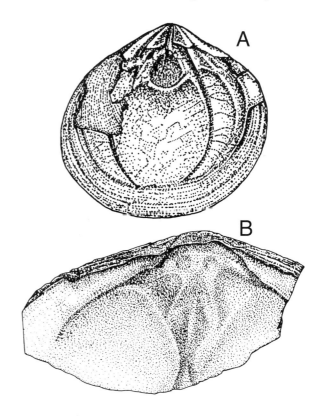

*Fig. 56.* Drawings of *Broeggeria salteri* (Holl); Kendyktas Range, Kazakhstan. □A. Ventral interior; based on RM Br 136000. ×8. □B. Dorsal interior; based on RM Br 135992. ×9.

*Occurrence.* – Upper Cambrian – Lower Ordovician (Tremadoc–Arenig); Kazakhstan, ?Bohemia, Scandinavia, Belgium, Britain, Argentina.

## *Broeggeria salteri* (Holl, 1865)

Figs. 45N, O, 54, 55C–Q, 56–61A–D

*Synonymy.* – □1865 *Obolella Salteri* sp. nov. – Holl, p. 102, Fig. 8. □1866 *Obolella Salteri* Holl – Davidson, p. 61, Pl. 4:28–29. □1882 *Obolus Salteri* Holl – Brøgger, p. 44, Pl. 10:10–11, 13. □1906 *Obolus (Brøggeria) Salteri* Holl – Moberg & Segerberg, p. 64, Pl. 1:27–30. □1909 *Obolus (Brøggeria) Salteri* Holl (var.?) – Westergård, p. 56, Pl. 2:17–19. □1909 *Obolus? inflatus* sp. nov. Westergård, p. 76, Pl. 5:25. □1912 *Obolus (Brøggeria) salteri* (Holl) – Walcott, p. 424, Pls. 13:1, 1a–n; 15:4, 4a–d [synonymy]. □1937 *Obolus (Brøggeria)* cf. *salteri* (Holl) – Harrington, p. 103, Pl. 5:4. □1938 *Obolus (Brøggeria) salteri* (Holl) – Harrington, p. 119, Pl. 1:4–5 [synonymy]. □?1938 *Obolus (Brøggeria?) elongatus* sp. nov. Harrington, p. 120, Pl. 1:6–10. □1952 *Brøggeria salteri* (Holl) – Waern, p. 234, Pl. 1:5. □1982 *Broeggeria salteri* (Holl) – Rushton & Bassett *in* Owens *et al.*, p. 23, Pl. 7a–g, i [synonymy].

*Fig. 55.* □A. *Broeggeria* sp.; Dorsal interior; olistolith (sample 7825-12.5) in Satpak Formation; Kujandy Section, northeastern Central Kazakhstan; RM Br 136009; ×8. □B. Detail of dorsal larval shell, showing pitted ornamentation; olistolith (sample 7822-5) in Satpak Formation; Kujandy Section, northeastern Central Kazakhstan; RM Br 135972; ×550. □C. *Broeggeria salteri*; Dorsal exterior; Selety Limestone (sample 325); Selety River, northeastern Central Kazakhstan; RM Br 135995; ×10. □D–O. *Broeggeria salteri* (Holl); Agalatas Formation (sample 564); Kendyktas Range, Kazakhstan. □D. Dorsal exterior; RM Br 136001; ×6. □E. Lateral view of D; ×12. □F. Dorsal exterior; RM Br 136002; ×12. □G. Surface ornamentation of F; ×88. □H. Ventral exterior; RM Br 136003; ×12. □I. Lateral view of H; ×18. □J. Ventral interior; RM Br 135993; ×27 □K. Oblique posterior view of J; ×35. □L. Lateral view of dorsal interior; RM Br 135994; ×14. □M. Dorsal interior; RM Br 135992; ×11. □N. Lateral view of M; ×16. □O. Detail of M, showing pseudointerarea and visceral field; ×100. □P. *Broeggeria salteri* (Holl); detail of juvenile ventral valve, showing pseudointerarea; Selety Limestone (sample 325); Selety River, Kazakhstan; RM Br 135991; ×50. □Q. Lateral view of P; ×28.

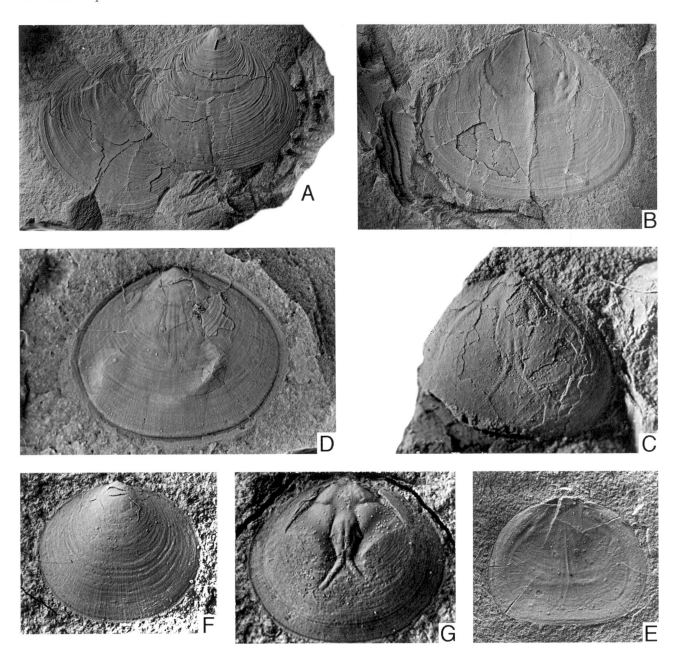

*Fig. 57. Broeggeria salteri* (Holl). □A. Ventral view of complete shell; Dictyonema/Ceratopyge Shale; St. Olofsgatan, Oslo; PMO 540; ×5. □B. Partly exfoliated ventral valve; figured by Walcott (1912, Pl. 13:1n); Dictyonema/Ceratopyge Shale; Oslo; PMO 20018; ×4. □C. Dorsal internal mould; figured by Brøgger (1882, Pl. 10:10) and Walcott (1912, Pl. 13:1m); PMO 1519; ×10. □D. Dorsal internal mould; Ceratopyge Shale; Ottenby, Öland; RM Br 133941; ×7. □E. Dorsal interior; figured by Waern (1952, Pl. 1:5); Ceratopyge Shale; Bödahamn core, Öland; PMU Öl 109; ×8. □F. Dorsal exterior; Dictyonema Shale; Sandby, Scania; LO 6572t (on the same slab as LO 1776t); ×7. □G. Dorsal internal mould; figured by Moberg & Segerberg (1906, Pl. 1:27); Dictyonema Shale; Sandby, Scania; LO 1776t; ×10.

*Fig. 58* (opposite page). *Broeggeria salteri* (Holl). □A. Ventral exterior; Bjørkåsholmen Limestone; Ottenby, Öland; RM Br 133940; ×6. □B. Detail of ornamentation of A; ×80. □C. Detail of micro-ornamentation of A, showing pits; ×800. □D. Dorsal exterior; Bjørkåsholmen Limestone (sample Öl-1); Ottenby, Öland; LO 6515t; ×20. □E. Detail of D, showing ornamentation and possible predatory boring; ×75. □F. Dorsal exterior; Bjørkåsholmen Limestone (sample Ng-1); Bjørkåsholmen, Oslo region; RM Br 136259; ×25. □G. Dorsal exterior; Bjørkåsholmen Limestone (sample Ng-1); Bjørkåsholmen, Oslo region; RM Br 136260; ×32. □H. Lateral view of G; ×35. □I. Detail of dorsal larval shell; Bjørkåsholmen Limestone (sample Ng-1); Bjørkåsholmen, Oslo region; RM Br 136263; ×120. □J. Lateral view of ventral valve; Bjørkåsholmen Limestone (sample Öl-1); Ottenby, Öland; LO 6516t; ×36. □K. Detail of ventral larval shell; Bjørkåsholmen Limestone (sample Ng-1); Bjørkåsholmen, Oslo region; RM Br 136312; ×75. □L. Detail of pitted ornamentation of ventral valve; Bjørkåsholmen Limestone (sample Öl-1); Ottenby, Öland; LO 6517t; ×560. □M. Detail of broken ventral valve edge, showing shell structure; Bjørkåsholmen Limestone (sample Öl-1); Ottenby, Öland; LO 6518t; ×125. □N. Detail of M, showing baculate shell structure; ×2700. □O. Detail of M, showing baculae; ×4125.

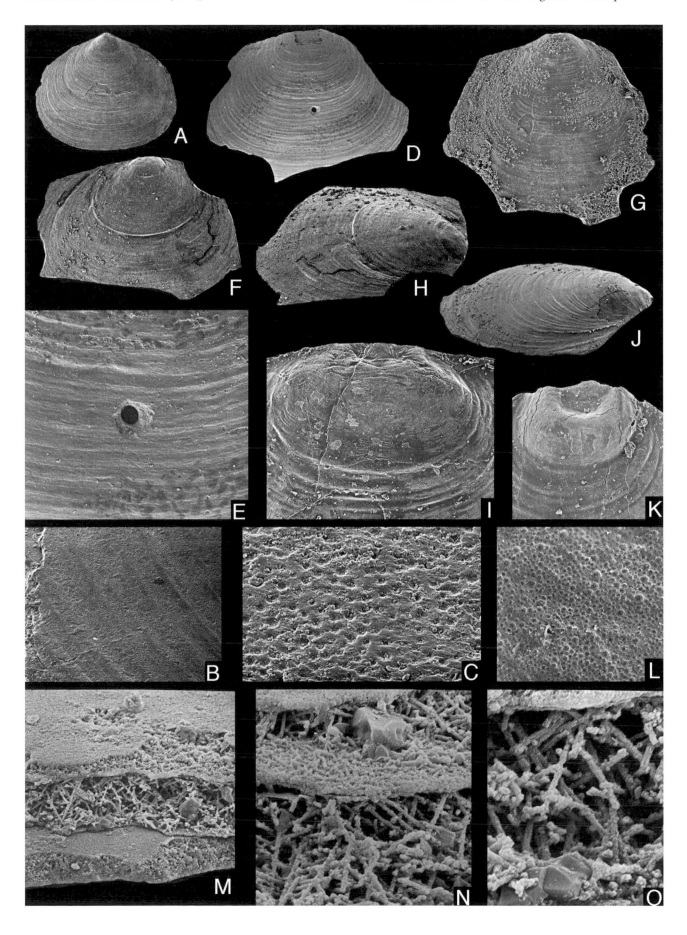

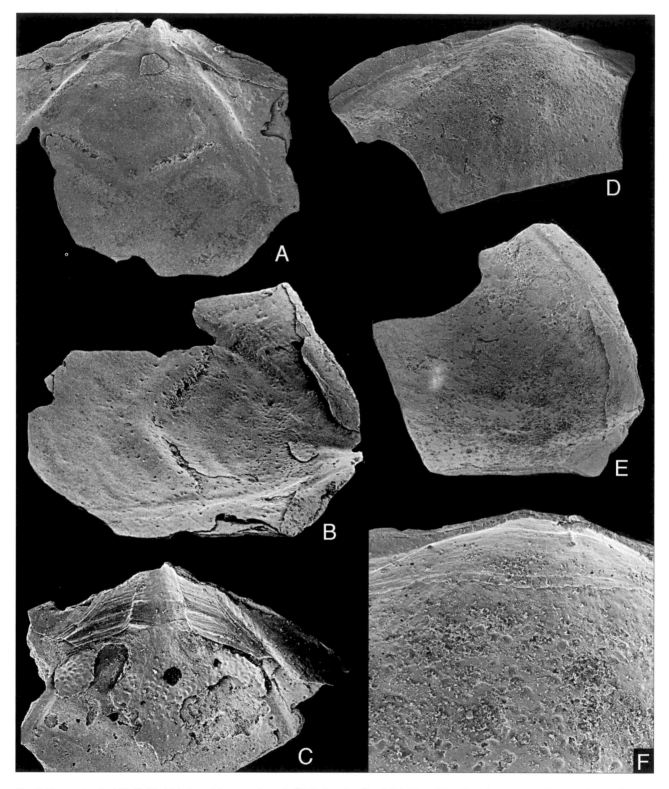

*Fig. 59. Broeggeria salteri* (Holl); Bjørkåsholmen Limestone (sample Öl-1); Ottenby, Öland. □A. Ventral interior; LO 6518t; ×20. □B. Lateral view of A; ×24. □C. Ventral interior; LO 6519t; ×30. □D. Dorsal interior; LO 6520t; ×20. □E. Lateral view of D; ×27. □F. Detail of pseudointerarea of D; ×60.

---

*Fig. 60* (opposite page). *Broeggeria salteri* (Holl). □A. Ventral internal mould; Tremadoc; Dolgelly, Wales; BM B 14373; ×7. □B. Ventral exterior; Portmadoc, Gwynedd, Wales; BM B 6013a; ×7. □C. Ventral internal mould; Portmadoc, Gwynedd, Wales; BM B 6013b; ×7. □D. Ventral internal mould; figured by Walcott (1912, Pl. 13:1d); Dictyonema Shale; Cape Breton, Nova Scotia; USNM 51671a; ×4. □E. Dorsal internal mould; Dictyonema Shale; Cape Breton, Nova Scotia; BM B 84102; ×7.

*Lectotype.* – Selected by Cocks (1978, p. 18); BM B4044; external mould of dorsal or ventral valve from the Upper Cambrian White-Leaved-Oak Shales; Malvern Hills, Herefordshire, South Wales (refigured by Rushton & Bassett *in* Owens *et al.*, Pl. 7b; see also Fig. 54F herein).

*Material.* – Figured from southern Kendyktas Range. Ventral valves: RM Br 135993; RM Br 135996; RM Br 136000 (L 6.5, W 6.5, LI 1, WI 3.8, ML 0.7, MG 0.24, LV 2.3, WV 3); RM Br 136003. Dorsal valves: RM Br 135992; RM Br 135994; RM Br 135997; RM Br 135998 (L 6, W 6.6, LI 0.3; WI 2.7; MG 1.6; LV 3.3; WV 2.5); RM Br 135999 (L 7.4, W 6.7, LI 0.4, WI 2.9, MG 1.4, LV 3.2, WV 3.4); RM Br 136001; RM Br 136002. Total of 13 complete shells, 69 ventral valves, and 73 dorsal valves.

Figured from northeastern Central Kazakhstan. Ventral valves: RM Br 135991; RM Br 135995; RM Br 136013. Total of 9 ventral and 4 dorsal valves.

Figured from South Urals: Dorsal valve: RM Br 136010. Total of 2 complete, 2 ventral valves, and 2 dorsal valves.

Figured from Scandinavia. Ventral valves: RM Br 133940 (L 6.5, W 7.1); RM Br 136312; PMO 540 (L 7.5, W 7); PMO 20018 (L 10.2, W 12.2); LO 5616t; LO 6517t; LO 6518t; LO 6519t. Dorsal valves: RM Br 133941 (L 7, W 8.3, LV 3.8); RM Br 136259; RM Br 136260; RM Br 136263; PMO 1519 (L 4.6, LV 2.2); LO 1776t (L 10, W 11.6, LV 4.6); LO 6515t; LO 6520t; LO 6572t; PMU Öl 109 (L 4.2, W 5.1, LI 0.3, LV 2.8).

Total of 30 ventral and 20 dorsal valves, and an undetermined number of fragmented valves and specimens in shales.

Figured from Britain. Ventral valves: BM B 143373; BM B 6013a; BM B 6013b; IGS GSM 85075.

Figured from Canada. Ventral valve: USNM 51671a. Dorsal valve: BM B 84102.

*Measurements.* – See Tables 13–14.

*Diagnosis.* – As for genus.

*Description of material from Scandinavia and southern Kendyktas Range.* – The shell is dorsi-biconvex, about 20% as thick as wide in adults, slightly inequivalved and subcircular. The ventral valve is gently and evenly convex in lateral profile, on average 97% as long as wide (*N* 8; specimens from the southern Kendyktas Range, Table 13,, where the largest recorded valve is 6.5 mm long and 6.8 mm wide; in Scandinavia, the recorded maximum size is considerably larger, 10.3 mm in length and 14.3 mm in width in one unfigured ventral valve from the Oslo region). The ventral pseudo-interarea is orthocline, on average 25% as long as wide (*N* 8; specimens from the Kendyktas; Table 13), occupying 48% of the valve width in the same sample (Table 13). The pedicle groove is deep, narrowly triangular, with steep lateral slopes; the ventral propareas are raised above the valve floor and have well-defined flexure lines (Figs. 55J, 59A–C). The ven-

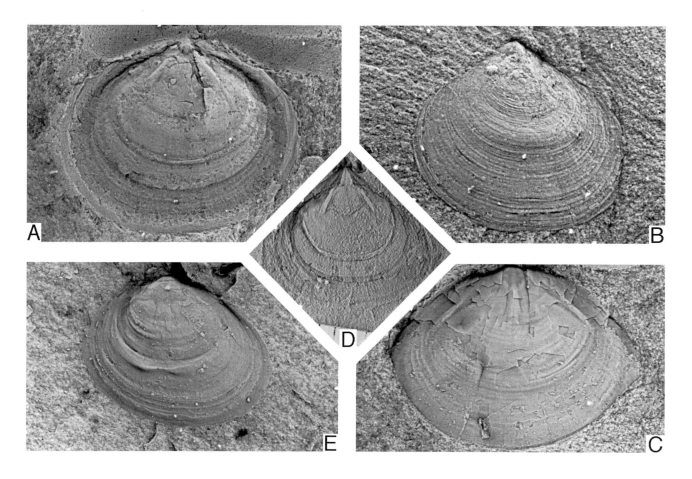

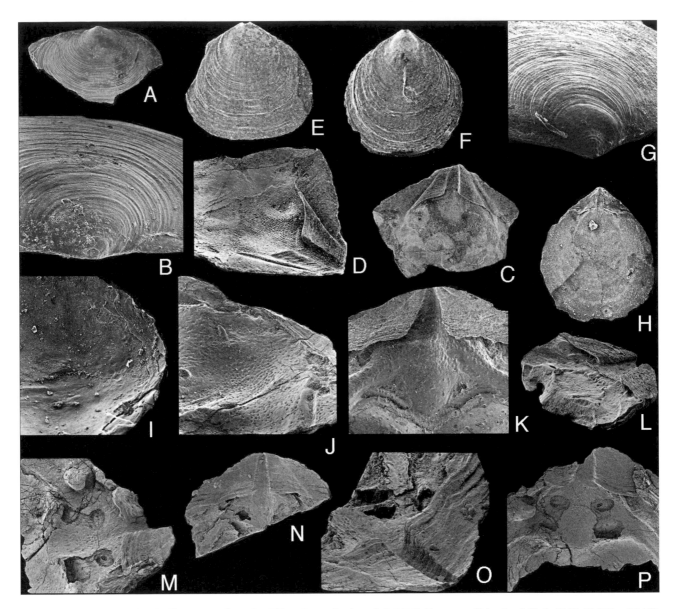

*Fig. 61.* □A–D. *Broeggeria salteri* (Holl). □A. Dorsal exterior; Kidryas Formation (sample B-768-1); Tyrmantau Ridge, South Urals; RM Br 136010; ×6.5. □B. Posterior view of A; ×45. □C. Ventral interior; Selety Limestone (sample 325); Selety River, northeastern Central Kazakhstan; RM Br 136013; ×13. □D. Lateral view of C; ×24. □E–P. *Broeggeria* sp.; Olistoliths within Satpak Formation; Kujandy Section, northeastern Central Kazakhstan. □E. Ventral exterior (sample 7822-5); RM Br 136011; ×22. □F. Ventral exterior (sample 7822-6); RM Br 136012; ×18. □G. Posterior view of F; ×50. □H. Ventral interior (sample 7822-6); RM Br 136016; ×25. □I. Lateral view of dorsal interior (sample 7822-6); RM Br 136014; ×50. □J. Lateral view of dorsal interior (sample 7825-12.5); RM Br 136310; ×25. □K. Ventral interior (sample 7825-12.5); RM Br 136015; ×20. □L. Lateral view of ventral interior (sample 7825-12.5); RM Br 136017; ×12. □M. Lateral view of ventral interior (sample 79103-b); RM Br 136018; ×17. □N. Ventral interior, (sample 79103-b); RM Br 136019; ×11. □M. Detail of ventral pseudointerarea of N; ×23. □P. Ventral interior (sample 79103-b); RM Br 136020; ×13.

tral visceral platform is well defined, on average occupying 48% of the total valve width (*N* 8; specimens from Kendyktas; Table 13), extending to 40% of the valve length (same sample; Fig. 54E; Table 13). The ventral visceral area is bisected by a pair of slightly divergent, median furrows (Fig. 59A–C). The ventral *vascula lateralia* are submarginal, arcuate, widely divergent proximally. In one particularly well-preserved interior (Figs. 54E, 56A), the main vascular trunks occupy up to about half the valve width, at the mid-length;

the distal ends of the trunks are strongly curved medially, extending for about 85% of the total valve length, and almost converge anteriorly; the minor canals bifurcate peripherally, up to a point approximately 1 mm from the valve edge, but continue medially without bifurcation to the centre of the valve (Figs. 54E, 56A).

The dorsal valve is gently and evenly convex in lateral profile, on average 94% as long as wide (*N* 7; specimens from Kendyktas, Table 14, where the maximum size is 8.2 mm in length and 8.6 mm in width; in Scandinavia the recorded maximum length and width is 10 and 11.6 mm, respectively; Fig. 57G). The dorsal pseudointerarea is apsacline, forming a wide and short, crescent-shaped strip bounding the posterior margin; it occupies on average 41% of the valve width (*N* 4; specimens from Kendyktas; Table 14). The median groove is wide, gently concave, occupying on average 20% of the pseudointerarea in the same sample (Figs. 55M–O, 58B, 59D–F; Table 14). The dorsal visceral area is raised above the valve floor, with a long median tongue extending anteriorly for 52% of the total length (*N* 6; Table 14); it is bisected by a fine median ridge (Fig. 56B). The dorsal central muscle scars are strongly elongate and thickened, bounded laterally by ridges; they are placed close to the anterior lateral muscle scars. The dorsal *vascula lateralia* are peripheral, arcuate in shape, and widely divergent; the *vascula media* are long, almost straight and widely divergent (Figs. 54C, D, 57N, 57G). In one particularly well-preserved dorsal internal mould (Fig. 57C; see also Walcott 1912, Pl. 13:1m), the *vascula lateralia* occupy up to about 75% of the total valve width, extending for about 60% of the valve length; the *vascula media* occupy up to about 30% of the valve width, extending for about 80% of the valve length; the minor canals are relatively poorly preserved (Fig. 57C). In a second well-preserved dorsal interior (Fig. 57E; see also Waern 1952, Pl. 1:5), the *vascula lateralia* and *vascula media* are unusually curved and long; the distal ends are almost touching at a point only half a millimetre from the anterior margin of the valve; the minor canals are also well shown, and seem to bifurcate mainly peripherally (Fig. 57E). The entire visceral areas of both valves have distinctive rounded, closely packed depression, up to about 60 µm across (Fig. 59A–F).

The larval shell is usually not preserved on adults; on one juvenile dorsal valve it is transversely oval, about 0.45 mm wide and 0.3 mm long, with an ornamentation of radially arranged, outwardly convex sets of 'drapes' of fine fila close to the anterior margin; there is a depressed area, 0.12 mm wide, at the mid-posterior margin (Fig. 58I). The larval shell of one juvenile ventral valve is damaged posteriorly but is of similar size as the dorsal one; it also has a depressed area at about the mid-posterior, which is somewhat larger and 0.2 mm wide (Fig. 58K). The postlarval shell is finely pitted with irregularly distributed circular pits of varying sizes, up to about 4 µm across (Fig. 58B, C, L). The postlarval shell is ornamented by fine, closely spaced fila, and a few more pronounced disturbances in growth at irregular intervals. The baculate shell structure is well preserved, with long slender baculae, around 500 nm thick (Fig. 58M–O).

*Discussion.* – The specimens of *B. salteri* from the Kendyktas Formation of the southern Kendyktas Range, Kazakhstan, are closely comparable with topotypes (Fig. 54F) and with the material from the Tremadoc of Carmarthen, South Wales (see also Fig. 60A–C), described by Rushton & Bassett (*in* Owens *et al.*, 1982), both in the size and main characters of the valves. Compared with *B. salteri* from Cape Breton,

Nova Scotia, the only difference is that the Kendyktas specimens have more strongly impressed internal characters. As noted by Walcott (1912, p. 425), some difference in the relative size of the vascular platforms between specimens may be due to exfoliation, compaction, and other preservational factors. The outline of the specimens preserved in shales is obviously also strongly dependent on *post mortem* deformation.

The specimens of *B. salteri* from Argentina (Harrington 1937, 1938) are poorly illustrated but do not appear to show any differences with our material.

One fragmentary dorsal valve has a possible circular predatory boring, about 70 µm across, with a countersunk rim (Fig. 58E), the only possible evidence of predation encountered.

The depressions on the visceral areas are obviously identical to those described from *Lingulella antiquissima* but somewhat larger; Walcott (1912, p. 424) also commented on the 'punctate interior' of *B. salteri*.

Williams & Holmer (1992) suggested that radial sets of 'drapes' identical to those described above are formed because of stresses in the outer mantle lobe induced by the setal muscles.

*Occurrence.* – The distribution of *B. salteri* and the *Broeggeria* assemblage is discussed above (see p. 31). In Scandinavia, it possibly occurs in the Upper Cambrian and is common in the Dictyonema and Ceratopyge shales and the Bjørkåsholmen Limestone of Öland, Scania, Västergötland, and in the Oslo region (Figs. 5–8; Appendix 1K). In northeastern Central Kazakhstan, it was found in the upper part of the Selety Limestone, at the Selety River (Appendix 1H), and in the Satpak Formation at the Kujandy (Fig. 25; Appendix 1G) and Aksak–Kujandy sections (Fig. 27; Appendix 1H). In the southern Kendyktas Range it is recorded from the Kendyktas Formation and Agalatas Formation (Fig. 35; Appendix 1J). In the South Urals, *B. salteri* is recorded from the lower part of the Kidryas Formation at Tyrmantau Ridge (Fig. 11; Appendix 1A).

## *Broeggeria* sp.

Figs. 55A, B, 61E–P

*Material.* – Figured. Ventral valves: RM Br 136011, RM Br 136012; RM Br 136015; RM Br 136016; RM Br 136017; RM Br 136018; RM Br 136019; RM Br 136020. Dorsal valves: RM Br 135972; RM Br 136009; RM Br 136014; RM Br 136310. Total of 44 ventral and 55 dorsal valves.

*Remarks.* – The valves mostly represent juvenile individuals, but some fragmentary adults also occur. The internal characters are closely comparable with *B. salteri*.

*Occurrence.* – The species is abundant in olistoliths within the Satpak Formation at the Aksak–Kujandy, Sasyksor Lake, and Kujandy sections (Fig. 25; Appendix 1G–H).

*Table 13. Broeggeria salteri,* average dimensions and ratios of ventral valves.

| | L | W | LI | WI | ML | MG | LV | WV | $L/W$ | $LI/WI$ | $MG/WI$ | $WI/W$ | $ML/L$ | $LV/L$ | $WV/W$ |
|---|---|---|---|---|---|---|---|---|---|---|---|---|---|---|---|
| Samples 551, 556, 556-1, 563 | | | | | | | | | | | | | | | |
| N | 10 | 8 | 9 | 9 | 9 | 9 | 9 | 9 | 8 | 9 | 9 | 8 | 9 | 9 | 8 |
| X | 5.04 | 5.54 | 0.70 | 2.81 | 0.40 | 0.24 | 1.93 | 2.50 | 97% | 25% | 9% | 48% | 8% | 40% | 48% |
| S | 1.15 | 1.22 | 0.25 | 0.76 | 0.15 | 0.06 | 0.36 | 0.54 | 0.07 | 0.04 | 0.02 | 0.06 | 0.02 | 0.10 | 0.06 |
| MIN | 3.12 | 3.04 | 0.32 | 1.40 | 0.24 | 0.16 | 1.36 | 1.60 | 87% | 18% | 6% | 39% | 4% | 25% | 39% |
| MAX | 6.48 | 6.80 | 1.04 | 3.76 | 0.72 | 0.36 | 2.32 | 3.08 | 108% | 30% | 12% | 55% | 11% | 57% | 55% |

*Table 14. Broeggeria salteri,* average dimensions and ratios of dorsal valves.

| | L | W | LI | WI | ML | MG | LV | WV | $L/W$ | $LI/WI$ | $ML/L$ | $MG/WI$ | $WI/W$ | $LV/L$ | $WV/W$ |
|---|---|---|---|---|---|---|---|---|---|---|---|---|---|---|---|
| Samples 551, 553, 554, 556, 556-1 | | | | | | | | | | | | | | | |
| N | 7 | 7 | 4 | 4 | 4 | 4 | 6 | 6 | 7 | 4 | 4 | 4 | 4 | 6 | 6 |
| X | 6.11 | 6.55 | 0.33 | 2.93 | 0.26 | 1.44 | 3.38 | 3.48 | 94% | 11% | 4% | 20% | 41% | 52% | 50% |
| S | 1.28 | 1.46 | 0.07 | 0.16 | 0.08 | 0.23 | 0.78 | 0.66 | 0.05 | 0.03 | 0.01 | 0.05 | 0.05 | 0.07 | 0.09 |
| MIN | 3.92 | 3.76 | 0.24 | 2.72 | 0.16 | 1.12 | 2.40 | 2.48 | 91% | 8% | 2% | 13% | 35% | 41% | 38% |
| MAX | 8.24 | 8.64 | 0.40 | 3.08 | 0.32 | 1.60 | 4.80 | 4.24 | 104% | 14% | 5% | 24% | 47% | 58% | 63% |

## Genus *Keskentassia* gen. nov.

*Name.* – After the type locality, the Keskentas Ridge.

*Type and only species.* – *Keskentassia multispinulosa* gen. et sp. nov.; Agalatas Formation.

*Diagnosis.* – Shell strongly biconvex, slightly inequivalved. Ventral pseudointerarea forming crescent-shaped strip; ventral propareas strongly elevated, flat, lacking flexure lines. Ventral *vascula lateralia* submedial. Dorsal pseudointerarea reduced, mainly consisting of wide median groove. Dorsal transmedian, outside lateral, and middle lateral muscle scars placed on separate muscle platforms along posterolateral margins of visceral area. Post-larval shell with finely pitted micro-ornament, and spinose valve margin.

*Discussion.* – The most distinctive features of the genus are: (1) the reduced dorsal pseudointerarea, (2) the median location of the ventral *vascula lateralia*, and (3) the presence of fine marginal spines that are otherwise lacking within the Family Elkaniidae. It differs from *Broeggeria* and *Elkania* also in having more strongly raised visceral platforms with deep median depressions in both valves. The convexity of the shell is similar to that of *Elkania*.

*Occurrence.* – At the type locality only.

## *Keskentassia multispinulosa* sp. nov.

Figs. 62A–R, 63

*Name.* – Latin *multus*, much, and *spinula*, small spine; alluding to the row of fine spines along the valve margin.

*Holotype.* – Fig. 62C; RM Br 136021; ventral valve (L 3.49, W 3.28, H 1, LV 1.28, WV 1.72); upper part of the Agalatas Formation (sample 564).

*Paratypes.* – Figured. Ventral valves: RM Br 136025; RM Br 136026. Dorsal valves: Br.136022 (L 3.32, W 3.96, H 1.48, LV 1.68, WV 1.44); RM Br 136023; RM Br 136024. Total of 24 ventral and 17 dorsal valves.

*Diagnosis.* – As for genus.

*Description.* – The shell is strongly biconvex, about 40–60% as thick as long, transversely oval to subcircular. The ventral valve is strongly and evenly convex. The ventral beak is strongly swollen, slightly incurved (Fig. 62B). The ventral pseudointerarea forms a wide and short strip bounding the posterior margin, bisected by a narrow, deep pedicle groove, with steep lateral slopes; the propareas are strongly raised, flattened, lacking well-developed flexure lines. The ventral visceral area is strongly thickened and raised, with a deep median depression. The posterolateral muscle fields are situated on separate platforms, bounded anteromedianly by deep, widely divergent furrows connected with the proximal parts of the vascular trunks; the *vascula lateralia* are submedian, straight, and slightly divergent (Fig. 62C–E, N, O).

The dorsal valve is strongly convex, with maximum height somewhat posterior to mid-length. The dorsal umbo is strongly swollen. The dorsal pseudointerarea is strongly reduced, represented mainly by a short, wide median groove. The dorsal visceral field is strongly thickened and raised, with a deep median depression; the median tongue is elongated, extending to mid-length. The dorsal central muscle scars are elongate and convergent anteriorly, situated close to the anterior lateral scars. The dorsal *vascula lateralia* are arcuate and submarginal; the *vascula media* are long and widely divergent (Fig. 62G–M, P–R). The larval and postlarval shell is finely pitted, but the ornamentation is poorly preserved. The valve margin has a row of fine, closely spaced spines, about 18–20 per mm (Fig. 62F, G).

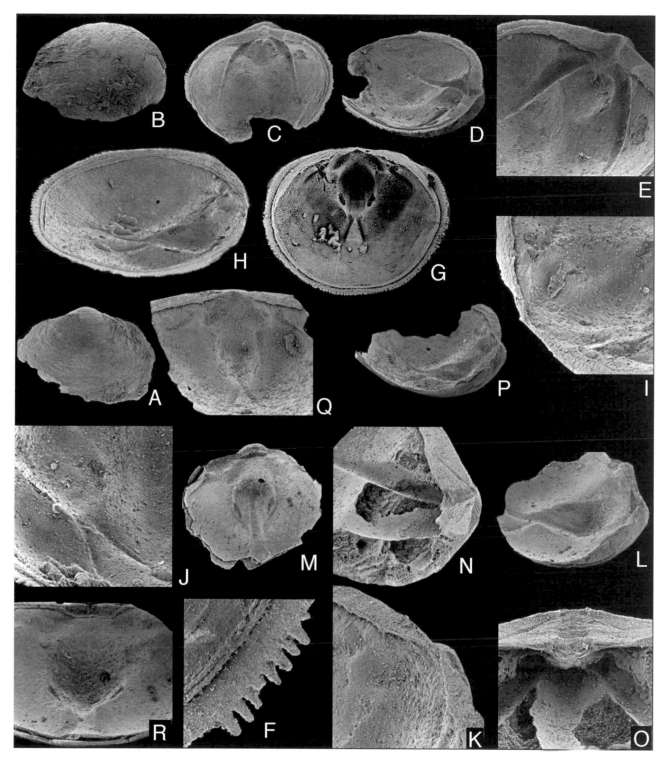

*Fig. 62. Keskentassia multispinulosa* gen. et sp. nov.; Agalatas Formation (sample 564); Kendyktas Range, Kazakhstan. □A. Ventral exterior; RM Br 136026; ×9. □B. Lateral view of A; ×11. □C. Holotype; ventral interior; RM Br 136021; ×10. □D. Lateral view of C; ×11. □E. Oblique lateral view of C; ×25. □F. Detail of C, showing valve margin with spines; ×80. □G. Dorsal interior; RM Br 136022; ×11. □H. Lateral view of G; ×17. □I. Detail of G, showing pseudointerarea; ×30. □J. Detail of G, showing visceral area and *vascula media*; ×32. □K. Detail of G, showing pseudointerarea and posterior part of the visceral field; ×32.5. □L. Lateral view of dorsal interior; RM Br 136023; ×16.5. □M. Plane view of L; ×13. □N. Lateral view of ventral interior; RM Br 136025; ×25. □O. Ventral pseudointrarea of N; ×37. □P. Lateral view of dorsal interior; RM Br 136024; ×15. □Q. Anterior view of P; ×15. □R. Anterior view of P; ×25.

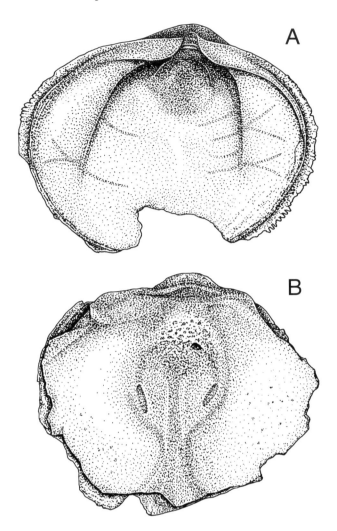

*Fig. 63.* Drawings of *Keskentassia multispinulosa* gen. et sp. nov. □A. Ventral interior; based on RM Br 136021. ×20. □B. Dorsal interior; based on RM Br 136023. ×26.

*Discussion.* – Marginal spines are quite unusual in obolids, otherwise known only from *Thysanotos*; however, in contrast to that genus there are no regularly spaced growth lamellae with spines. In *Thysanotos*, these represent temporary interruptions in growth that perhaps are related to spawning, but both in this genus and in *Keskentassia* the spines obviously were formed (and subsequently covered) continuously along the shell margin.

*Occurrence.* – At the type locality only (Fig. 35; Appendix 1J).

## Genus *Lamanskya* Moberg & Segerberg, 1906

*Type species.* – By monotypy; *Lamanskya splendens* Moberg & Segerberg, 1906; Bjørkåsholmen Limestone, Ottenby, Öland.

*Diagnosis.* – See Holmer (1993, p. 153).

*Remarks.* – Holmer (1993) recently reviewed the records of the genus and this information is therefore not repeated here.

## *Lamanskya splendens* Moberg & Segerberg, 1906

*Synonymy.* – □1906 *Lamanskya splendens* gen. et sp. nov. – Moberg & Segerberg, p. 71, Pl. 3:17. □1993 *Lamanskya splendens* Moberg & Segerberg – Holmer, p. 153, Figs. 3–7, 8a.

*Holotype.* – By monotypy; LO1795T; exfoliated dorsal valve (W 6.8, L 6, H 2.5); Bjørkåsholmen Limestone; Ottenby.

*Material.* – Unfigured: fragmentary valves and the specimens illustrated by Holmer (1993).

*Remarks.* – Fragments of this species can be recognized by their distinctive ornamentation, but the only more complete specimens have already been described by Holmer (1993); the species occurs in the Bjørkåsholmen Limestone at Ottenby, Mossebo, and Stora Backor (Figs. 5–7; Appendix 1K). It is also recorded from the Koagash River (Appendix 1F).

## Family Lingulellotretidae Koneva & Popov, 1983

*Diagnosis.* – Shell elongate oval or subtriangular. Larval shell smooth. Pseuointerareas of both valves well developed. Ventral pseudointerarea with elongate oval pedicle foramen. Ventral interior with internal pedicle tube; ventral *vascula lateralia* straight, divergent in posterior half. Dorsal *vascula lateralia* peripheral.

*Genera included.* – *Lingulellotreta* Koneva, 1983; *Aboriginella* Koneva, 1983; *Mirilingula* Popov, 1983.

*Occurrence.* – Lower Cambrian (Botomian) – Lower Ordovician (Tremadoc).

## Genus *Mirilingula* Popov, 1983

*Type species.* – Original designation by Popov (*in* Koneva & Popov 1983, p. 116); *Mirilingula mutabilis* Popov, 1983; Lower Ordovician; Malyj Karatau, southern Kazakhstan.

*Diagnosis.* – Shell elongate oval, moderately to strongly biconvex, with shallow sulcus on both valves. Ventral pseudointerarea lacking flexure lines. Dorsal pseudointerarea strongly elevated, flattened and undivided. Visceral areas poorly defined in both valves. Dorsal interior with low median ridge.

*Occurrence.* – Upper Cambrian – Lower Ordovician (Tremadoc); Kazakhstan.

## *Mirilingula* sp.

Fig. 64

*Synonymy.* – □1956 *Lingulella* cf. *mosia* (Hall) – Nikitin, p. 11, Pl. 1:2–4. □1983 *Mirilingula* sp. – Koneva & Popov, p. 121, Pls. 28:1–3; 29:7–10.

*Material.* – Figured. Ventral valves: RM Br 136027; RM Br 136028; RM Br 136032 (L 1.8, W 1.28). Dorsal valves: RM Br 136029; RM Br 136030; RM Br 136033 (L 2.76, W 2.2); RM Br 136034; RM Br 136035 (L 1.28, W 1.16). Total of 9 ventral and 10 dorsal valves.

*Description.* – The shell is subequally biconvex, elongate oval. The ventral valve is about 1.4–1.5 times as wide as long. The ventral pseudointerarea is apsacline, with a deep, narrowly triangular pedicle groove; the propareas are strongly elevated and flattened. The ventral visceral area is poorly defined and does not extend to the centre of the valve; it is bisected by two fine, slightly divergent ridges (Fig. 64A–D).

The dorsal valve is about 1.1–1.3 times as long as wide. The dorsal pseudointerarea is undivided, flattened, and orthocline, strongly elevated above the valve floor (Fig. 64I, J). The dorsal visceral area is poorly defined, extending somewhat anterior to mid-length, bisected by a fine, long median ridge. The muscle scars of both valves are generally poorly defined, but the dorsal central muscle scars are elongate oval, placed somewhat posterior to the centre of the valve.

*Discussion.* – All specimens in our collection lack the pedicle foramen that is present in adults of all lingulellotretids; our specimens therefore probably represent juveniles. They are closely comparable with the specimens of *Mirilingula* sp. described by Koneva & Popov (1983, p. 121) from the Late Cambrian Batyrbai Section of Malyj Karatau, Southern Kazakhstan. Lingulids, described by Nikitin (1956) under the name of *Lingulella* cf. *mosia*, were collected from the same locality and are probably also synonymous with *Mirilingula* sp.

*Fig. 64. Mirilingula* sp.; Selety Limestone (sample 325); Selety River, northeastern Central Kazakhstan. □A. Ventral interior; RM Br 136027; ×45. □B. Lateral view of B; ×28. □C. Ventral interior; RM Br 136028; ×18. □D. Lateral view of D; ×45. □E. Posterior view of dorsal valve; RM Br 136029; ×18. □F. Posterior view of dorsal valve; RM Br 136030; ×11. □G. Dorsal exterior; RM Br 136032; ×19. □H. Dorsal exterior; RM Br 136033; ×12. □I. Dorsal interior; RM Br 136034, ×28. □J. Dorsal interior of juvenile specimen; RM Br 136035, ×50.

*Occurrence.* – It occurs only in the Selety Limestone, at the Selety River (Appendix 1H).

# Family Zhanatellidae Koneva, 1986

*Diagnosis.* – Shell subcircular to elongate suboval. Larval and postlarval shell with finely pitted micro-ornament. Ventral valve with flattened pseudointerarea, bisected by deep pedicle groove; flexure lines variably developed; beak with semicircular emarginature. Dorsal pseudointerarea divided by median groove or undivided.

*Genera included.* – *Zhanatella* Koneva, 1986; *Fossuliella* Popov & Ushatinskaya, 1992; *Rowellella* Wright, 1963.

*Discussion.* – The original definition of the family included *Zhanatella* Koneva and *Kyrshabaktella* Koneva. Popov & Ushatinskaya (1992, p. 65) recognized the morphological similarities between *Zhanatella* and *Rowellella* in terms of ornamentation and the presence of a deep ventral emarginature, and included also the new genus *Fossuliella*. Ushatinskaya (1992, p. 83) noted that *Kyrshabaktella* lacks pitted micro-ornament, and they referred it to a separate family.

*Occurrence.* – Middle Cambrian – Upper Ordovician.

# Genus *Zhanatella* Koneva, 1986

*Type species.* – Original designation by Koneva (1986, p. 50); *Zhanatella rotunda*; Upper Cambrian; Malyj Karatau, southern Kazakhstan.

*Diagnosis.* – Shell subequally biconvex, circular; ornamented by evenly spaced, concentric rugellae. Ventral pseudointerarea wide, with deep, narrow pedicle groove; propareas flattened, elevated with well-defined flexure lines. Ventral visceral field slightly thickened anteriorly, not extending to mid-length; ventral *vascula lateralia* submedianly placed, widely divergent in posterior half, becoming arcuate anteriorly. Dorsal pseudointerarea with wide, lens-shaped median groove. Dorsal visceral field small, with narrow median tongue, not extending to mid-length, bordered laterally by low ridges; dorsal *vascula lateralia* subperipheral, arcuate; *vascula media* long, divergent.

*Species included.* – *Zhanatella rotunda* Koneva, 1986; *Zhanatella* sp. Henderson (*in* Henderson *et al.*), 1992.

*Discussion.* – Henderson (*in* Henderson *et al.* 1992, Pls. 2:15, 16; 3:7, 8, 10) described fragmentary *Zhanatella* sp. from the Upper Cambrian of West Antarctica. The genus is also represented in the Upper Cambrian of North America (Popov & Holmer, unpublished).

*Occurrence.* – Upper Cambrian; Kazakhstan, Antarctica, North America.

# *Zhanatella rotunda* Koneva, 1986
Figs. 40, 65

*Synonymy.* – □1986 *Zhanatella rotunda* sp. nov. – Koneva, p. 50, Pl. 5:1–12. □1992 *Zhanatella rotunda* Koneva – Popov & Ushatinskaya, Pls. 2:5, 6; 3:1–3.

*Holotype.* – GA 427/259; ventral valve (L 0.39, W 0.42); Aksai Stage (*Pseudagnostus pseudangustilobus* Biozone); Kyrshabakty River, Malyj Karatau, southern Kazakhstan.

*Material.* – Figured. Ventral valves: RM Br 136041 (L 4.64, W 4.96, LI 0.56, WI 3.28, ML 0.44, WG 0.48, LV 1.4, WV 2.8); RM Br 136044 (L 2, W 2.72, LI 0.32, WI 1.44, ML 0.16, WG 0.12, LV 0.88); RM Br 136045; RM Br 136048 (L 3.36, W 3.52, LI 0.48, WI 2.24, ML 0.16, WG 0.32, LV 1.28); RM Br 136049. Dorsal valves: RM Br 136042; RM Br 136043; RM Br 136046; RM Br 136047 (L 3.68, W 3.24); RM Br 136050. Total of 52 ventral and 62 dorsal valves.

*Diagnosis.* – As for genus.

*Description.* – The shell is gently ventri-biconvex and subcircular; the maximum length is placed somewhat anterior to the mid-length. The ventral valve is about 74–95% as long as wide and gently convex, with the maximum height placed about one third of the valve length from the umbo. The ventral pseudointerarea is orthocline, about 17–31% as long as wide, occupying about 55–66% of the total valve width, with a deep pedicle groove forming the emarginature (Fig. 65U). The ventral propareas are flattened, slightly raised above the valve floor, and bisected by flexure lines (Fig. 65K, Q, U). The ventral visceral area is slightly raised anteriorly and does not extend to the centre of the valve; the anterolateral muscle field is subtriangular in outline and has six pairs of scars representing the central, middle, and out-

*Fig. 65.* *Zhanatella rotunda* Koneva; Kujandy Formation; northeastern Central Kazakhstan. □A. Ventral exterior; Satpak Syncline (sample 7843); RM Br 136041; ×10. □B. Lateral view of A; ×12. □C. Posterior view of A; ×30. □D. Dorsal interior; Satpak Syncline (sample 7843); RM Br 136042; ×15. □E. Ventral exterior; olistolith within Satpak Formation (sample 7844-2); RM Br 136044; ×13. □F. Lateral view of E; ×16. □G. Detail of E, showing apical area and emarginature; ×100. □H. Detail of E; ×215. □I. Detail of E, showing pitted micro-ornament; ×650. □J. Ventral interior; Satpak Syncline (sample 7835); RM Br 136045; ×12. □K. Lateral view of J; ×24. □L. Lateral view of dorsal interior; olistolith within Satpak Formation (sample 7844-2); RM Br 136043; ×33 □M. Detail of pseudointerarea of L; ×55. □N. Detail of broken shell edge of L, showing baculate shell ultrastructure; ×1000. □O. Detail of shell ultrastructure L; ×3250. P□. Lateral view of ventral interior; olistolith within Satpak Formation (sample 7844-2); RM Br 136048; ×11. □Q. Detail of P, showing pseudointerarea; ×40. □R. Lateral view of dorsal interior; Satpak Syncline (sample 7843); RM Br 136046; ×30. □S. Oblique posterior view of dorsal exterior; Satpak Syncline (sample 7843); RM Br 136047; ×16. □T. Detail of S, showing larval shell; ×50. □U. Oblique lateral view of ventral interior; Satpak Syncline (sample 7835); RM Br 136049; ×33. □V. Detail of ornamentation of dorsal valve; RM Br 136050; ×275. □W. Detail of V; ×1100.

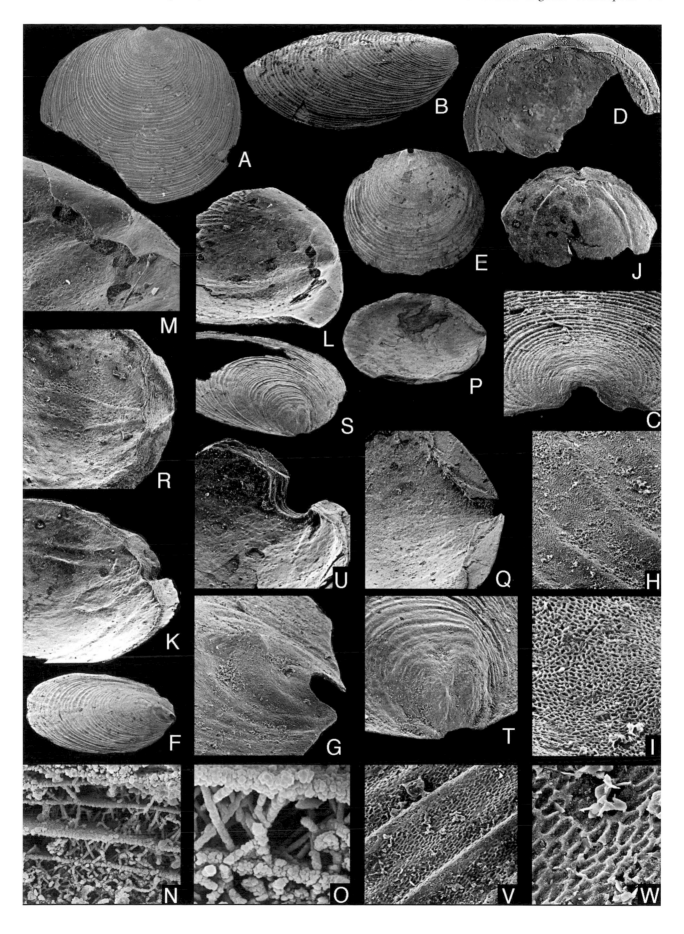

side lateral muscles. The posterolateral muscle fields are elongate triangular and widely divergent. The mantle canal system of both valves is baculate; with ventral, submarginal *vascula lateralia* widely divergent in their proximal parts. The ventral umbonal muscle scars are placed directly anterolateral to the anterior margin of the pedicle groove (Fig. 65J, K).

The dorsal valve is gently and evenly convex. The dorsal pseudointerarea is low, orthocline, and has a widely triangular median groove that is not raised above the valve floor; the propareas are narrow and slightly raised, with flexure lines (Fig. 65L, M). The dorsal visceral field has a narrow median tongue, bisected by a fine median ridge and extending to centre of the valve; the anterior oblique muscle scars are placed close to the suboval central muscle scars. The dorsal *vascula lateralia* are peripherally placed and long; the *vascula media* are divergent (Fig. 65L, M, R).

The ventral larval shell is rounded, around 0.4 mm across, and has a rounded emarginature, about 0.1 – 0.15 mm across; there is a raised, short ridge at the anterior margin of the shell (Fig. 65G). The dorsal larval shell is bisected by two low ridges (Fig. 65T). The larval and postlarval shell is finely pitted; the larval pits are irregularly distributed, rounded, and of varying size, up to about 2 μm across, while those on the postlarval shell are elongate oval, up to about 7 μm across, and defined by elevated ridges. Close to the larval shell they are arranged with the longest dimensions in varying directions (Fig. 65I), but laterally the elongate pits are arranged with the longest dimension oriented perpendicular to the direction of growth (Fig. 65V, W). The postlarval shell is ornamented by regularly spaced rugellae, about 0.1 mm apart. The baculate shell structure in the secondary layer is well developed with thin, slender baculae, up to about 1 μm thick, consisting of granular apatite (Fig. 65N, O).

*Discussion.* – The described material is completely identical in all morphological details to that described by Koneva (1986). However, our material is better preserved and allows a description of the micro-ornamentation, as well as the musculature. The dorsal and ventral larval shells are somewhat similar to those of *Tropidoglossa modesta* (Walcott) described by Rowell (1966), in having a double and single ridge, respectively. The pitted postlarval micro-ornamentation of *Z. rotunda* is closely similar to that of *T. modesta* in having elongate pits arranged in varying directions (Holmer & Popov, unpublished). However, *T. modesta* is not similar to *Z. rotunda* in any other respect.

*Occurrence.* – *Z. rotunda* occurs in the middle part of the Kujandy Formation, at the Satpak Syncline (Fig. 29; Appendix 1I), and at the Kujandy Section (Fig. 25; Appendix 1G). It is also found in olistoliths within the Satpak and Erzhan formations at the Aksak–Kujandy Section (Fig. 27; Appendix 1H).

## Genus *Fossuliella* Popov & Ushatinskaya, 1992

*Type species.* – By original designation; *Lingulella linguata* Pelman, 1977; Middle Cambrian Mayaktakh Formation (Maya Stage); northeastern Central Siberia.

*Diagnosis.* – Shell elongate oval. Ventral umbo and pseudointerarea perforated by semicircular to elongate oval emarginature; ventral propareas short, with well-defined flexure lines. Ventral visceral field not extending to mid-length; ventral *vascula lateralia* submarginal and straight, slightly divergent. Dorsal pseudointerarea high, orthocline; median groove wide, subtriangular; propareas reduced. Dorsal visceral field extending somewhat anterior to mid-length; proximal parts of dorsal *vascula lateralia* straight, slightly divergent; *vascula media* long, divergent. Larval and postlarval shell with pitted micro-ornament.

*Species assigned.* – *Lingulella linguata* Pelman, 1977; *Fossuliella konevae* sp. nov.

*Discussion.* – *Fossuliella* is distinguished from other Cambrian lingulids in having a finely pitted larval and postlarval shell and a deep emarginature perforating the ventral umbo. These feature suggest that the genus is related to *Zhanatella*. *Fossuliella* differs however in having (1) an elongate oval shell, (2) submarginal ventral *vascula lateralia*, (3) a high triangular dorsal pseudointerarea with reduced propareas, and (4) ornamentation with finer rugellae.

## *Fossuliella konevae* sp. nov.
Figs. 66–67

*Synonymy.* – □1992 *Disorystus* sp. [*sic*] – Popov & Ushatinskaya, Pl. 3:5–7.

*Name.* – In honour of Svetlana P. Koneva.

*Holotype.* – Fig. 66A–B; RM Br 136029; dorsal valve (L 1.62, W 1.2, LI 0.32, WI 0.82, ML 0.16, WG 0.44, LS 0.34); from an olistolith (sample 79101) within Erzhan Formation; Aksak–Kujandy Section.

*Paratypes.* – Figured. Complete shells: RM Br 136038 (L 1, W 0.95, T 0.1); RM Br 136039 (L 0.84, W 0.59); RM Br 136040. Ventral valves: RM Br 136032; RM Br 136034; RM Br 136036; RM Br 136037. Dorsal valves: RM Br 136030; RM Br 136031 (L 1.82, W 1.42, LI 36, WI 0.92, ML 0.2, WG 0.56, LS 0.38); RM Br 136035. Total of 21 complete shells, 15 ventral valves, and 85 dorsal valves.

*Measurements.* – See Table 15.

*Diagnosis.* – Ventral beak perforated by wide, deep emarginature; ventral propareas reduced and short. Dorsal pseudointerarea orthocline, mainly consisting of wide,

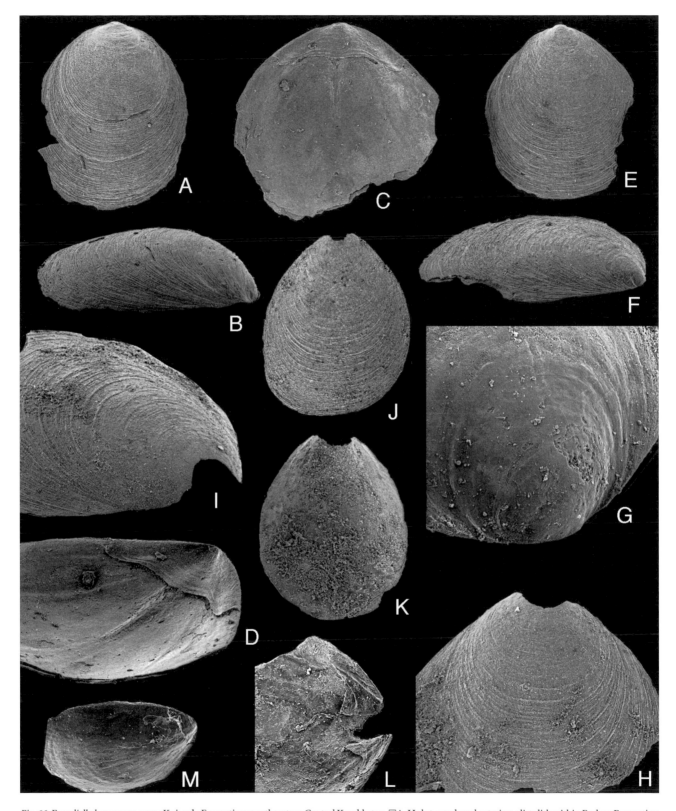

*Fig. 66. Fossuliella konevae* sp. nov.; Kujandy Formation; northeastern Central Kazakhstan. □A. Holotype; dorsal exterior; olistolith within Erzhan Formation (sample 79101); Aksak–Kujandy Section; RM Br 136029; ×30. □B. Lateral view of A; ×33. □C. Dorsal interior; olistolith within Erzhan Formation (sample 79101); Aksak–Kujandy Section; RM Br 136030; ×32. □D. Oblique lateral view of C; ×60. □E. Dorsal exterior; olistolith within Erzhan Formation (sample 79101); Aksak–Kujandy Section; RM Br 136031; ×26. □F. Lateral view of E; ×32. □G. Larval shell of E; ×180. □I I. Ventral exterior; olistolith within Erzhan Formation (sample 79101); Aksak–Kujandy Section; RM Br 136037; ×38. □I. Oblique posterior view of H; ×50. □J. Ventral exterior; Kujandy Section (sample 7823-4); Aksak–Kujandy Section; RM Br 136032; ×18. □K. Ventral interior; olistolith within Erzhan Formation (sample 79101); Aksak–Kujandy Section; RM Br 136034; ×30. □L. Pseudointerarea and emarginature of ventral valve; olistolith within Erzhan Formation (sample 79101); Aksak–Kujandy Section; RM Br 136036; ×38. □M. Lateral view of interior of juvenile dorsal valve; Satpak Syncline (sample 7836a); RM Br 136035; ×55.

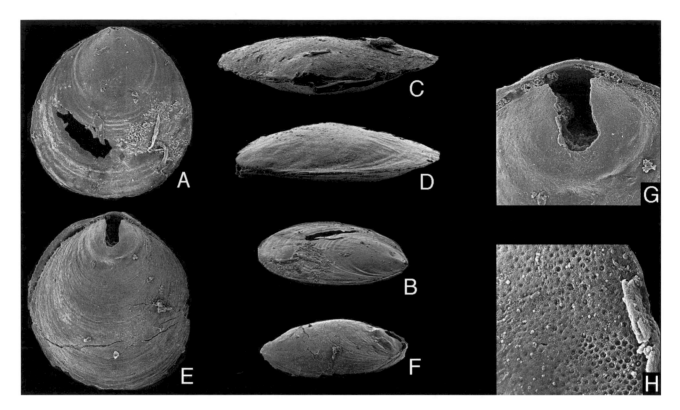

*Fig. 67. Fossuliella konevae* sp. nov.; Kujandy Formation (sample 7827); Aksak–Kujandy Section, northeastern Central Kazakhstan. □A. Ventral view of complete juvenile shell; RM Br 136038; ×44. □B. Lateral view of A; ×40. □C. Posterior view of A; ×75. □D. Lateral view of complete juvenile shell; RM Br 136040; ×41. □E. Ventral view of complete juvenile shell; RM Br 136039; ×67. □F. Lateral view of E; ×30. □G. Detail of E, showing ventral larval shell with emarginature; ×165. □H. Detail of ventral larval shell of E, showing pitted micro-ornamentation; ×500.

subtriangular median groove. Dorsal interior with short median ridge, extending to central muscle scars.

*Description.* – The shell is somewhat dorsi-biconvex and elongate oval. The ventral valve is gently and evenly convex, about 1.4 times as long as wide. The ventral pseudointerarea is orthocline and has a deep, semicircular to elongate oval emarginature that perforates the umbo of the valve; the propareas are short and concave in profile, with flexure lines (Fig. 66I–L). The ventral interior visceral area is poorly defined and subtriangular, occypying the posterior half of the valve (Fig. 66L).

The dorsal valve is moderately and evenly convex, on average 1.3 times as long as wide (Table 15). The dorsal

pseudointerarea is relatively high and orthocline, on average 40% as wide as long and occupying 55% of the valve width (Table 15); the median groove is widely triangular and deeply concave (Fig. 66C, D). The dorsal visceral field is weakly impressed and has a narrow median tongue that extends, on average, 56% of the valve length and occupies 64% of the maximum valve width (Table 14). A short dorsal median ridge originates directly anterior to the median groove and extends to the central muscle scars (Fig. 66C, D, M). The proximal parts of the dorsal *vascula lateralia* are straight and divergent; the *vascula media* are long and widely divergent.

The postlarval shell has an ornamentation of fine, closely spaced rugellae, up to about 80 μm apart, with finer growth

*Table 15. Fossuliella konevae*, average dimensions and ratios of dorsal valves.

|  | L | W | LI | WI | ML | MG | LS | LV | WV | $L/W$ | $LI/WI$ | $ML/L$ | $MG/WI$ | $WI/W$ | $LV/L$ | WV/W |
|---|---|---|---|---|---|---|---|---|---|---|---|---|---|---|---|---|
| Sample 79101 | | | | | | | | | | | | | | | | |
| N | 9 | 9 | 9 | 9 | 9 | 9 | 9 | 4 | 7 | 9 | 9 | 9 | 9 | 9 | 4 | 7 |
| X | 1.85 | 1.43 | 0.30 | 0.80 | 0.19 | 0.53 | 0.49 | 1.15 | 0.95 | 129% | 40% | 10% | 72% | 55% | 56% | 64% |
| S | 0.54 | 0.37 | 0.13 | 0.32 | 0.07 | 0.16 | 0.21 | 0.36 | 0.28 | 0.12 | 0.15 | 0.03 | 0.24 | 0.13 | 0.08 | 0.10 |
| MIN | 1.22 | 1.04 | 0.16 | 0.30 | 0.08 | 0.36 | 0.24 | 0.92 | 0.68 | 113% | 24% | 7% | 46% | 24% | 46% | 49% |
| MAX | 2.84 | 2.22 | 0.54 | 1.40 | 0.32 | 0.86 | 0.88 | 1.68 | 1.30 | 145% | 73% | 17% | 127% | 65% | 65% | 81% |

lines in between; it also has a pitted postlarval micro-orna-mentation that is, however, poorly preserved in this material. The larval shells are usually poorly defined on the adults (Fig. 66G, I), but they can be observed on some juvenile complete shells (Fig. 67A–H). In two specimens, the larval shells of both valves are close to circular, about 0.5 mm across; the ventral larval shell has a minute pedicle groove (Fig. 66A–D). In the third complete shell, the larval shells of both valves are less well defined and smaller, about 0.3 mm across, and pitted with closely spaced circular pits, of somewhat varying size, up to 4 μm across (Fig. 67H); the ventral larval shell has a long slit, which almost reaches the anterior margin, and which may represent the beginning of the emarginature (Fig. 67E–G).

*Discussion.* – *F. konevae* differs from the type species, *F. linguata*, in having (1) a larger and deeper emarginature, (2) shorter and more reduced ventral propareas, and (3) a shorter dorsal median ridge. Moreover, the postlarval rugel-lose ornamentation of *F. linguata* is more subdued; the finer details of the postlarval pitting is much better preserved in *F. linguata*, described by Popov & Ushatinskaya (1992, Pl. 2:1), and cannot be compared in detail with that of *F. konevae*. Popov & Ushatinskaya (1992, Fig. 10) published a recon-struction of the dorsal musculature and mantle canals of *F. konevae* [erroneously referred to *Dysoristus*].

*Occurrence.* – This species occurs in the Kujandy Formation at the Kujandy Section (Fig. 25; Appendix 1G) and the Aksak–Kujandy Section (Fig. 27; Appendix 1H). It is also present in olistoliths within the Erzhan Formation at the Aksak–Kujandy Section.

# Genus *Rowellella* Wright, 1963

*Type species.* – Original designation by Wright (1963, p. 233); *Rowellella minuta* Wright, 1963; Upper Ordovician (Ashgill, Cautleyan) Portrane Limestone; Portrane, Ireland.

*Diagnosis.* – Shell dorsi-biconvex, elongate oval to subrect-angular; ornamented by concentric rugellae, becoming lamellose peripherally. Ventral pseudointerarea with wide pedicle groove and narrow, elevated propareas, lacking flex-ure lines. Ventral visceral field slightly thickened anteriorly, not extending to mid-length. Dorsal valve geniculate ven-trally, with low, undivided pseudointerarea. Dorsal visceral field with anterior platform bearing central and anterior lateral muscle scars, extending far anterior to mid-length; *vascula lateralia* of both valves peripheral; *vascula media* short, divergent.

*Remarks.* – Holmer (1989, p. 76) gave a brief review of the genus and a list of included species.

*Occurrence.* – Ordovician (Tremadoc–Ashgill); Ireland, Bo-hemia, Baltoscandia, Poland, Kazakhstan, North America.

## *Rowellella* sp.
Fig. 68.

*Synonymy.* – □1973 *Rowellella* sp. 2 – Biernat, p. 60, Pl. 3:2–4. □?1986 *Rowellella parallela* sp. nov. – Bednarczyk, p. 411, Pl. 1:1, 5. □?1986 *Rowellella multilamellata* sp. nov. – Bed-narczyk, p. 411, Pl. 1:2, 4.

*Material.* – Figured. Indeterminate valves: LO 6521t; LO 6522t; LO 6523t. Total of 6 valves.

*Remarks.* – The complete outline of neither valve is pre-served, but it appears to have been elongate suboval to subtriangular. The posterior parts of the valves are invariably broken and characters of the pseudointerareas and visceral areas cannot be observed. The surface ornamentation con-sists of rugellae superposed on growth lamellae (Fig. 68A–C). Similarly poorly preserved specimens of *Rowellella* were de-scribed by Bednarczyk (1986) as *R. parallela* and *R. multila-mellata* from lower Arenig beds in northern Poland. How-ever, the fragmentary preservation of these specimens makes it difficult to compare them in detail with any other species. It is possible that the specimens of *Rowellella* sp. 2 described by Biernat (1973) from the Tremadoc chalcedonites of the Holy Cross Mountains also belong to the same species.

*Occurrence.* – *R.* sp. occurs in the Bjørkåsholmen Limestone at Ottenby, Fågelsång, Scania, and at Stora Backor (Figs. 5–6; Appendix 1K).

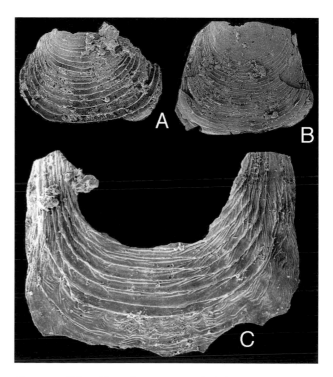

*Fig. 68. Rowellella* sp. □A. Valve exterior; Bjørkåsholmen Limestone (sam-ple Sk-3); Fågelsång, Scania; LO 6521t; ×40. □B. Valve exterior; Bjørkåsh-olmen Limestone (sample Sk-3); Fågelsång, Scania; LO 6522t; ×32. □C. Valve exterior; Bjørkåsholmen Limestone (sample Öl-1); Ottenby, Öland; LO 6523t; ×82.

# Family Dysoristidae Popov & Ushatinskaya, 1992

*Diagnosis.* – Shell subcircular or elongate oval. Larval and postlarval shell ornamented by fine pits. Ventral valve with circular foramen, becoming enlarged anteriorly through resorption, and forming pedicle track; posterior part of pedicle track closed by plate. Ventral pseudointerarea flat, undivided. Dorsal pseudointerarea with median groove and well-developed propareas. Dorsal visceral field with median tongue, extending to mid-length.

*Genera included.* – *Dysoristus* Bell, 1944; *Ferrobolus* Havlíček, 1982.

*Discussion.* – *Dysoristus* was previously placed within the siphonotretids (Rowell 1962, p. 148). Havlíček (1982, p. 71) described *Ferrobolus* and included both genera within the Subfamily Schizamboninae. However, as noted by Popov & Ushatinskaya (1992, p. 66), the baculate shell structure (Fig. 70L–M) and the pitted micro-ornamentation of the larval (Fig. 73L, M) and postlarval shells (Fig. 74E) of species of *Dysoristus* and *Ferrobolus* indicate that they are more closely related to the lingulids. Moreover, both genera lack hollow spines, the most distinctive feature of siphonotretids.

The Family Dysoristidae is most similar to the lingulid Family Zhanatellidae Koneva, 1986, which is also characterized by having a finely pitted larval and postlarval micro-ornamentation. Dysoristids differ from zhanatellids in having a pedicle foramen.

*Occurrence.* – Upper Cambrian – Lower Ordovician (Arenig).

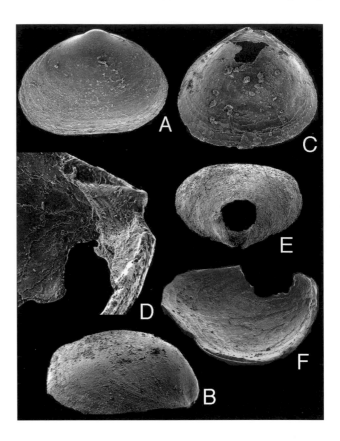

*Fig. 69. Dysoristus lochmanae* Bell; Dresbachian; Nevada. □A. Dorsal exterior; RM Br 136265; ×21. □B. Lateral view of A; ×30. □C. Ventral interior; RM Br 136266; ×21. □D. Detail of C, showing pseudointerarea; ×75. □E. Posterior view of ventral exterior; RM Br 136267; ×21. □F. Lateral view of dorsal interior; RM Br 136268; ×21.

# Genus *Dysoristus* Bell, 1944

*Type species.* – Original designation by Bell (1944, p. 146); *Dysoristus lochmanae* Bell, 1944; Upper Cambrian *Cedaria* Biozone; Montana (refigured in Fig. 69 herein).

*Diagnosis.* – Shell biconvex, subtriangular to elongate oval. Pedicle track widely subtriangular. Dorsal visceral field with narrow anterior tongue, extending anterior to mid-length, and bisected by weak median ridge. *Vascula lateralia* of both valves marginal, arcuate; *vascula media* divergent.

*Species included.* – *Dysoristus lochmanae* Bell, 1944; *Linnarssonella transversa* Walcott, 1908; *Dysoristus orientalis* sp. nov; ?*Schizambon* sp. Henderson (*in* Henderson *et al.*), 1992.

*Discussion.* – Pelman (1977, p. 34) described *Dysoristus belli* from the Upper Lower Cambrian – Middle Cambrian of Siberia. However, this species lacks an enclosed pedicle foramen, and Koneva (1986, p. 52) referred it to the obolide *Kyrshabaktella* Koneva.

*Fig. 70* (opposite page). □A–S. *Dysoristus orientalis* sp. nov.; Kujandy Formation; northeastern Central Kazakhstan. □A. Dorsal exterior; olistolith within Erzhan Formation (sample 79101); Aksak–Kujandy Section; RM Br 136060; ×24. □B. Dorsal interior; olistolith within Erzhan Formation (sample 79101); Aksak–Kujandy Section; RM Br 136061; ×24. □C. Lateral view of B; ×60. □D. Holotype; ventral exterior; olistolith within Erzhan Formation (sample 79101); Aksak–Kujandy Section; RM Br 136051; ×16. □E. Lateral view of D; ×18. □F. Posterior view of D; ×45. □G. Lateral view of dorsal interior, showing pseudointerarea; olistolith within Erzhan Formation (sample 79101); Aksak–Kujandy Section; RM Br 136052; ×50. □H. Dorsal interior; olistolith within Erzhan Formation (sample 79101); Aksak–Kujandy Section; RM Br 136053; ×18. □I. Lateral view of dorsal interior; olistolith within Erzhan Formation (sample 79101); Aksak–Kujandy Section; RM Br 136055; ×40. □J. Anterior view of ventral valve; Kujandy Section (sample 7823-4); RM Br 136054; ×16. □K. Detail of J, showing pseudointerarea and internal pedicle foramen; ×60. □L. Shell structure of G; ×400. □M. Baculate shell structure of G; ×1800. □N. Lateral view of ventral pseudointerarea; Satpak Syncline (sample 7836-a); RM Br 136056; ×33. □O. Lateral view of ventral beak with pedicle foramen; olistolith within Erzhan Formation (sample 79101); Aksak–Kujandy Section; RM Br 136057; ×65. □P. Lateral view of ventral beak with pedicle foramen; Kujandy Section (sample 7823-4); RM Br 136058; ×33. □Q. Lateral view of ventral beak with pedicle foramen; Satpak Section (sample 7836-a); RM Br 136313; ×28. □R. Posterior view of Q; ×45. □S. Oblique anterior view of ventral interior; olistolith within Erzhan Formation (sample 79101); Aksak–Kujandy Section; RM Br 136059; ×38.

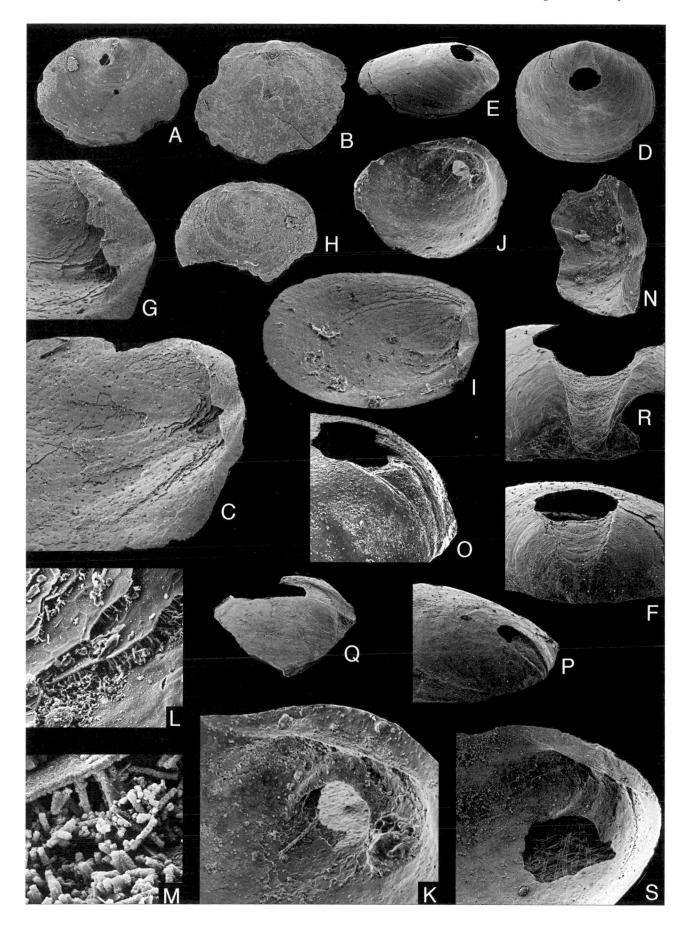

Henderson (*in* Henderson *et al.* 1992, Pl. 2:13–14) described a ventral valve [referred to *Schizambon* sp.] from the Upper Cambrian of West Antarctica. It is comparable with *Dysoristus*, in the shape of the pedicle opening and the apparent lack of spines.

*Occurrence.* – Upper Cambrian; North America, Kazakhstan, ?Antarctica.

## *Dysoristus orientalis* sp. nov.

Fig. 70A–S

*Name.* – Latin *orientalis*, of the east.

*Synonymy.* – □1992 *Disorystus* sp. [*sic*] – Popov & Ushatinskaya, Pl. 3:4 [not 3:5–7].

*Holotype.* – Fig. 70D–F; RM Br 136051; ventral valve (L 1.9, W 1.92, LI 0.38, WI 1.46, ML 0.16); from an olistolith (sample 79101) within Erzhan Formation; Aksak–Kujandy Section.

*Paratypes.* – Figured. Ventral valves: RM Br 136054 (L 2.08, 1.82, LI 0.24, WI 1.08, ML 0.16); RM Br 126056; RM Br 136057 (L 1.44, W 1.64, LI 0.24, WI 0.84, ML 0.14); RM Br 136058 (W 2.02, LI 0.38, WI 1.46, ML 0.14); RM Br 126059; RM Br 136313. Dorsal valves: RM Br 136052; RM Br 136053; RM Br 13655; RM Br 136060; RM Br 136061 (L 1.36, W 1.48, LI 0.16, WI 0.88, ML 0.1, WG 0.48, LV 0.82). Total of 18 ventral and 46 dorsal valves.

*Measurements.* – See Tables 16–17.

*Diagnosis.* – Shell ventri-biconvex, elongate suboval to subcircular. Ventral pseudointerarea orthocline to slightly anacline and undivided; pedicle foramen large, subcircular;

pedicle track elongate subtriangular. Dorsal pseudointerarea wide, orthocline to slightly apsacline, with poorly defined, narrow median groove. Dorsal interior with umbonal muscle scars divided by short ridge.

*Description.* – The shell is dorsi-biconvex and inequivalved, elongate suboval to subcircular. The ventral valve is on average 98% as long as wide (Table 16); in lateral profile it is moderately and unevenly convex, with the maximum height in the posterior third of the valve. The ventral pseudointerarea is orthocline to slightly anacline, flat and undivided, about as long as wide, and occupies on average 60% of the maximum valve width (Fig. 70J, K, N, S; Table 16). The pedicle foramen is large and subcircular, up to about 0.4 mm wide, placed at the end of a triangular, tear-shaped pedicle track, totally up to about 0.8 mm long, and occupying about 40% of the total valve length, covered posteriorly by a triangular, slightly concave plate (Fig. 70D–F, O–R). The ventral visceral area is poorly defined, extending to the centre of the valve. The ventral *vascula lateralia* are arcuate and submarginal. The dorsal valve is flattened, only slightly convex, subcircular, on average 92% as long as wide (Table 17). The dorsal pseudointerarea is low and orthocline to slightly apsacline; it occupies on average 65% of the maximum valve width (Table 17). The median groove is narrow and poorly defined, supported anteriorly by a short ridge that divides the umbonal muscle scar. The dorsal interior has a pair of large, subtriangular central muscle scars at about the centre of the valve; the anterior lateral muscle scars are small and placed anteromedianly to the central scars, somewhat anterior to the mid-length (Fig. 70B, C, G–I). The dorsal *vascula lateralia* are submedianly placed and widely divergent.

The larval and postlarval shells are covered by fine circular pits, but the ornamentation is not well preserved in this material. The baculate shell structure has thin, slender baculae, usually somewhat more than 1 μm thick, consisting of granular apatite (Fig. 70L, M).

*Discussion.* – *D. orientalis* differs from *D. transversus* (Walcott) and *D. lochmanae* Bell (Fig. 69), in having (1) a subcircular outline, (2) a flattened dorsal valve, (3) a wider dorsal pseudointerarea, and (4) a very short dorsal median ridge. Popov & Ushatinskaya (1992, Fig. 10) published a reconstruction of the musculature and mantle canals of *D. orientalis*; however, only the figured ventral valve represent this species, and the dorsal valve belongs to *Fossuliella konevae* sp. nov.

*Table 16. Dysoristus orientalis*, average dimensions and ratios of ventral valves.

| | L | W | LI | WI | ML | $L/_W$ | $WI/_W$ |
|---|---|---|---|---|---|---|---|
| Samples 7823-4, 79101 | | | | | | | |
| N | 4 | 5 | 5 | 5 | 5 | 4 | 5 |
| X | 1.80 | 1.86 | 0.27 | 1.13 | 0.18 | 98% | 60% |
| S | 0.33 | 0.21 | 0.07 | 0.31 | 0.05 | 0.08 | 0.13 |
| MIN | 1.44 | 1.64 | 0.18 | 0.84 | 0.14 | 88% | 51% |
| MAX | 2.08 | 2.04 | 0.38 | 1.46 | 0.24 | 114% | 76% |

*Table 17. Dysoristus orientalis*, average dimensions and ratios of dorsal valves.

| | L | W | LI | WI | ML | MG | LS | LV | WV | $L/_W$ | $LI/_{WI}$ | $MG/_{WI}$ | $WI/_W$ | $LS/_L$ | $LV/_L$ | WV/W |
|---|---|---|---|---|---|---|---|---|---|---|---|---|---|---|---|---|
| Samples 7823-4, 79101 | | | | | | | | | | | | | | | | |
| N | 4 | 3 | 5 | 4 | 5 | 5 | 4 | 4 | 2 | 2 | 4 | 4 | 3 | 3 | 3 | 3 |
| X | 1.48 | 1.31 | 0.18 | 0.94 | 0.12 | 0.37 | 0.30 | 0.87 | 0.80 | 92% | 20% | 44% | 65% | 23% | 63% | 27% |
| S | 0.16 | 0.25 | 0.07 | 0.22 | 0.04 | 0.14 | 0.11 | 0.22 | 0.25 | 0.00 | 0.04 | 0.09 | 0.05 | 0.05 | 0.04 | 0.05 |
| MIN | 1.32 | 1.02 | 0.10 | 0.68 | 0.06 | 0.22 | 0.14 | 0.58 | 0.62 | 92% | 15% | 35% | 59% | 20% | 60% | 24% |
| MAX | 1.64 | 1.48 | 0.28 | 1.20 | 0.18 | 0.56 | 0.38 | 1.06 | 0.98 | 92% | 23% | 55% | 68% | 29% | 67% | 32% |

*Occurrence.* – The species occurs in the Kujandy Formation in the Kujandy Section (Fig. 25; Appendix 1G) and in the Satpak Syncline (Fig. 29; Appendix 1I). It is also present in olistoliths within the Erzhan Formation in the Aksak–Kujandy Section (Fig. 27; Appendix 1H).

## Genus *Ferrobolus* Havlíček, 1982

*Type species.* – Original designation by Havlíček (1982, p. 71); *Ferrobolus catharinus* Havlíček, 1982; Třenice Formation; Holoubkov, Bohemia.

*Diagnosis.* – Shell flattened, subcircular to elongate suboval. Pedicle track elongate, narrowly triangular, with rounded pedicle foramen continued as interior tube. Ventral visceral area small, not extending to mid-length; ventral *vascula lateralia* submedial.

*Species included.* – *Ferrobolus catharinus* Havlíček, 1982; *Ferrobolus concavus* sp. nov.; *Ferrobolus fragilis* sp. nov.

*Discussion.* – *Ferrobolus* differs mainly from *Dysoristus* in having a more flattened subcircular to elongate suboval shell and a more elongated and narrow pedicle track; an interior pedicle tube is present in *Ferrobolus* but poorly developed to absent in *Dysoristus*. Moreover, the larval shells of both valves are well defined in *Ferrobolus*, sometimes forming an acrotretoid-like, circular, pitted shell with raised boundaries (Figs. 71L, M, 74E, F, L, M).

*Occurrence.* – Upper Cambrian – Lower Ordovician (Arenig); Bohemia, Kazakhstan.

## *Ferrobolus concavus* sp. nov.

Fig. 71

*Name.* – Latin *concavus*, concave; alluding to the concave dorsal valve.

*Holotype.* – Fig. 71E–G; RM Br 136068; ventral valve (W 3, LI 0.18, WI 0.4); from an olistolith (sample 7825-12.5) within the Satpak Formation; Kujandy Section.

*Paratypes.* – Figured. Dorsal valves: RM Br 136067; RM Br 136069 (L 1.68, W 1.54); RM Br 136070 (L 2.14, W 1.86), RM Br 136071 (L 1.18 W 1.39). Total of 3 ventral and 6 dorsal valves.

*Diagnosis.* – Shell planoconvex to slightly concavo-convex; subcircular to elongate suboval. Pedicle track covered posteriorly by broadly triangular, concave plate. Dorsal valve gently concave in lateral profile. Dorsal pseudointerarea triangular, anacline, wide, occupying more than half of valve width; median groove broad and shallow. Dorsal interior with short median ridge and large, long, slightly thickened posterolateral muscle fields extending anterior to mid-length.

*Description.* – The shell is planoconvex to slightly concavo-convex and elongate suboval. The ventral valve is gently convex in lateral profile with the maximum height placed near the beak. The ventral pseudointerarea is orthocline, slightly concave and undivided, 45% as long as wide in the holotype. The pedicle foramen is small, about 0.18 mm wide and long in the holotype and placed at the end of a short pedicle track, totally 0.36 mm long; it is covered posteriorly by a short, broadly triangular and deeply concave plate (Fig. 71E–G); the foramen is continued as a short internal pedicle tube. The ventral visceral area is only slightly thickened and do not extend to the centre of the valve. The ventral mantle canals and muscle scars are not visible.

The dorsal valve is concave in lateral profile (Fig. 71B), on average 97% (OR 83–115%; $N$ 4) as wide as long. The dorsal pseudointerarea is anacline, high and strongly convex in profile, on average 20% (OR 13–25%; $N$ 3) as long as wide and occupying 62% (OR 47–73%; $N$ 3) of the total valve width; the median groove is poorly defined, broadly triangular and flattened (Fig. 71H, I). The dorsal interior has a poorly defined median ridge, extending for about 36–50% of the valve length (Fig. 71C–D); the posterolateral muscle fields are large, and slightly thickened, extending for about 46% (OR 41–54%; $N$ 3) of the valve length (Fig. 71H).

The valves are ornamented with fine concentric rugellae, numbering about 40–50 per mm. The dorsal larval shell is well defined, close to circular, about 0.3 mm across, and has two distinct nodes at about the centre (Fig. 71L); the larval pits are of varying size, up to about 2 μm across (Fig. 71M). The ventral larval shell could not be identified on the material (Fig. 71G), and the postlarval pits are also poorly preserved.

*Discussion.* – *F. concavus* differs from other species of the genus in having (1) a concave dorsal profile of the dorsal valve, (2) a dorsal median ridge, and (3) an unusually large dorsal pseudointerarea, with (4) a broad, very shallow median groove. In addition it differs from the type species, *F. catharinus*, in having a smaller maximum size, smaller pedicle track and opening, and larger dorsal posterolateral muscle fields that extends to the middle of the valve.

*Occurrence.* – At the type locality only (Appendix 1G).

## *Ferrobolus fragilis* sp. nov.

Figs. 72

*Name.* – Latin *fragilis*, fragile.

*Holotype.* – Fig. 72B–F; RM Br 136066; ventral valve (L 0.88, W 1.02, LI 0.04, WI 0.32); Olenty Formation (sample 601); Sasyksor Lake.

*Paratypes.* – Figured. Ventral valves; RM Br 136062 (L 1.23, W 1.85); RM Br 136064. Dorsal valves: RM Br 136063; RM Br 136147. Total of 5 ventral and 4 dorsal valves.

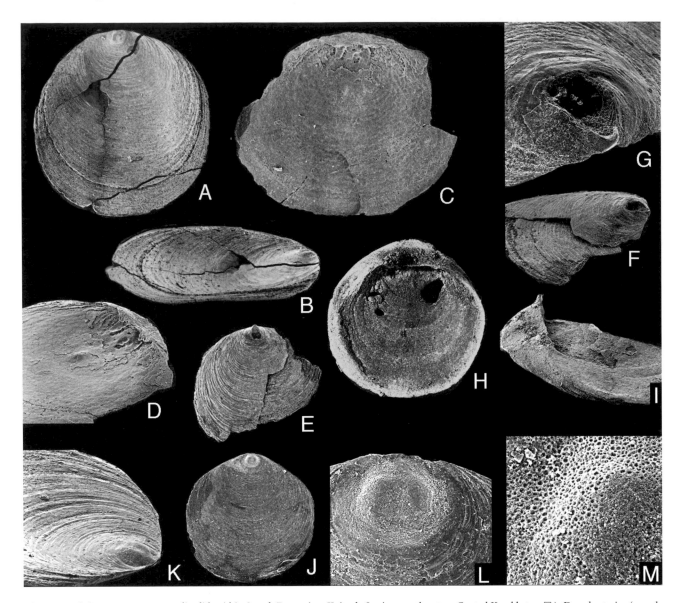

*Fig. 71. Ferrobolus concavus* sp. nov.; olistolith within Satpak Formation; Kujandy Section, northeastern Central Kazakhstan. □A. Dorsal exterior (sample 7822-4); RM Br 136070; ×20. □B. Lateral view of A; ×24. □C. Dorsal interior (sample 7822-4); RM Br 136071; ×39. □D. Lateral view of B; ×55. □E. Holotype; ventral exterior (sample 7825-12.5); RM Br 136068; ×11. □F. Lateral view of E; ×24. □G. Ventral beak with pedicle track and foramen of E; ×60. □H. Dorsal interior of A; ×15. □I. Lateral view of dorsal interior (sample 7822-4); RM Br 136067; ×39. □J. Dorsal exterior (sample 7822-4); RM Br 136069; ×22. □K. Lateral view of J; ×60. □L. Detail of J, showing larval shell; ×135. □M. Detail of J, showing pitted micro-ornament of larval shell; ×500.

*Diagnosis.* – Shell small and thin, circular. Pedicle track narrow, elongated. Dorsal pseudointerarea with median groove not elevated above valve floor. Visceral areas and muscle scars of both valves poorly defined.

*Description.* – The shell is about equally biconvex and circular. The ventral valve is gently convex; the ventral pseudointerarea is flattened, orthocline, triangular, about 10–12% as long as wide, and occupying about 46% ($N\,2$) of the total valve width (Fig. 72N). The pedicle foramen is small, rounded, and about 0.14×0.1 mm, placed at the end of an elongate, narrow pedicle track, up to about 0.4 mm long,

extending for about 20–25% of the valve length in the largest valves (Fig. 72J), but shorter in younger ones (Fig. 72B); it is closed by an elongate triangular, concave plate, extending for about 18% (Fig. 72J–M); the foramen is continued as a low, internal pedicle tube (Fig. 72N). The visceral field and muscle scars of both valves are poorly defined.

The dorsal valve is gently convex; the dorsal pseudointerarea is low and orthocline, about 10–12% ($N\,2$) as long as wide; the median groove is shallow, occupying about 30–45% ($N\,2$) of the width of the pseudointerarea (Fig. 72G).

The larval shell is close to circular, around 0.25 mm wide, poorly defined on the dorsal valve (Fig. 72I), but on the

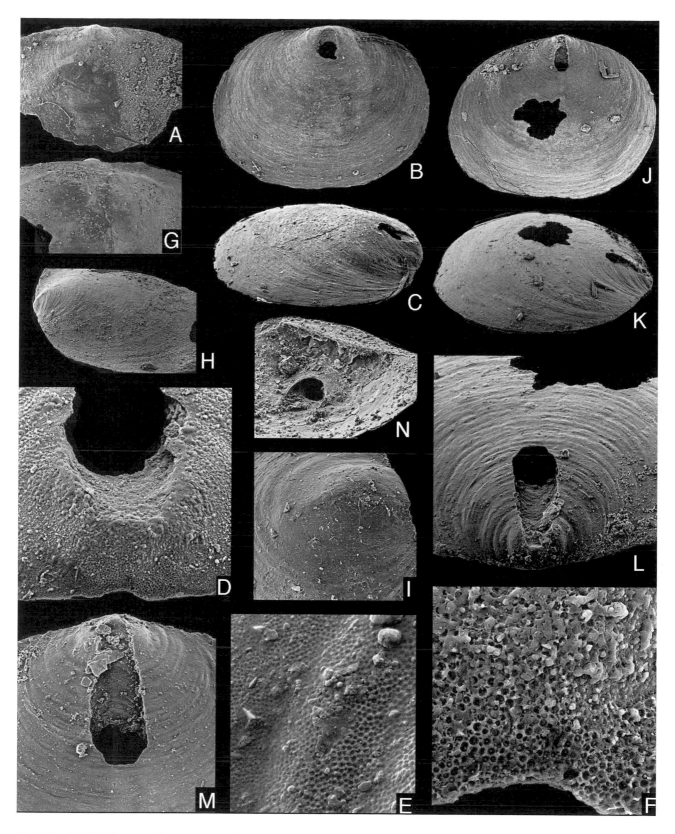

*Fig. 72. Ferrobolus fragilis* sp. nov.; Olenty Formation (sample 601); Sasyksor Lake, northeastern Central Kazakhstan. □A. Dorsal exterior, RM Br 136147; ×38. □B. Holotype; ventral exterior; RM Br 136066; ×43. □C. Lateral view of B; ×54. □D. Detail of B, showing ventral beak with pedicle track and foramen; ×300. □E. Pitted micro-ornament of postlarval shell of B; ×750. □F. Pitted micro-ornament of larval shell of B; ×1025. □G. Dorsal interior; RM Br 136063; ×40. □H. Lateral view of dorsal exterior G; ×32. □I. Larval shell of G; ×180. □J. Ventral exterior; RM Br 136062; ×30. □K. Lateral view of J; ×37. □L. Posterior view of J; ×97. □M.Ventral beak with pedicle track and foramen of J; ×105. □N. Lateral view of ventral interior; RM Br 136064; ×75.

ventral one it is defined by a slightly raised rim, which is bisected by the pedicle track (Fig. 72M); the larval pits are of slightly varying size, up to 2 μm across (Fig. 72F). The postlarval shell surface is also finely pitted, with closely packed pits of more uniform size, around 2 μm (Fig. 72E).

*Discussion.* – *F. fragilis* sp. nov. is comparable only with the type species; it differs in having (1) a smaller maximum size, (2) a circular outline, and (3) a dorsal pseudointerarea, which is not raised above the inner surface of the valve. It also differs from *F. concavus* sp. nov. in having a longer, narrower pedicle track with a longer posterior plate, and a flat ventral pseudointerarea. The dorsal larval shell of *F. fragilis* also appears to differ from those of *F. concavus*, in that the nodes are lacking.

*Occurrence.* – This species occurs in the Olenty Formation, at the Sasyksor Lake (Appendix 1H), and in the Koagash Formation at the Koagash River (Appendix 1F).

# Order Siphonotretida Kuhn, 1949

## Superfamily Siphonotretoidea Kutorga, 1848

## Family Siphonotretidae Kutorga, 1848

## Subfamily Siphonotretinae Kutorga, 1848

## Genus *Siphonobolus* Havlíček, 1982

*Type species.* – Original designation by Havlíček (1982, p. 61); *Siphonotreta simulans* Růžička, 1927; Třenice Formation; Holoubkov, Bohemia.

*Diagnosis.* – Shell ventri-biconvex, elongate oval. Growth lamellae with rows of evenly spaced, small spines of one size. Pedicle foramen moderately large, subcircular; ventral pseudointerarea poorly defined, low, undivided. Dorsal pseudointerarea large, orthocline, divided by broad, poorly defined median groove; ventral interior with long pedicle tube and slightly thickened visceral field extending to valve centre. Dorsal visceral field with long, broad median tongue, extending almost to anterior margin and bisected by low median ridge. *Vascula lateralia* of both valves marginal, arcuate.

*Species included.* – *Siphonotreta simulans* Růžička, 1927; *Siphonotreta uralensis* Lermontova, 1933.

*Occurrence.* – Lower Ordovician (Tremadoc–Arenig); Bohemia, South Urals.

## *Siphonobolus uralensis* (Lermontova, 1933)
Fig. 73

*Synonymy.* – ☐1933 *Siphonotreta uralensis* sp. nov. Lermontova *in* Lermontova & Razumovskij, p. 202, Pl. 1:7–13.

*Lectotype.* – Selected here; CNIGR 40/6565; ventral valve (Fig. 73G herein); Kidryas Formation; east of Novo-Dmitrievka (sample 564 of Lermontova & Razumovskij 1933).

*Material.* – Figured. Ventral valves: RM Br 136074 (L 4.24, W 4.16, H 1.04); RM Br 136077; RM Br 136080; RM Br 136079. Dorsal valves: RM Br 136072; RM Br 136073; RM Br 130675; RM Br 136076; RM Br 136081. Total of 156 ventral and 74 dorsal valves.

*Measurements.* – See Tables 18–19.

*Diagnosis.* – Shell subcircular. Ventral valve about as long as wide, with maximum height slightly anterior to beak. Ventral pseudointerarea low, apsacline, poorly defined laterally. Ventral interior with long, cylindrical pedicle tube. Dorsal valve gently to moderately convex, with maximum height in posterior third. Dorsal pseudointerarea high, anacline. Shell surface finely pustulose, with fine, hollow spines of equal size.

*Description.* – The shell is ventri-biconvex and subcircular. The ventral valve is on average as long as wide and 28% as high as long (Table 18), moderately and unevenly convex in lateral profile, with the maximum height placed anterior to the beak (Fig. 73E, L). The ventral pseudointerarea is low, apsacline and poorly defined laterally (Fig. 73F, Q). The pedicle foramen is subcircular, about 0.16–0.24 mm wide (Fig. 73P)). The ventral interior has a long, cylindrical, internal pedicle tube and arcuate submarginal *vascula lateralia*; the visceral area is slightly raised anteriorly and extends to the mid-length (Fig. 73M–O).

The dorsal valve is on average 98% as long as wide (Table 19), gently to moderately convex with the maximum height in the posterior third of the valve length (Fig. 73H). The dorsal pseudointerarea is undivided, slightly anacline and strongly raised above the valve floor. The dorsal interior lacks characters, with the exception of a pair of large, but poorly defined, central muscle scars, and arcuate, submarginal *vascula lateralia* (Fig. 73B, C).

The shell surface is finely pustulose, with rows of thin and hollow spines of about equal size (Fig. 73I). The larval shells of both valves are poorly defined (Fig. 73K, P).

*Discussion.* – This species is very similar to the type species, *S. simulans* (Růžička), in size and general shape. However, it differs in having (1) a long internal pedicle tube, (2) finely pustulose ornamentation, and (3) no median ridge in either valve.

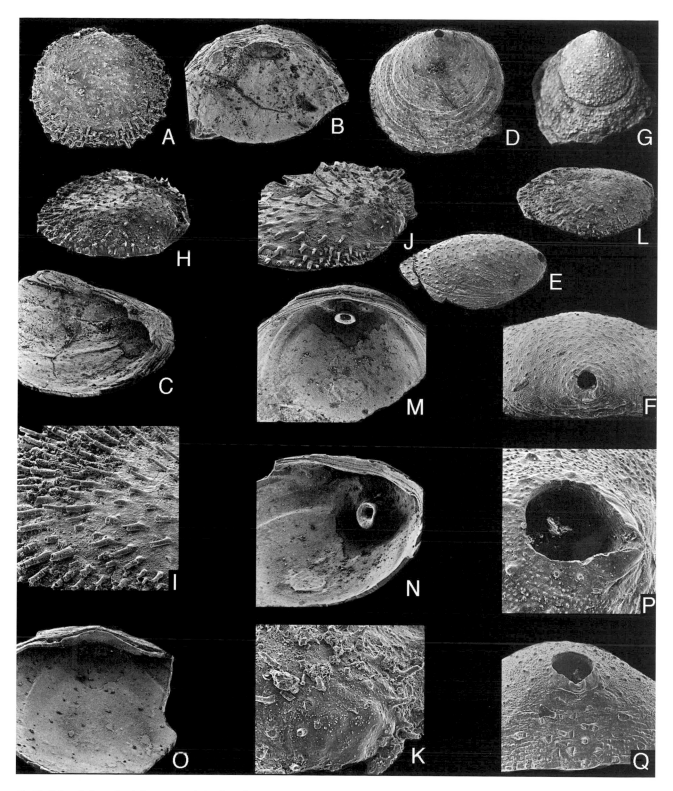

*Fig. 73. Siphonobolus uralensis* (Lermontova); South Urals. □A. Dorsal exterior; Malaya Kayala River (sample G-113); RM Br 136072; ×14. □B. Dorsal interior; Akbulaksai Formation (sample B-578-2); Alimbet Farm; RM Br 136073; ×10. □C. Lateral view of B; ×13. □D. Ventral exterior; Akbulaksai Formation (sample B-578-2); Alimbet Farm; RM Br 136074; ×8. □E. Lateral view of D; ×9. □F. Posterior view of D; ×15. □G. Lectotype; ventral exterior; Kidryas Formation; Novo-Dmitrievka; CNIGR 40/6565; ×8. □H. Lateral view of dorsal exterior; Malaya Kayala River (sample G-113); RM Br 136075; ×13. □I. Detail of ornamentation of H; ×40. □J. Lateral view of dorsal exterior; Malaya Kayala River (sample G-113); RM Br 136076; ×24. □K. Detail of larval shell of J; ×100. □L. Lateral view of ventral exterior; Malaya Kayala River (sample G-113); RM Br 136077; ×24. □M. Ventral interior; Akbulaksai Formation (sample B-578-2); Alimbet Farm; RM Br 136080; ×14. □N. Oblique lateral view of M; ×15. □O. Dorsal interior; Alimbet Formation (sample B-578-1); Alimbet Farm; RM Br 136081; ×16. □P. Detail of pedicle foramen; Akbulaksai Formation (sample B-578-2); Alimbet Farm; RM Br 136079; ×75. □Q. Posterior view of P; ×33.

*Occurrence.* – *S. uralensis* is one of the most abundant species in the upper parts of the Alimbet, Kidryas, and Akbulaksai formations, where it occurs at the following localities: Tyrmantau Ridge, Alimbet Farm, Alimbet River, Bolsjaya and Malaya Kayala rivers, and the Kosistek River (Figs. 11, 13, 16; Appendix 1B, C, F).

*Table 18. Siphonobolus uralensis*, average dimensions and ratios of ventral valves.

| | L | W | H | $L/W$ | $H/L$ |
|---|---|---|---|---|---|
| Samples B-69-1, B-578-5, B-780 | | | | | |
| N | 6 | 6 | 3 | 6 | 3 |
| X | 3.17 | 3.13 | 0.63 | 100% | 28% |
| S | 1.36 | 1.18 | 0.24 | 0.07 | 0.10 |
| MIN | 1.50 | 1.64 | 0.46 | 91% | 16% |
| MAX | 5.30 | 4.80 | 1.04 | 110% | 35% |

*Table 19. Siphonobolus uralensis*, average dimensions and ratios of dorsal valves.

| | L | W | H | ML | WI | $L/W$ |
|---|---|---|---|---|---|---|
| Samples B-69-1, B-578-5, B-780 | | | | | | |
| N | 6 | 6 | 1 | 1 | 1 | 6 |
| X | 3.02 | 3.08 | 0.32 | 0.06 | 1.12 | 98% |
| S | 0.94 | 0.93 | | | | 0.09 |
| MIN | 1.32 | 1.40 | 0.32 | 0.06 | 1.12 | 86% |
| MAX | 4.20 | 4.02 | 0.32 | 0.06 | 1.12 | 112% |

# Genus *Siphonotretella* gen. nov.

*Name.* – Greek *siphon*, pipe, and *tretos*, perforated.

*Type species.* – *Siphonotretella jani* gen. et sp. nov.; Bjørkåsholmen Limestone; Ottenby.

*Diagnosis.* – Shell ventri-biconvex and subcircular. Ventral valve subconical. Ventral pseudointerarea procline to slightly apsacline, poorly defined laterally; pedicle foramen apical, small, rounded, not becoming enlarged by resorption. Internal characters of both valves poorly known. Dorsal valve gently convex, with maximum height somewhat anterior to apex. Dorsal pseudointerarea extremely reduced, mainly consisting of median groove. Shell surface covered by fine hollow spines of about equal size.

*Species included.* – *Siphonotretella jani* sp. nov.; '*Siphonotreta acrotretomorpha* Gorjansky' – Biernat, 1973.

*Discussion.* – *Siphonotretella* is similar to *Siphonobolus* in having a ventri-biconvex shell and growth lamellae with rows of evenly spaced, small spines of one size. However, *Siphonotretella* differs in having a much smaller pedicle foramen that is formed completely without resorption, and an almost completely reduced dorsal pseudointerarea. The new genus also differs from *Siphonobolus* in lacking an internal pedicle tube.

The dorsal valve of *Siphonotretella* is also somewhat similar to that of *Acanthambonia* Cooper, which also has a pedicle foramen not enlarged by resorption; however, *Acanthambonia* differs in having (1) exclusively hemiperipheral growth of both valves, (2) a small but well-developed dorsal pseudointerarea, and (3) an internal pedicle tube along the valve floor.

## *Siphonotretella jani* sp. nov.
Fig. 74A–I

*Name.* – Alluding to the Långe Jan ('Long John') Light House, close to the type locality.

*Holotype.* – Fig. 74C–E; LO 6525T; ventral valve (L 0.72, W 0.72, H 0.3); Bjørkåsholmen Limestone (sample Öl-1); Ottenby.

*Paratypes.* – Figured. Dorsal valves: RM Br 129097 (L 0.56, W 0.6); RM Br 136261 (L 1.08, W 1.23); LO 6524t; LO 6526t; LO 6527t (L 0.64, W 0.74, H 0.2); LO 6528t. Total of 9 ventral and 18 dorsal valves.

*Diagnosis.* – Ventral valve subconical, low, with maximum height somewhat anterior to beak. Ventral pseudointerarea poorly defined, apsacline. Dorsal pseudointerarea extremely reduced, crescent-shaped.

*Description.* – The shell is ventri-biconvex, subcircular to circular. The ventral valve is conical, about as long as wide and 40% as high as long in one valve, with the maximum height somewhat anterior to the beak (Fig. 74C–D). The ventral pseudointerarea is apsacline and poorly defined laterally. The pedicle foramen is situated at the apex; it is small and rounded and does not become enlarged by resorption (Fig. 74E). The internal characters of both valves are poorly known. The dorsal valve is 86–93% (N3) as long as wide, and 30% as high as long in one valve. In lateral profile, it is gently convex with the maximum height placed somewhat anterior to the umbo (Fig. 74G). The dorsal pseudointerarea is reduced to a thin plate, occupied mainly by the median groove (Fig. 74B). The larval shell is circular, around 0.2–0.3 mm across (Fig. 74E, F). The shell surface is covered by very fine hollow spines of about equal size (Fig. 74H).

*Occurrence.* – *S. jani* is restricted to the Bjørkåsholmen Limestone in Scandinavia, where it has been recorded at the following localities: Ottenby, Flagabro, Fågelsång, Stora Backor, Mossebo (Figs. 5–7; Appendix 1K).

## *Siphonotretella* sp.
Fig. 75

*Synonymy.* – □?1973 *Siphonotreta acrotretomorpha* Gorjansky – Biernat, p. 105, Pls. 27:4–10; 28; 29; 30:6–8.

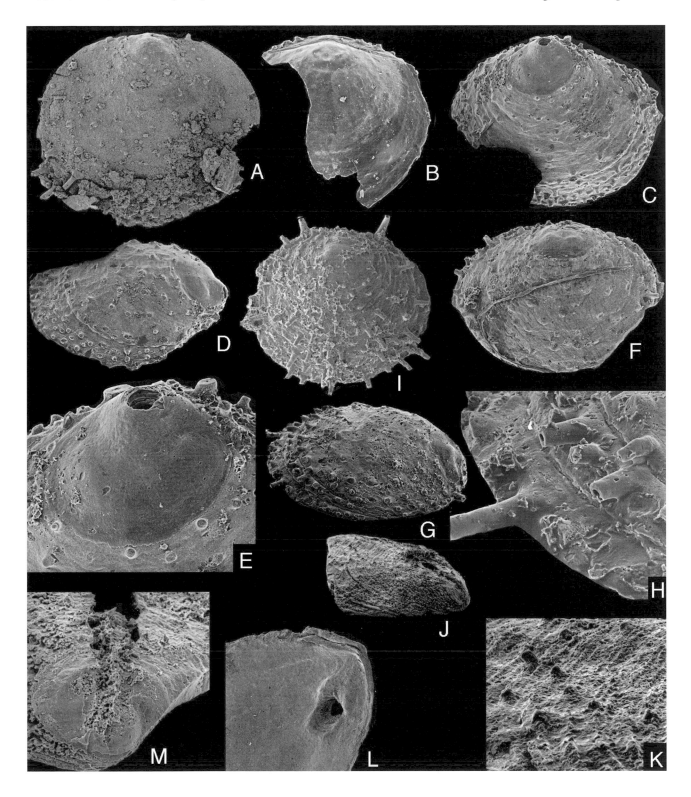

*Fig. 74.* □A–I. *Siphonotretella jani* gen. et sp. nov.; Bjørkåsholmen Limestone; Scandinavia. □A. Dorsal exterior; Galgberget (coll. G. Holm), Oslo; RM Br 136261; ×49. □B. Dorsal interior; Ottenby (sample Öl-1), Öland; LO 6524t; ×75. □C. Holotype; ventral exterior; Ottenby (sample Öl-1), Öland; LO 6525T; ×75. □D. Lateral view of C; ×75. □E. Larval shell of C; ×200. □F. Dorsal exterior; Ottenby (sample Öl-1), Öland; LO 6526t; ×75. □G. Lateral view of dorsal valve; Ottenby (sample Öl-1), Öland; LO 6527t; ×75. □H. Detail of spinose ornamentation of valve fragment; Ottenby (sample Öl-1), Öland; LO 6528t; ×320. □I. Dorsal exterior; Flagabro (sample Sk-1), Scania; RM Br 129097; ×75. □J–M. *Schizambon* sp.; Agalatas Formation (sample 564); Kendyktas Range, Kazakhstan; RM Br 136149. □J. Lateral view of ventral valve; ×30. □K. Detail of ornamentation of J; ×135. □L. Oblique lateral view of interior of J; ×45. □M. Detail of pedicle opening of J; ×120.

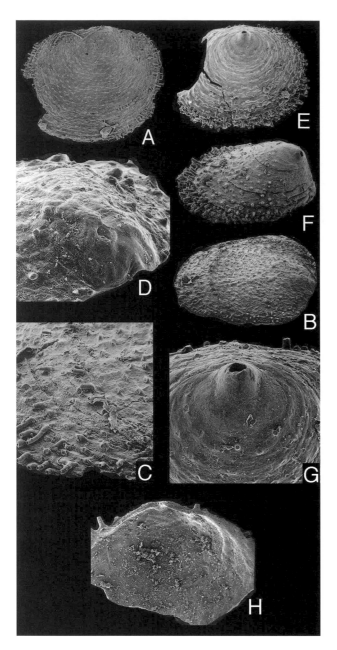

*Fig. 75.* Siphonotretella sp.; Koagash Formation; South Urals. □A. Dorsal exterior; Koagash River (sample K-458); RM Br 136082; ×15. □B. Side view of A; ×60. □C. Detail of ornamentation of A; ×120. □D. Detail of larval shell of A; ×100. □E. Ventral exterior; Karabutak River (sample 1163); RM Br 136083; ×30. □F. Lateral view of E; ×37. □G. Detail of larval shell E; ×100. □H. Dorsal interior; Koagash River (sample K-458); RM Br 136084; ×45.

*Material.* – Figured. Ventral valve: RM Br 136083 (L 0.96, W 1.15, H 0.48). Dorsal valves: RM Br 136082 (L 2, W 2.05); RM Br 136084. Total of 1 ventral and 3 dorsal valves.

*Description.* – The shell is ventri-biconvex, subcircular. The ventral valve is subconical, about 83% as long as wide and about 50% as high as long in one valve (Fig. 75E, F). The ventral pseudointerarea is procline, poorly defined laterally (Fig. 75E). The pedicle foramen is situated at the apex; it is small and rounded and does not become enlarged by resorption (Fig. 75G). The internal characters of both valves are poorly known.

The dorsal valve is gently convex, with the maximum height placed somewhat anterior to the umbo; it is about 79% as long as wide in one specimen (Fig. 75B). The dorsal pseudointerarea is reduced to a thin plate, occupied mainly by the median groove (Fig. 75H). The shell surface is covered by very fine hollow spines of about equal size (Fig. 75C).

*Discussion.* – The external characters of the ventral valve and the shape of the pedicle foramen are closely similar to the siphonotretid species described by Biernat (1973) as '*Siphonotreta acrotretomorpha* Gorjansky'; from comparative studies of the type material of Gorjansky's species from the Leetse beds in Estonia, it is evident that *S. acrotretomorpha* is not conspecific with the species described by Biernat (1973) from the Holy Cross Mountains. However, Biernat's species might be conspecific with *Siphonotretella* sp. described here.

'*Siphonotreta*' *acrotretomorpha* (Gorjansky, 1969) differs mainly in having a pedicle foramen which is enlarged anteriorly by resorption.

*S.* sp. differs from *S. jani* in having a comparatively higher conical ventral valve with maximum height at about the beak, and a procline pseudointerarea.

*Occurrence.* – This rare species occurs in the Koagash Formation at the Koagash and Karabutak rivers (Appendix 1E–G). In Kazakhstan, it is known from the Olenty Formation at the Sasyksor Lake (Appendix 1H).

## Subfamily Schizamboninae Havlíček, 1982

*Diagnosis.* – Shell ventri-biconvex, ornamented with spines, usually of more than one size. Pedicle foramen large, subtriangular to circular, usually extending forward by resorption, to produce elongate track, which may be closed posteriorly by concave plate; pedicle tube variably developed.

*Genera included.* – *Schizambon* Walcott, 1884; *Multispinula* Rowell, 1962; *Cyrbasiotreta* Williams & Curry, 1985; *Nushbiella* Popov, 1986.

*Occurrence.* – Middle Cambrian – Upper Ordovician.

## Genus *Schizambon* Walcott, 1884

*Type species.* – Subsequent designation by Oehlert (*in* Fischer 1887, p. 1266); *Schizambon typicalis* Walcott, 1884; Lower Ordovician Pogonip Limestone, Hamburg Ridge, Eureka district, Nevada.

*Diagnosis.* – Shell dorsi-biconvex; ornamented by short discontinuous costellae and evenly spaced, fine spines. Ventral pseudointerarea low, undivided, apsacline; pedicle track elongate triangular, posterior part covered by plate. Dorsal pseudointerarea low, with poorly defined median groove; visceral areas of both valves slightly thickened. Dorsal visceral area with median tongue extending anterior to centre. *Vascula lateralia* of both valves marginal, arcuate.

*Occurrence.* – Upper Cambrian – Middle Ordovician; North America, Europe, Asia.

## *Schizambon* sp.

Figs. 74J–M

*Material.* – Figured. Ventral valve: RM Br 136149 (the single available specimen).

*Description.* – The valve is flattened, only slightly convex with the maximum height somewhat anterior to the umbo; the beak is marginal (Fig. 74J). The pedicle foramen is small, subcircular, about 0.1 mm across, situated at the anterior end of a long, narrow, subtriangular pedicle track, which is covered by a thick, slightly concave plate (Fig. 74L). The ventral interior has a slightly thickened visceral area; the composite muscle scars of the transmedian, middle lateral and outside lateral muscles are situated on the posterolateral sides of the visceral area. The *vascula lateralia* are straight and slightly divergent in their proximal parts (Fig. 74L). The shell surface is ornamented by fine spines of about equal size (Fig. 74K).

*Discussion.* – *Schizambon* sp. is similar to the type species, *S. typicalis* Walcott (Rowell 1962) in general shell shape and ornamentation, as well as in the morphology of the pedicle opening.

*Occurrence.* – The species occurs in the upper part of the Agalatas Formation (sample 564; Appendix 1J).

# Order Acrotretida Kuhn, 1949

# Superfamily Acrotretoidea Schuchert, 1893

# Family Acrotretidae Schuchert, 1893

*Diagnosis.* – Ventral valve conical, more rarely convex. Ventral pseudointerarea commonly bisected by intertrough or interridge; pedicle foramen circular or elongate oval, apical or immediately posterior to beak; foramen completely or partly within larval shell. Apical process usually present, but variable. Dorsal interior with median septum or ridge; median buttress commonly present; cardinal muscle fields usually thickened, slightly divergent, surrounded by raised ridge.

*Occurrence.* – Lower Cambrian – ?Silurian.

# Genus *Acrotreta* Kutorga, 1848

*Type species.* – Subsequent designation by Davidson (1853, p. 133); *Acrotreta subconica* Kutorga, 1848; Lower Ordovician (Billingen Stage) Päite beds; Tosna River, Ingria.

*Diagnosis.* – Shell finely pustulose with wide posterior margin. Ventral valve highly conical; ventral pseudointerarea well defined, apsacline to catacline or slightly procline; pedicle foramen never completely enclosed within larval shell. Dorsal pseudointerarea with wide median groove and propareas. Ventral interior with apical process forming a high septum and muscular platform, and enclosing internal pedicle tube, usually occupying more than half of valve height; three pairs of strongly thickened pinnate mantle canals. Dorsal interior with triangular median septum, and median buttress.

*Species included.* – See list of Holmer & Popov (1994).

*Occurrence.* – Lower–Middle Ordovician (Tremadoc?; Arenig–Llandeilo); South Ural Mountains, Ingria, northern Estonia, Poland, Bohemia, Sweden, North America.

*Discussion.* – *Acrotreta* was revised recently by Holmer & Popov (1994), who gave a review and comparative discussion of the genus as well as a select list of the species currently included. It would seem that most species are of Arenig or younger age; the only possible Tremadoc record is *Acrotreta dissimilis* (Biernat, 1973) from chalcedonites in the Holy Cross Mountains, Poland. This species is currently being redescribed together with the other rich lingulate fauna from these beds (Biernat & Holmer, unpublished).

*Acrotreta* also occurs in the Třenice Formation (here referred to the Arenig) of Bohemia, where *A. grandis* (Klouček, 1919) [=? *Conotreta obesa* Havlíček, 1980] has been recorded (Havlíček 1980).

Bednarczyk (1959a, b) described seven species (that he referred to *Conotreta*) from the 'Upper Tremadoc' at a single locality in the Holy Cross Mountains, Poland (see also Bednarczyk, 1964); these were referred to a single species, *Acrotreta kozlowskii* (Bednarczyk, 1959a) by Holmer & Popov (1994).

## *Acrotreta korynevskii* Holmer & Popov, 1994

Fig. 76

*Synonymy.* – □1994 *Acrotreta korynevskii* sp. nov. Holmer & Popov, p. 441, Fig. 9:1–12.

*Holotype.* – RM Br 133926; ventral valve; Koagash Formation; Koagash River (sample B-523).

*Paratypes.* – Figured. Ventral valves: RM Br 136087; RM Br 136088; RM Br 136093; RM Br 136095. Dorsal valves: RM Br 136085; RM Br 136086; RM Br 136089; RM Br 136091; RM Br 136092; RM 136094.

*Measurements.* – See Holmer & Popov (1994).

*Diagnosis.*– Ventral valve moderately conical, up to ⁴/₅ as high as wide; procline to catacline. Ventral pseudointerarea flattened, poorly divided. Ventral interior with extremely wide and high apical process, occupying well over half the height of the valve; pinnate mantle canals exceptionally strongly thickened. Dorsal valve convex, almost lacking median sulcus. Dorsal interior with narrow, triangular median septum, reaching its highest point at mid-valve.

*Description.*– See Holmer & Popov (1994). The ventral valve in this material has characteristic profile with an umbo that is somewhat curved backwards (Fig. 76F, H); the ventral pseudointerarea is wide, flattened and poorly divided (Fig. 76F, G, S).

*Discussion.* – *A. korynevskii* differs from the type species mainly in having a less convex ventral valve, poorly developed pustulose ornamentation, higher and thicker ventral apical process, and a higher dorsal median septum.

The material illustrated here is less fragmented than the specimens figured by Holmer & Popov (1994). A detailed comparative discussion was provided by the same authors.

*Distribution.*– This species is known from the Akbulaksai and Koagash formations at the following localitites: Alimbet (Fig. 16), Akbulaksai (Fig. 18; Appendix 1D), Karabutak, and Koagash rivers (Appendix 1E).

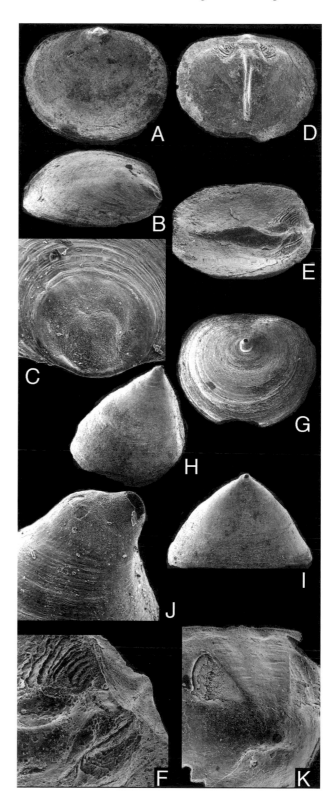

*Fig. 76* (opposite page). *Acrotreta korynevskii* Holmer & Popov; South Urals. □A. Dorsal exterior; Koagash Formation; Koagash River (sample B-523); RM Br 136085; ×32. □B. Dorsal interior; Akbulaksai Formation; Akbulaksai River (sample B-607-15); RM Br 136086; ×15. □C. Lateral view of B; ×18. □D. Detail of B, showing pseudointerarea and cardinal muscle fields; ×33. □E. Ventral exterior; Koagash Formation; Koagash River (sample B-523); RM Br 136087; ×16. □F. Lateral view of E; ×16. □G. Posterior view of E; ×18. □H. Lateral view of E; ×18. □I. Ventral interior; Koagash Formation; Koagash River (sample B-523); RM Br 136088; ×13. □J. Lateral view of K; ×65. □K. Lateral view of dorsal interior; Koagash Formation; Koagash River (sample B-523); RM Br 136091; ×25. □L. Plane view of K; ×16. □M. Lateral view of dorsal exterior; Akbulaksai Formation; Akbulaksai River (sample B-607-15); RM Br 136089; ×15. □N. Larval shell of M; ×65. □O. Lateral view of dorsal interior; Koagash Formation; Koagash River (sample B-523); RM Br 136094; ×33. □P. Dorsal pseudointerarea and cardinal muscle scars of O; ×28. □Q. Ventral interior; Koagash Formation; Koagash River (sample B-523); RM Br 136093; ×32. □R. Oblique lateral view of juvenile dorsal exterior; Koagash Formation; Koagash River (sample B-523); RM Br 136092; ×32. □S. Detail of apex of ventral exterior; Koagash Formation; Koagash River (sample B-523); RM Br 136095; ×50. □T. Lateral view of R; ×33.

*Fig. 77. Conotreta shidertensis* sp. nov.; Olenty Formation; Sasyksor Lake (sample 601). □A. Dorsal exterior; RM Br 136225; ×15. □B. Lateral view of A; ×21. □C. Larval shell of A; ×100. □D. Dorsal interior; RM Br 136226; ×14. □E. Lateral view of D; ×18. □F. Dorsal pseudointerarea and cardinal muscle fields of D; ×28. □G. Holotype; ventral exterior; RM Br 136227; ×15. □H. Lateral view of G; ×15. □I. Larval shell of G; ×100. □J. Posterior view of G. □K. Ventral interior; RM Br 136228; ×50.

# Genus *Conotreta* Walcott, 1889

*Type species.* – Original designation by Walcott (1889, p. 365); *Conotreta rusti* from the Upper Ordovician Trenton Group ('Trenton limestone'); Trenton Falls, New York.

*Discussion.* – As noted by numerous previous authors, the type species is in need of revision. This work is in preparation by one of us (Holmer) but is outside the scope of the present paper; the syntypes and some possible topotypes indicate that *Conotreta* differs from *Acrotreta* in lacking a ventral median septum. In addition, the pedicle opening of *Conotreta* is invariably placed within the larval shell, while that of *Acrotreta* is partly outside.

Holmer (1989, p. 81) proposed that *C. multisinuata* Cooper, 1956, might serve as a kind of 'provisional standard of reference', and comparisons with new topotypes of the type species seem to indicate that this species illustrate characters typical of *Conotreta*.

*Occurrence.* – Ordovician (Arenig–Caradoc); South Urals, Ingria, Estonia, Poland, Bohemia, Sweden, North America.

## *Conotreta shidertensis* sp. nov.
Fig. 77

*Name.* – After the Shiderty River, close to the type locality.

*Holotype.* – Fig. 77G–J; RM Br 136227; ventral valve (L 1.92, W 2.32 H 1.34); Olenty Formation (sample 601); Sasyksor Lake.

*Paratypes.* – Figured. Ventral valve: RM Br 136228. Dorsal valves: RM Br 136225; RM Br 136226 (L 2.14, W 2.52, LI 0.15, WI 1.36, WG 0.48, LM1 0.54, WM1 1.28, LS 1.64). Total of 11 ventral and 21 dorsal valves.

*Measurements.* – See Tables 20–21.

*Diagnosis.* – Ventral valve widely conical. Ventral pseudo-interarea almost catacline, with narrow, poorly defined interridge. Ventral interior with elongate, subtriangular, thickened cardinal muscle fields; apical process elongate subtriangular, situated anterior to pedicle opening. Dorsal valve with low and wide, slightly anacline pseudointerarea, occupying about half of maximum valve width. Dorsal interior with low triangular median septum, lacking septal rods; dorsal cardinal muscle scars large, transversley oval, thickened.

*Description.* – The shell is subcircular. The ventral valve is widely conical, on average 86% as long as wide and 64% as high as long (Fig. 77G–I; Table 20). The ventral pseudointerarea is almost catacline, with a narrow, poorly defined interridge (Fig. 77I). The pedicle foramen is small, rounded, situated on the top of a low external pedicle tube (Fig. 77J). The lateral sides of the ventral valve are gently convex in transverse profile (Fig. 77I). The ventral interior has large, elongate subtriangular and strongly thickened cardinal muscle fields, and a high, triangular apical process, situated directly anterior to the interior pedicle opening (Fig. 77K).

The dorsal valve is gently and evenly convex in lateral profile, on average 85% as long as wide (Fig. 77B; Table 21). The dorsal pseudointerarea is orthocline to slightly anacline, on average 19% as long as wide and occupying 48% of the valve width; the median groove is gently concave, lens-like (Fig. 77F). The dorsal interior has a triangular median septum, extended on average 81% of the valve length, with the maximum height placed on average 52% of the valve length from the posterior margin (Fig. 77D; Table 21). The cardinal muscle scars are large, transversely suboval in outline, strongly thickened, occupying on average 47% of the valve width and 30% of the length (Fig. 77D, F; Table 21).

The valves are ornamented by poorly developed fila. The larval shells of both valves are well defined, circular, about 0.3 mm across; the dorsal one has three distinct nodes (Fig. 77C).

*Discussion.* – *C. shidertensis* is closely comparable with *C. mica* Gorjansky (see also Holmer 1989) and *C. siljanensis* Holmer, 1989, from the Middle Ordovician of the northern

*Table 20. Conotreta shidertensis,* average dimensions and ratios of ventral valves.

|  | L | W | H | L/W | H/L |
|---|---|---|---|---|---|
| Sample 601 | | | | | |
| N | 4 | 4 | 4 | 4 | 4 |
| X | 1.27 | 1.49 | 0.83 | 86% | 64% |
| S | 0.44 | 0.56 | 0.34 | 0.07 | 0.05 |
| MIN | 0.96 | 1.14 | 0.64 | 83% | 58% |
| MAX | 1.92 | 2.32 | 1.34 | 96% | 70% |

*Table 21. Conotreta shidertensis,* average dimensions and ratios of dorsal valves.

|  | L | W | LI | WI | ML | MG | LM1 | WM1 | LS | PHS | BS | L/W | LI/WI | WI/W | MG/WI | LM1/L | WM1/W | LS/L | PHS/L | BS/L |
|---|---|---|---|---|---|---|---|---|---|---|---|---|---|---|---|---|---|---|---|---|
| Sample 601 | | | | | | | | | | | | | | | | | | | | |
| N | 10 | 10 | 10 | 10 | 10 | 10 | 10 | 10 | 10 | 10 | 10 | 10 | 10 | 10 | 10 | 10 | 10 | 10 | 10 | 10 |
| X | 1.60 | 1.88 | 0.17 | 0.90 | 0.15 | 0.33 | 0.48 | 0.88 | 1.28 | 0.82 | 0.33 | 85% | 19% | 48% | 37% | 30% | 47% | 81% | 52% | 21% |
| S | 0.34 | 0.39 | 0.03 | 0.24 | 0.03 | 0.09 | 0.09 | 0.20 | 0.23 | 0.16 | 0.07 | 0.05 | 0.03 | 0.05 | 0.06 | 0.04 | 0.05 | 0.06 | 0.07 | 0.02 |
| MIN | 1.14 | 1.40 | 0.12 | 0.64 | 0.10 | 0.22 | 0.32 | 0.56 | 0.88 | 0.50 | 0.22 | 78% | 16% | 40% | 25% | 21% | 40% | 72% | 44% | 19% |
| MAX | 2.18 | 2.68 | 0.20 | 1.28 | 0.18 | 0.52 | 0.64 | 1.22 | 1.64 | 1.04 | 0.44 | 91% | 25% | 57% | 45% | 35% | 56% | 88% | 68% | 24% |

East Baltic and Sweden, both in external characters and in the possession of a robust apical process. *C. shidertensis* differs from these two species in having a less conical ventral valve and a relatively lower dorsal median septum that lacks septal rods.

*Occurrence.* – At the type locality only (Appendix 1H).

# Genus *Dactylotreta* Rowell & Henderson, 1978

*Type species.* – Original designation by Rowell & Henderson (1978, p. 3); *Dactylotreta redunca* Rowell & Henderson, 1978; Upper Cambrian (Idamean Stage) Georgina Limestone; Glenormiston district, western Queensland, Australia.

*Diagnosis.* – Shell subcircular, with convex posterior margin. Ventral valve highly conical; ventral pseudointerarea catacline to somewhat procline, poorly defined laterally, and bisected by narrow intertrough; pedicle foramen within larval shell. Ventral interior with large apical process completely occluding apex, with apical pits posterolaterally to internal foramen. Dorsal pseudointerarea long, relatively high; median groove well developed, with anacline to orthocline propareas. Dorsal interior with low triangular median septum and median buttress. Larval shell with pits of one size.

*Species included.* – *Dactylotreta redunca* Rowell & Henderson, 1978; *Dactylotreta solitaria* Popov, 1984; *Dactylotreta patriella* Zell & Rowell, 1988; *Dactylotreta batkanensis* sp. nov.; *Dactylotreta pharus* sp. nov; *Dactylotreta* sp. Henderson (*in* Henderson *et al.*), 1992.

*Occurrence.* – Middle Cambrian (*Lejopyge laevigata* Biozone) – Lower Ordovician (Tremadoc); North America, Greenland, Asia, Australia, Antarctica.

## *Dactylotreta batkanensis* sp. nov.
Fig. 78

*Name.* – After the Batkan River, close to the type locality.

*Holotype.* – Fig. 78E–H; RM Br 136099; dorsal valve (L 2.88, W 2.92, LI 0.4, WI 1.44, WG 0.96, LS 2.2); Kidryas Formation; Tyrmantau Ridge.

*Paratypes.* – Figured: Ventral valves: RM Br 136100 (L 1.84, W 2.04, H 1.76); RM Br 136096; RM Br 136097; RM Br 136101. Dorsal valve: RM Br 136098 (L 1.72, W 1.88, LI 0.16, WI 1.12, WG 0.6, LS 1.24). Total of 18 ventral and 20 dorsal valves.

*Measurements.* – See Tables 22–23.

*Diagnosis.* – Shell large for genus; subcircular. Ventral valve as high as long; anterior and posterior slopes straight in lateral profile. Ventral pseudointerarea procline to slightly catacline, with poorly defined intertrough. Apical process with poorly defined median ridge. Dorsal valve almost flat, with strongly thickened, low triangular median septum.

*Description.* – The valves are subcircular. The ventral valve is widely conical, on average 88% as long as wide and 86% as high as long (Table 22). In lateral profile, the anterior and posterior slopes of the ventral valve are straight (Fig. 78J). The ventral pseudointerarea is procline to slightly catacline, with a narrow, poorly defined intertrough (Fig. 78K). The pedicle foramen is small, rounded, and enclosed within the larval shell (Fig. 78M). The apical process completely fills the apex; it has a poorly defined median ridge, with apical pits situated posterolaterally to the internal pedicle opening. The internal pedicle tube is completely embedded within the apical process (Fig. 78N).

The dorsal valve is flattened in lateral profile, on average 90% as long as wide (Table 23); there is a shallow median sulcus arising a short distance in front of the beak (Fig. 78A–C). The pseudointerarea is slightly anacline and occupies on average 51% of the total valve width; the median groove is concave and broadly triangular (Fig. 78H). The cardinal muscle fields are elongate suboval and slightly thickened

*Table 22. Dactylotreta batkanensis,* average dimensions and ratios of ventral valves.

|            | L    | W    | H    | L/W  | H/L  |
|------------|------|------|------|------|------|
| **Sample B-768-1** |      |      |      |      |      |
| N          | 8    | 8    | 7    | 8    | 7    |
| X          | 1.39 | 1.57 | 1.20 | 88%  | 86%  |
| S          | 0.58 | 0.65 | 0.62 | 0.03 | 0.10 |
| MIN        | 0.63 | 0.74 | 0.51 | 84%  | 74%  |
| MAX        | 2.10 | 2.30 | 1.90 | 94%  | 104% |

*Table 23. Dactylotreta batkanensis,* average dimensions and ratios of dorsal valves.

|            | L    | W    | LI   | WI   | MG   | LM1  | WM1  | LS   | PHS  | BS   | L/W  | WI/W | MG/WI | LM1/L | WM1/W | LS/L | PHS/L | BS/L |
|------------|------|------|------|------|------|------|------|------|------|------|------|------|-------|-------|-------|------|-------|------|
| **Sample B-768-1** |      |      |      |      |      |      |      |      |      |      |      |      |       |       |       |      |       |      |
| N          | 13   | 13   | 13   | 13   | 13   | 12   | 12   | 13   | 11   | 13   | 13   | 13   | 13    | 13    | 12    | 12   | 13    | 11   | 13 |
| X          | 1.49 | 1.65 | 0.14 | 0.84 | 0.53 | 0.56 | 1.02 | 1.13 | 0.97 | 0.30 | 90%  | 51%  | 64%   | 37%   | 60%   | 76%  | 66%   | 20%  |
| S          | 0.32 | 0.32 | 0.03 | 0.20 | 0.11 | 0.13 | 0.19 | 0.35 | 0.32 | 0.10 | 0.05 | 0.06 | 0.06  | 0.05  | 0.05  | 0.16 | 0.24  | 0.04 |
| MIN        | 0.87 | 0.94 | 0.09 | 0.44 | 0.29 | 0.30 | 0.59 | 0.40 | 0.57 | 0.14 | 82%  | 40%  | 55%   | 31%   | 53%   | 25%  | 50%   | 10%  |
| MAX        | 1.87 | 1.91 | 0.21 | 1.17 | 0.64 | 0.77 | 1.27 | 1.57 | 1.86 | 0.47 | 98%  | 66%  | 77%   | 49%   | 72%   | 91%  | 137%  | 26%  |

*Fig. 78. Dactylotreta batkanensis* sp. nov.; Kidryas Formation; Tyrmantau Ridge, South Urals. □A. Dorsal exterior (sample B-768-3); RM Br 136098; ×25. □B. Same view as A; ×18. □C. Lateral view of A; ×24. □D. Larval shell of A; ×165. □E. Holotype; dorsal interior (sample B-768-1); RM Br 136099; ×25. □F. Same view as E; ×30. □G. Lateral view of E; ×39. □H. Dorsal pseudointerarea and cardinal muscle fields of E; ×60. □I. Ventral exterior (sample B-768-1); RM Br 136096; ×15. □J. Lateral view of ventral exterior; (sample B-768-1); RM Br 136100; ×20. □K. Posterior view of J; ×18. □L. Lateral view of ventral exterior (sample B-768-1); RM Br 136097; ×8. □M. Larval shell of I; ×70. □N. Ventral interior (sample B-768-3); RM Br 136101; ×24.

*Fig. 79* (opposite page). *Dactylotreta pharus* sp. nov.; Bjørkåsholmen Limestone (sample Öl-1); Ottenby, Öland. □A. Dorsal exterior; LO 6530; ×29. □B. Lateral view of A; ×32. □C. Larval shell of A; ×135. □D. Detail of edge of larval shell of A; ×410. □E. Ornamentation of A; ×165. □F. Detail of pustulose ornamentation of A; ×750. □G. Dorsal interior; LO 6531t; ×45. □H. Lateral view of G; ×45. □I. Pseudointerarea of G; ×120. □J. Holotype; ventral exterior; LO 6532T; ×29. □K. Posterior view of J; ×22. □L. Lateral view of J; ×22. □M. Larval shell of J; ×150. □N. Lateral view of ventral valve; LO 6533t; ×37. □O. Ventral interior; LO 6534t; ×24.

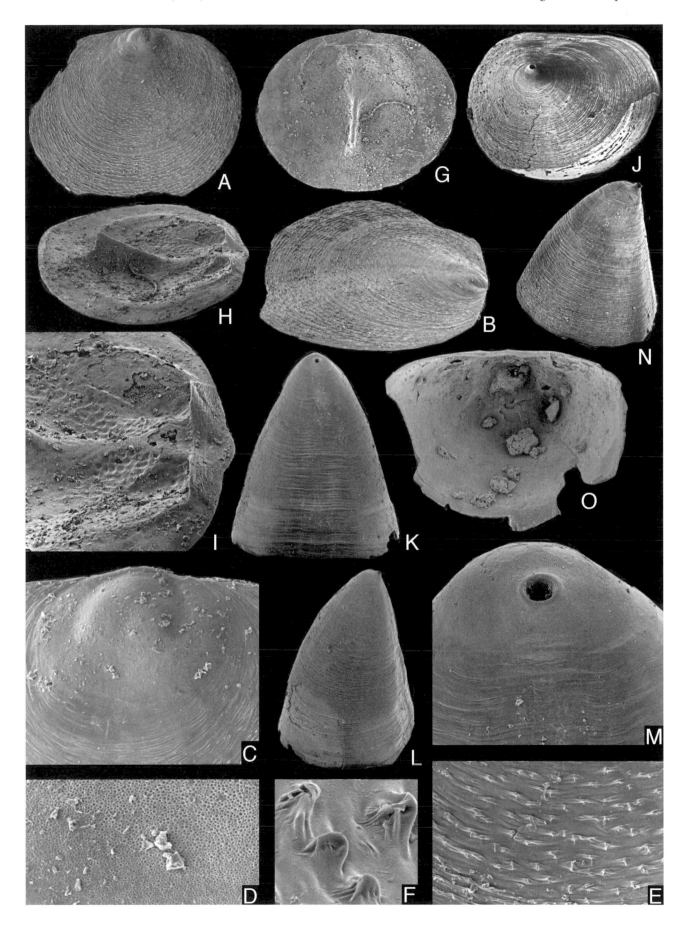

anteriorly; they occupy on average 60% of the valve width and 37% of the length (Table 23); the median buttress is narrow and triangular. The median septum is low, thickened, and triangular in profile, occupying on average 76% of the valve length (Fig. 78E–G). Both valves are ornamented by fine concentric rugellae. The larval shells of both valves are poorly defined (Fig. 78D, M).

*Discussion. – D. batkanensis* is distinguished from all other species of the genus in having a ventral valve with straight anterior and posterior slopes, as well as large, slightly thickened dorsal cardinal muscle fields. It differs from the type species, *D. redunca*, in having a thickened dorsal median septum and from *D. patriella* Zell & Rowell in having a broadly triangular dorsal pseudointerarea.

*Occurrence.* – At the type locality only (Fig. 11; Appendix 1A).

## Dactylotreta pharus sp. nov.

Figs. 79–80

*Synonymy.* – □1992 'Acrotretoid C'– Williams & Holmer, p. 659, Pl. 3:5.

*Name.* – Proposed by G. Holm, in an unpublished set of drawings (Fig. 80), probably alluding to the shape of the ventral valve that may be reminiscent of the high lighthouse that once stood on the island of Pharus outside Alexandria.

*Holotype.* – Fig. 79J–M; Lo 6532t; ventral valve (L 1.7, W 1.8, H 2.2); Bjørkåsholmen Limestone (sample Öl-1); Ottenby.

*Paratypes.* – Figured. Ventral valves: RM Br 133948 (L 1.8, W 1.8; H 2.4); RM Br 133949 (L 1.2, W 1.3, H 1.3); LO 6533t (L

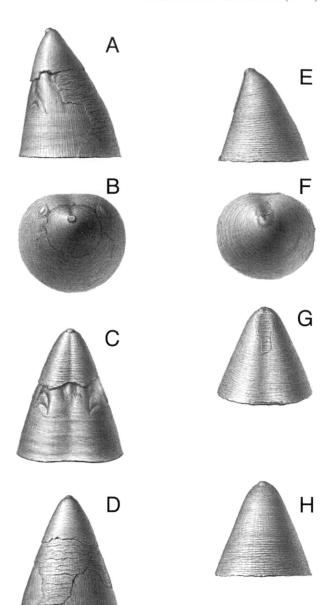

*Fig. 80. Dactylotreta pharus* sp. nov.; Camera-lucida drawings prepared by G. Holm around 1900. □A. Partly exfoliated ventral exterior; Bjørkåsholmen Limestone; Tyrifjord (detailed locality data lacking), Norway; RM Br 133948; ×12. □B. Lateral view of A; ×12. □C. Posterior view of A; ×12. □D. Anterior view of A; ×12. □E. Ventral exterior; Bjørkåsholmen Limestone; Ottenby, Öland; RM Br 133949; ×16. □F. Lateral view of E; ×16. □G. Posterior view of E; ×16. □H. Anterior view of E; ×16.

*Table 24. Dactyolotreta pharus,* average dimensions and ratios of ventral valves.

|         | L    | W    | H    | $L/W$ | $H/L$ |
|---------|------|------|------|-------|-------|
| Sample Öl-1 | | | | | |
| N       | 5    | 5    | 5    | 5     | 5     |
| X       | 1.20 | 1.43 | 1.50 | 85%   | 120%  |
| S       | 0.30 | 0.42 | 0.69 | 0.04  | 0.26  |
| MIN     | 0.84 | 0.98 | 0.84 | 80%   | 95%   |
| MAX     | 1.54 | 1.92 | 2.36 | 89%   | 153%  |

*Table 25. Dactyolotreta pharus,* average dimensions and ratios of dorsal valves.

|         | L    | W    | LI   | MG   | WI   | LM1  | WM1  | LS   | PHS  | BS   | $L/W$ | $LI/WI$ | $MG/W$ | $WI/W$ | $LM1/L$ | $WM1/W$ | $LS/L$ | $PHS/L$ | $BS/L$ |
|---------|------|------|------|------|------|------|------|------|------|------|-------|---------|--------|--------|---------|---------|--------|---------|--------|
| Sample Öl-1 | | | | | | | | | | | | | | | | | | | |
| N       | 15   | 15   | 15   | 15   | 15   | 15   | 15   | 15   | 8    | 15   | 15    | 15      | 15     | 15     | 15      | 15      | 15     | 8       | 15     |
| X       | 1.32 | 1.40 | 0.15 | 0.32 | 0.69 | 0.45 | 0.79 | 1.03 | 0.74 | 0.29 | 95%   | 23%     | 23%    | 49%    | 34%     | 56%     | 78%    | 54%     | 22%    |
| S       | 0.21 | 0.26 | 0.03 | 0.07 | 0.16 | 0.09 | 0.21 | 0.19 | 0.19 | 0.04 | 0.05  | 0.03    | 0.03   | 0.05   | 0.04    | 0.07    | 0.05   | 0.09    | 0.03   |
| MIN     | 1.06 | 1.06 | 0.10 | 0.22 | 0.50 | 0.32 | 0.48 | 0.76 | 0.46 | 0.20 | 85%   | 18%     | 17%    | 40%    | 28%     | 45%     | 70%    | 43%     | 17%    |
| MAX     | 1.64 | 1.84 | 0.24 | 0.48 | 0.96 | 0.62 | 1.18 | 1.32 | 1.00 | 0.38 | 100%  | 28%     | 27%    | 58%    | 40%     | 67%     | 86%    | 68%     | 29%    |

1.1, W 1.2, H 1); LO 6534t. Dorsal valves: LO 6530t; LO 6531t. Total of 29 ventral and 78 dorsal valves.

*Measurements.* – See Tables 24–25.

*Diagnosis.* – Ventral valve highly conical, generally around 1.2–1.5 times as high as long. Ventral pseudointerarea procline to slightly cataline, with intertrough; pedicle foramen forming short tube. Apical process with well-defined median ridge. Dorsal valve flat, only slightly convex. Dorsal median septum moderately high, subtriangular, extending about ¾ of valve length. Ornamentation of minute rugellae with distinctive pustules.

*Description.* – The ventral valve is highly conical, on average 85% as long as wide and 120% as high as long (Figs. 79K, L, 80; Table 24). The ventral pseudointerarea is procline to slightly cataline and poorly defined laterally; and has an intertrough (Figs. 79K, 80C, G). The pedicle foramen is small, rounded, placed within the larval shell and forming a short external pedicle tube (Fig. 79M). The ventral interior is completely filled by the apical process that has a well-developed median ridge; the apical pits are situated posterolaterally to the internal pedicle opening (Fig. 79O).

The dorsal valve is flattened, only slightly convex, on average 95% as long as wide (Table 25); there is a broad and very shallow sulcus originating near the beak (Fig. 79A, B). The dorsal pseudointerarea is wide, subtriangular, slightly anacline, on average 23% as long as wide and occupying 49% of the valve width. The median groove is wide, subtriangular (Fig. 79I). The dorsal interior has a triangular median septum, extending on average 78% of the valve length (Table 25). The median buttress is elongate subtriangular, widening posteriorly. The cardinal muscle fields are elongate suboval in outline, slightly thickened; they occupy on average 56% of the total valve width and extend for 34% of the length (Fig. 79G, H; Table 25).

The shell is ornamented by fine, closely spaced, concentric rugellae with distinctive pustules (Fig. 79E, F). The larval shells of both valves are somewhat transversely oval, around 0.3×0.25 mm (Fig. 79C); there is usually no raised rim around the shell, and there is a gradual decrease in the size of the larval pits at the transition to the postlarval shell (Fig. 79D). The outer rim of the larval shell has unusually well-developed disturbances in the growth lines forming up to 12–15 sets of radial 'drapes' (Fig. 79C).

*Discussion.* – *D. phurus* is distinguished from all other species of the genus in having a short external pedicle tube. It also differs from the approximately contemporaneous *D. batkanensis* in having (1) a smaller maximum size, (2) significantly higher ventral valve, and (3) a slightly higher, but not so strongly thickened dorsal median septum. The ornamentation of pustules is also distinctive; Williams & Holmer (1992 p. 659, Pl. 3:5) described the micro-ornamentation of this species and its significance. They suggested that the disturbances (so-called 'nick points') at the outer rim of the larval shell were formed due to stresses in the outer mantle lobe induced by the setal muscles (Williams & Holmer 1992, Text-fig. 5).

*Occurrence.* – The species is restricted to the Bjørkåsholmen Limestone in Scandinavia; it is recorded at the following localities: Ottenby, Flagabro, Stora Backor (Figs. 5–7; Appendix 1K); and it is also known from the Oslo region.

## Genus *Eurytreta* Rowell, 1966

*Type species.* – Original designation by Rowell (1966, p. 9); *Acrotreta curvata* Walcott, 1902 (topotypes figured here in Fig. 81); Lower Ordovician 'Pogonip Limestone'; Eureka district (Walcott 1912, loc. 203a), Nevada.

*Diagnosis.* – Shell with short, slightly convex posterior margin. Ventral valve strongly convex to subconical; pseudointerarea apsacline to cataline, poorly defined laterally, and divided by poorly defined intertrough or interridge. Pedicle foramen small, apical, within larval shell, and usually forming short tube. Ventral interior with subtriangular apical process, placed anterior to pedicle foramen, and bearing semicircular depression. Dorsal valve slightly convex to deeply convex, sulcate; pseudointerarea short, divided by wide median groove. Dorsal interior with triangular median ridge or septum; dorsal cardinal muscle fields small, rounded, placed relatively close together, and not extending far anteriorly; median buttress wide; dorsal central muscle scars usually well defined, small, rounded.

*Species included.* – *Acrotreta curvata* Walcott, 1902; *Obolella sabrinae* Callaway, 1877; *Acrotreta bisecta* Matthew, 1901; *Acrothyra? chabakovi* Lermontova, 1933; *Eurytreta minor* Biernat, 1973; *Eurytreta campaniformis* Krause & Rowell, 1975; *Eurytreta kendyktassica* Popov, 1980; *Eurytreta discors* Popov, 1988; *Eurytreta sublata* Popov, 1988; *Eurytreta evanda* Popov, 1988.

*Discussion.* – *Eurytreta* is interpreted within wide morphological limits; it is a taxonomically difficult genus, and some species are so variable in morphology that it is sometimes difficult to define them in a meaningful way. In addition, it is sometimes difficult to distinguish small adults (like *E. minor* described below) from the juveniles of larger species of *Eurytreta* (like *E.* cf *sabrinae* and *E.* sp. a described below); juveniles of *Longipegma thulensis* gen. et sp. nov. can also sometimes resemble those of *Eurytreta*, and fragmentary ventral valves of all these species are sometimes impossible to tell apart.

A further complication has been the varying states of preservation of the various species. For instance, the problems of an exact comparison between *E. sabriane* and *E. bisecta* was commented on by Rushton & Bassett (*in* Owens *et al.* 1982), and here we also find it difficult to compare these species, preserved in shales, with those from the etched

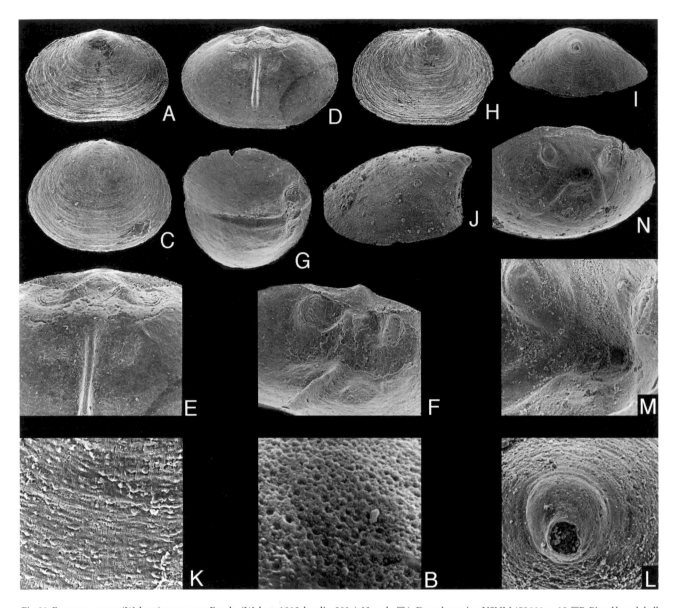

*Fig. 81. Eurytreta curvata* (Walcott); topotypes; Eureka (Walcott 1912, locality 203a), Nevada. □A. Dorsal exterior; USNM 459664a; ×13. □B. Pitted larval shell of A; ×600. □C. Dorsal exterior; USNM 459664b; ×19. □D. Dorsal interior; USNM 459664c; ×15. □E. Pseudointerarea and cardinal muscle field of D; ×32. □F. Oblique lateral view of D; ×30. □G. Lateral view of dorsal interior; USNM 459664d; ×18. □H. Ventral exterior; USNM 459664e; ×19. □I. Posterior view of H; ×19. □J. Lateral view of ventral exterior; USNM 459664f; ×25. □K. Larval shell of J; ×135. □L. Detail of pustulose ornamentation of J; ×135. □M. Ventral interior; USNM 459664g; ×23. □N. Apical process and cardinal muscle field of M; ×60.

material; thus the use of open nomenclature is used frequently below. In general, the lingulate fauna of the Ceratopyge and Dictyonema shales needs further study, based on much more material than was available for this study.

*Occurrence.* – Upper Cambrian (*Cambroistodus minutus* Biozone) – Middle Ordovician (Llanvirn); North America, Europe, Kazakhstan.

## *Eurytreta* aff. *curvata* (Walcott, 1902)
Fig. 82

*Material.* – Figured. Ventral valves: RM Br 136105; RM Br 136107; RM Br 136108. Dorsal valves: RM Br 136102; RM Br 136103; RM Br 136104; RM Br 136106. Total of 24 ventral and 39 dorsal valves.

*Measurements.* – See Table 26.

*Description.* – The shell is transversely oval. The ventral valve is subconical with the maximum height slightly anterior to the beak; in lateral profile it is moderately and evenly convex

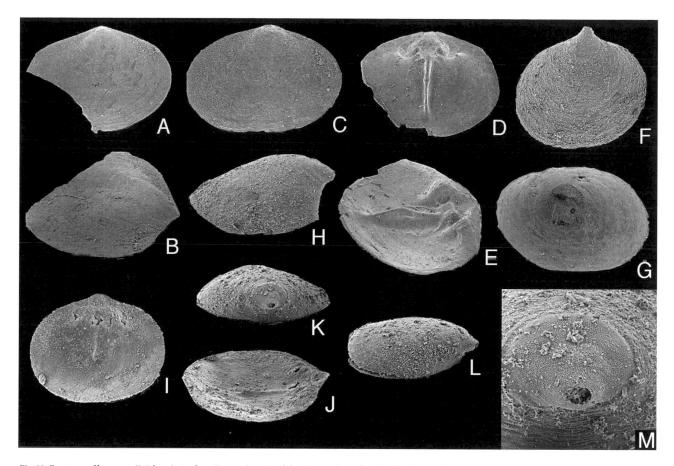

*Fig. 82. Eurytreta* aff. *curvata* (Walcott); Agalatas Formation; Kendyktas Range (sample 564), Kazakhstan. □A. Dorsal exterior; RM Br 136102; ×16. □B. Lateral view of A; ×22. □C. Dorsal exterior; RM Br 136103; ×15. □D. Dorsal interior; RM Br 136104; ×13. □E. Lateral view of D; ×18. □F. Ventral exterior; RM Br 136105; ×50. □G. Ventral interior; RM Br 136108; ×18. □H. Lateral view of ventral exterior; RM Br 136107; ×65. □I. Dorsal interior; RM Br 136106; ×30. □J. Lateral view of I; ×38. □K. Posterior view of H; ×55. □L. Lateral view of H; ×60. □M. Larval shell of H; ×195.

(Fig. 82F, H, L). The ventral pseudointerarea is recurved, apsacline in juvenile specimens and becomes procline or catacline in adults; the intertrough is wide and shallow (Fig. 82H, K, L). The pedicle foramen is placed within the larval shell and forms a short external pedicle tube (Fig. 82M). The ventral interior has a subtriangular apical process, placed in front of the internal pedicle opening; the cardinal muscle fields are elongate subtriangular, placed on the posterior slope of the valve (Fig. 82G).

The dorsal valve is gently convex, on average 85% as long as wide (Table 26), with a broad and very shallow sulcus originating near the beak (Fig. 82B). The dorsal pseudointerarea is somewhat anacline, occupying on average 53% of the maximum valve width. The dorsal interior has a low triangular median septum, extending on average 81% of the valve length (Table 26). The dorsal cardinal muscle fields are small, subcircular in outline, thickened, and closely spaced; they occupy on average 50% of the maximum valve width and

*Table 26. Eurytreta* aff. *curvata*, average dimensions and ratios of dorsal valves.

| | L | W | LI | WI | ML | MG | LM1 | WM1 | LS | PHS | BS | $L/W$ | $LI/WI$ | $ML/L$ | $WI/W$ | $MG/WI$ | $LM1/L$ | $WM1/W$ | $LS/L$ | $PHS/L$ | $BS/L$ |
|---|---|---|---|---|---|---|---|---|---|---|---|---|---|---|---|---|---|---|---|---|---|
| Sample 564 | | | | | | | | | | | | | | | | | | | | | |
| N | 5 | 5 | 4 | 5 | 4 | 4 | 5 | 5 | 5 | 5 | 5 | 5 | 4 | 4 | 5 | 4 | 5 | 5 | 5 | 5 | 5 |
| X | 1.71 | 2.04 | 0.21 | 1.08 | 0.20 | 0.36 | 0.54 | 0.99 | 1.40 | 0.95 | 0.44 | 85% | 21% | 12% | 53% | 35% | 32% | 50% | 81% | 55% | 26% |
| S | 0.20 | 0.36 | 0.03 | 0.20 | 0.02 | 0.08 | 0.07 | 0.09 | 0.24 | 0.14 | 0.07 | 0.09 | 0.03 | 0.02 | 0.04 | 0.05 | 0.03 | 0.07 | 0.06 | 0.03 | 0.02 |
| MIN | 1.94 | 2.48 | 0.24 | 1.32 | 0.22 | 0.46 | 0.64 | 1.12 | 1.74 | 1.16 | 0.56 | 98% | 25% | 14% | 56% | 41% | 36% | 56% | 90% | 60% | 29% |
| MAX | 1.44 | 1.64 | 0.18 | 0.88 | 0.18 | 0.26 | 0.46 | 0.88 | 1.16 | 0.82 | 0.38 | 76% | 18% | 10% | 47% | 28% | 27% | 39% | 75% | 51% | 24% |

extend for 32% of the length (Table 26). The median buttress is subrectangular, slightly raised posteriorly (Fig. 82D, E, I, J). The larval shells of both valves are somewhat transversely suboval, 0.15–0.2 mm wide and 0.12–0.15 mm long (Fig. 82M); the larval pitting is identical to that of *E. minor*.

*Discussion.* – The shells from the Agalatas Formation are very similar in the shape of shell and internal characters of the dorsal valve to *E. curvata*, as redescribed by Rowell (1966; see Fig. 81 herein). *E.* aff. *curvata* differs mainly in having (1) a less transversely oval shell; (2) a less strongly convex dorsal valve; (3) dorsal cardinal muscle scars that extend anteriorly for more than one-fourth of the valve length; and (4) a longer and somewhat higher dorsal median ridge. Our species also lack the pustulose ornamentation that appears to be developed in the type species (Fig. 81K).

*Occurrence.* – This species has been found only in the upper part of the Agalatas Formation (Appendix 1J).

## *Eurytreta chabakovi* (Lermontova, 1933)
Figs. 83

*Synonymy.* – □1933 *Acrothyra? chabakovi* sp. nov. – Lermontova *in* Lermontova & Razumovskij, p. 197, Pl. 1:1, 2, 5, 6 [not 3, 4].

*Material.* – Figured. Ventral valves: RM Br 136238; RM Br 136239; RM Br 136240 (L 0.7, W 0.84, Lmax 0.76, H 0.48); RM Br 136241; RM Br 136242; RM Br 136243. Dorsal valves: RM Br 136236 (L 1.84, W 2.52, LI 0.15, WI 1.36, WG 0.34, LM1 0.88, WM11.24, LS 1.26); RM Br 136237 (L 1.39, W 1.68, LI 0.08, WI 0.66, WG 0.36, LM1 0.36, WM1 0.82, LS 1.18); RM Br 136244; RM Br 136245; RM Br 136246; RM Br 136247. Coquina with numerous valves: RM Br 136248. Total of 340 ventral and 174 dorsal valves.

*Measurements.* – See Tables 27–28.

*Diagnosis.* – Shell large for genus with unusually high, strongly apsaconical ventral valve, up to more than $^4/_5$ as high as long. Ventral pseudointerarea with poorly defined intertrough. Dorsal valve gently and evenly convex, somewhat sulcate. Dorsal median ridge, low, triangular, extending

*Table 27. Eurytreta chabakovi,* average dimensions and ratios of ventral valves.

|  | L | W | $^L/_W$ | H | $^H/_W$ | max L | F |
|---|---|---|---|---|---|---|---|
| Samples G-113, H-381 | | | | | | | |
| N | 11 | 11 | 11 | 11 | 11 | 11 | 11 |
| X | 1.92 | 2.38 | 84% | 1.40 | 61% | 2.29 | 0.37 |
| S | 0.42 | 0.67 | 0.18 | 0.36 | 0.12 | 0.64 | 0.43 |
| MIN | 0.95 | 1.10 | 55% | 0.67 | 41% | 1.03 | 0.08 |
| MAX | 2.36 | 3.43 | 128% | 1.92 | 85% | 3.43 | 1.53 |

about $^4/_5$ of valve length; dorsal cardinal muscle fields small, subcircular, strongly thickened and extending about one-third of valve length.

*Description.* – The ventral valve is widely conical, on average 84% as long as wide and 61% as high as long (Table 27). The ventral pseudointerarea is apsacline with a narrow and shallow intertrough; the posterior surface is gently concave in lateral profile. The ventral beak extends on average 0.37 mm beyond the posterior margin (Fig. 83E, F, H, L; Table 27). The pedicle foramen is rounded and placed within the larval shell (Fig. 83I). The anterior slope of the valve is evenly convex the lateral profile; the lateral slopes are nearly straight (Fig. 83H). The ventral interior has a low, elongate subtriangular apical process placed anterior to the internal foramen. The ventral cardinal muscle fields are placed posterolaterally and slightly thickened (Fig. 83G). The ventral *vascula lateralia* are baculate, slightly divergent anteriorly.

The dorsal valve is on average 84% as long as wide and 19% as high as long (Table 28), gently and evenly convex in lateral profile, with a shallow sulcus arising near the beak (Fig. 83A, N, R). The dorsal pseudointerarea is low, anacline, occupying on average 44% of the valve width. The median groove is broadly triangular and bounded laterally by low ridges (Fig. 83C, K). The dorsal cardinal muscle fields are strongly thickened, subcircular and closely spaced, occupying on average 54% of the valve width and extending for 41% of the length (Table 28). The median buttress is well developed. The dorsal median septum is low, triangular, extending for on average 81% of the valve length (Table 28). The dorsal scars of the anterior lateral and central muscles are strongly thickened, forming muscle fieldes directly lateral to the median septum and extended to the centre of the valve (Fig. 83B–D, J, K). The larval shells of both valves are well defined, somewhat transversely suboval, around 0.2 mm wide and 0.17 mm long (Fig. 83I, Q); the larval pitting is identical to that of *E. minor*.

*Fig. 83. Eurytreta chabakovi* (Lermontova); Akbulaksai Formation; South Urals. □A. Dorsal exterior; exact locality unknown (sample 1007); RM Br 136236; ×15. □B. Dorsal interior; Alimbet River (sample B-780-1); RM Br 136237; ×11. □C. Pseudointerarea and cardinal muscle fields of B; ×24. □D. Lateral view of B; ×15. □E. Lateral view of ventral exterior; Alimbet River (sample B-780-1); RM Br 136238; 12. □F. Ventral exterior; Alimbet Farm (sample B-578-2); RM Br 136239; ×40. □G. Ventral interior; Alimbet Section (sample B-780-1); RM Br 136242; ×18. □H. Lateral view of ventral exterior; exact locality unknown (sample 1007); RM Br 136243; ×25. □I. Larval shell and ventral pseudointerarea of H; ×110. □J. Lateral view of dorsal interior; Alimbet Farm (sample B-578-2); RM Br 136244; ×38. □K. Dorsal pseudointerarea and cardinal muscle fields of J; ×110. □L. Lateral view of ventral exterior; Alimbet Farm (sample B-578-2); RM Br 136240; ×45. □M. Larval shell of L; ×135. □N. Dorsal exterior; exact locality unknown (sample 1007); RM Br 136245; ×30. □O. Ventral exterior, posterior view; Alimbet Farm (sample B-578-2); RM Br 136241; ×28. □P. Posterior view of dorsal exterior; Alimbet Farm (sample B-578-2); RM Br 136246; ×33. □Q. Dorsal larvel shell of P; ×196. □R. Dorsal exterior; Alimbet Farm (sample B-578-2); RM Br 136247; ×12. □S. Acrotretid coquina; Alimbet Farm (sample 2081-6b of B.M. Keller and K. S. Rozman); RM Br 136248; ×4.

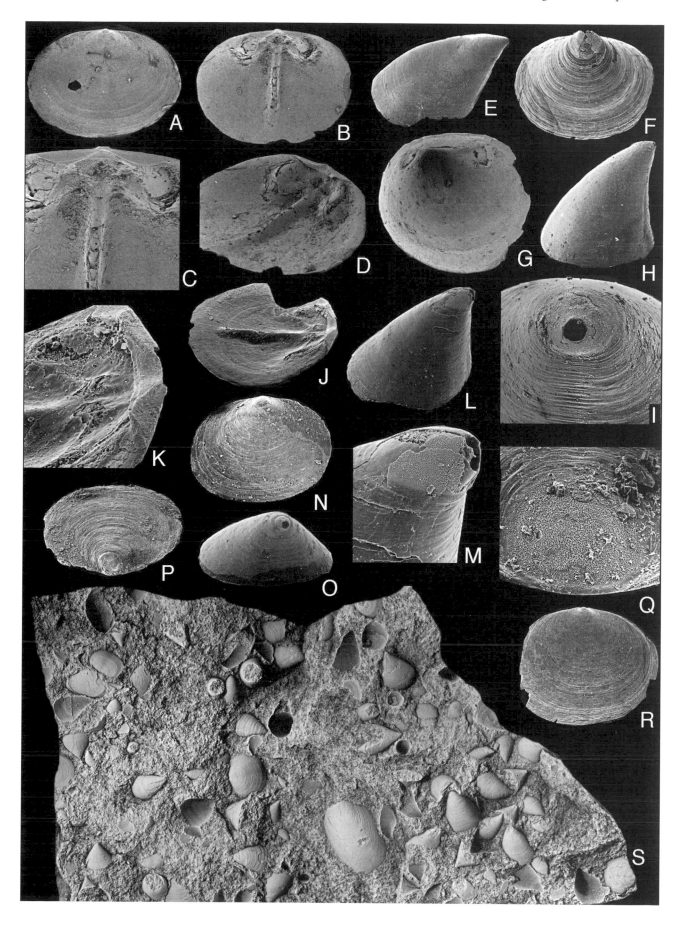

*Table 28. Eurytreta chabakovi*, average dimensions and ratios of dorsal valves.

| | L | W | H | LI | WI | LM1 | WM1 | LS | PHS | BS | L/W | H/L | WI/W | LM1/L | WM1/W | LS/L | PHS/L | BS/L |
|---|---|---|---|---|---|---|---|---|---|---|---|---|---|---|---|---|---|---|
| Samples N-381, B-780-1 | | | | | | | | | | | | | | | | | | |
| N | 20 | 20 | 18 | 20 | 20 | 18 | 18 | 18 | 14 | 18 | 20 | 18 | 20 | 18 | 18 | 18 | 14 | 18 |
| X | 1.54 | 1.86 | 0.29 | 0.17 | 0.82 | 0.66 | 1.08 | 1.34 | 0.90 | 0.29 | 84% | 19% | 44% | 41% | 54% | 81% | 53% | 18% |
| S | 0.70 | 0.84 | 0.12 | 0.08 | 0.47 | 0.27 | 0.51 | 0.54 | 0.31 | 0.10 | 0.06 | 0.04 | 0.11 | 0.14 | 0.09 | 0.09 | 0.09 | 0.04 |
| MIN | 0.47 | 0.51 | 0.07 | 0.06 | 0.26 | 0.19 | 0.41 | 0.54 | 0.43 | 0.17 | 72% | 11% | 23% | 21% | 36% | 58% | 41% | 10% |
| MAX | 2.85 | 3.35 | 0.55 | 0.40 | 2.15 | 1.33 | 2.23 | 2.15 | 1.45 | 0.53 | 94% | 28% | 64% | 90% | 72% | 90% | 76% | 24% |

*Discussion.* – *E. chabakovi* is distinguished from all other species of the genus by its significantly larger size and highly apsaconical ventral valve.

*Occurrence.* – The species is common in the Akbulaksai and Alimbet formations at the following localities: Alimbet Farm (Fig. 13; Appendix 1B); Alimbet River (Fig. 16; Appendix 1C); Akbulaksai River (Fig. 18; Appendix 1D); Bolshaya and Malaya Kayala rivers (Appendix 1E); and at Koagash River (Appendix 1F).

## *Eurytreta minor* Biernat, 1973

Fig. 84

*Synonymy.* – □1973 *Eurytreta minor* sp. nov. – Biernat, p. 74, Pl. 9:1–6; Fig. 26. □not ?1986 *Eurytreta minor* Biernat – Bednarczyk, p. 413, Pls. 1:3; 5:1–3.

*Holotype.* – Bp XV/16n; ventral valve; from Tremadoc chalcedonites; Wysoczki, Holy Cross Mountains, Poland.

*Material.* – Figured. Ventral valves: LO 6536t (L 0.48, W 0.6, H 0.33); LO 6537t (L 0.34, W 0.4, H 0.22). Dorsal valves: RM Br 136269 (L 0.6, W 0.8); RM Br 136277 (L 0.66, W 0.76, LI 0.1, WI 0.3, WG 0.16); LO 6535t; LO 6538t (L 0.55, W 0.63, LI 0.06, WI 0.22, WG 0.12, BS 0.17, LS 0.33); LO 6539t; LO 6540t. Total of 249 ventral and 125 dorsal valves.

*Diagnosis.* – Shell small for genus. Ventral valve with very short external pedicle tube; ventral pseudointerarea invariably procline, with poorly defined intertrough. Dorsal valve only slightly convex; dorsal pseudointerarea narrow and short, occupying about one-third of valve width. Dorsal median ridge extremely low and poorly visible; muscle fields of both valves poorly defined.

*Description.* – The ventral valve is low conical, on average 82% as long as wide and 68% as high as long (Table 29). The ventral pseudointerarea is invariably procline with a poorly defined intertrough; the posterior surface is gently concave in lateral profile. The anterior slope of the valve is evenly convex in lateral profile (Fig. 84D–F). The pedicle foramen is rounded, forming a very short pedicle tube, placed within the larval shell (Fig. 84G, I). The ventral interior has a poorly defined apical process placed anterior to the internal foramen.

The dorsal valve is on average 83% as long as wide (Table 30) and only slightly convex in lateral profile. The dorsal pseudointerarea is very narrow and short, occupying on average 36% of the valve width. The median groove is broadly triangular and the propareas are small (Fig. 84C, L). The muscle fields of both valves are poorly defined; the dorsal cardinal muscle fields occupy on average 36% of the valve width and 35% of the length (Table 30). The median buttress is poorly developed and usually not visible. The dorsal median ridge is extremely low and poorly visible; it originates at a short distance, around 0.15 mm, from the pseudointerarea and extends for, on average, 60% of the valve length (Fig. 84B, H, J; Table 30). The larval shells of both valves are well defined, transversely suboval, around 0.2 mm wide and 0.15 mm long (Fig. 84G, N); the larval pits are of two sizes: the larger, around 2 μm across, are surrounded by smaller ones, around 900 nm across (Fig. 84O).

*Table 29. Eurytreta minor*, average dimensions and ratios of ventral valves.

| | L | W | H | L/W | H/L |
|---|---|---|---|---|---|
| Samples Sk-1, Sk-2 | | | | | |
| N | 4 | 4 | 4 | 4 | 4 |
| X | 0.61 | 0.74 | 0.42 | 82% | 68% |
| S | 0.18 | 0.23 | 0.13 | 0.03 | 0.05 |
| MIN | 0.34 | 0.40 | 0.22 | 78% | 63% |
| MAX | 0.70 | 0.90 | 0.50 | 85% | 74% |

*Fig. 84. Eurytreta minor* Biernat.; Bjørkåsholmen Limestone; Scandinavia. □A. Dorsal exterior; Bjørkåsholmen (sample Ng-1), Oslo region; RM Br 136269; ×75. □B. Dorsal interior; Fågelsång (sample Sk-2), Scania; LO 6538t; ×90. □C. Pseudointerarea of B; ×225. □D. Ventral exterior; Fågelsång (sample Sk-2), Scania, LO 6536t; ×90. □E. Lateral view of D; ×112. □F. Posterior view of D; ×97. □G. Larval shell of D; ×200. □H. Lateral view of dorsal interior; Fågelsång (sample Sk-2), Scania; LO 6540t; ×65. □I. Lateral view of ventral exterior; Fågelsång (sample Sk-2), Scania; LO 6537t; ×165. □J. Lateral view of dorsal interior; Fågelsång (sample Sk-2), Scania; LO 6539t; ×75. □K. Dorsal interior; Mossebo (Vg-2), Västergötland; RM Br 136277; ×75. □L. Pseudointerarea of K; ×200. □M. Dorsal exterior; Fågelsång (sample Sk-2), Scania; LO 6535t; ×75. □N. Larval shell of M; ×225. □O. Larval pitting of M, ×900.

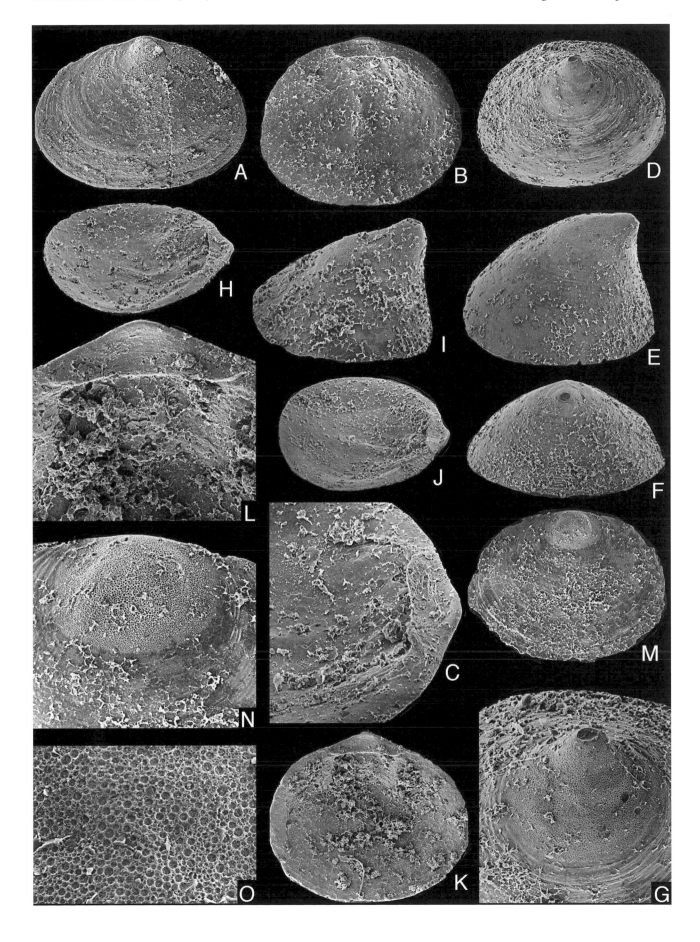

*Table 30. Eurytreta minor*, average dimensions and ratios of dorsal valves.

| | L | W | LI | WI | WG | LM1 | WM1 | LS | $L/W$ | $LI/WI$ | $LI/L$ | $WI/W$ | $LM1/L$ | $WM1/W$ | $LS/L$ |
|---|---|---|---|---|---|---|---|---|---|---|---|---|---|---|---|
| Samples Sk-1, Sk-2 | | | | | | | | | | | | | | | |
| N | 5 | 5 | 5 | 5 | 5 | 5 | 5 | 5 | 5 | 5 | 5 | 5 | 5 | 5 | 5 |
| X | 0.58 | 0.70 | 0.06 | 0.26 | 0.09 | 0.20 | 0.31 | 0.36 | 83% | 25% | 11% | 36% | 35% | 36% | 60% |
| S | 0.15 | 0.17 | 0.02 | 0.07 | 0.01 | 0.06 | 0.11 | 0.14 | 0.03 | 0.06 | 0.03 | 0.03 | 0.08 | 0.03 | 0.08 |
| MIN | 0.48 | 0.56 | 0.04 | 0.20 | 0.08 | 0.14 | 0.22 | 0.28 | 77% | 18% | 7% | 31% | 26% | 31% | 52% |
| MAX | 0.84 | 1.00 | 0.08 | 0.38 | 0.10 | 0.28 | 0.50 | 0.60 | 86% | 31% | 15% | 41% | 48% | 41% | 71% |

*Discussion.* – The Scandinavian material is similar to the Polish types in: (1) the small maximum size of the shell, (2) the outline and convexity of both valves, (3) the relative size and shape of the dorsal pseudointerarea, and (4) the extremely low and small median ridge that starts at a short distance from the pseudointerarea. As noted above, it is very difficult to distinguish *E. minor* from juveniles of the larger species, *E.* cf *sabrinae* and *E.* sp. a, sometimes occurring in the same samples; this problem needs further investigation on a larger material than was available for this study.

*Occurrence.* – *E. minor* was described from the Holy Cross Mountains; it is here recorded from the Bjørkåsholmen

Limestone in the following localities in Scandinavia: Ottenby, Flagabro, Fågelsång, Stora Backor, Mossebo, and at Bjørkåsholmen (Figs. 5–8; Appendix 1K).

## *Eurytreta sabrinae* (Callaway, 1877)

Figs. 85, 86N–P

*Synonymy.* – □1877 *Obolella sabrinae* sp. nov. – Callaway, p. 669, Pl. 24:12. □1912 *Acrotreta sabrinae* (Callaway) – Walcott, p. 702, Pl. 73:5, 5a–d [synonymy]. □1978 *Eurytreta? sabrinae* (Callaway) – Cocks, p. 23 [synonymy]. □1982 *Eu-*

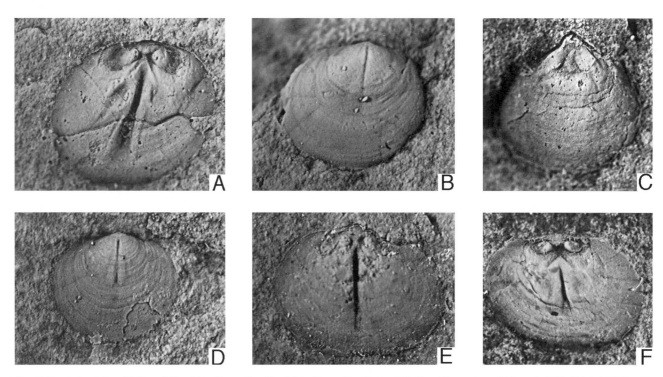

*Fig. 85. Eurytreta sabrinae* (Callaway). Kendyktas Formation; Kendyktas Range, Kazakhstan. □A. Dorsal internal mould (sample 554); RM Br 136109; ×20. □B. Dorsal exterior (sample 554); RM Br 136110; ×20. □C. ventral internal mould (sample 556-1); RM Br 136111, ×20. □D. Dorsal exterior (sample 554); RM Br 136112; ×20. E. dorsal internal mould (sample 556-1); RM Br 136113, ×20. □E. Dorsal internal mould (sample 554); RM Br 136114; ×20.

*Fig. 86* (opposite page). □A–F. *Ottenbyella carinata*? (Moberg & Segerberg); Scandinavia. □A. Dorsal exterior; figured by Moberg & Segerberg (1906, Pl. 1:26) as *Obolella sagittalis* Salter; Bjørkåsholmen Limestone; Fågelsång, Scania; LO 1775t; ×12. □B. Partial dorsal internal mould; figured by Waern (1952, Pl. 1:7) as *Acrotreta conula* Walcott; Ceratopyge Shale; Bödahamn core, Öland; PMO Öl 110; ×20. □C. Dorsal exterior; Ceratopyge Shale; Borgholm, Öland; RM Br 21265; ×15. □D. Dorsal internal mould; figured by Brøgger (1882, Pl. 10:8) as *Obolella sagittalis* Salter; ?Tøyen Shale; Krekling, Norway; PMO 19107; ×20. □E. Ventral exterior; figured by Brøgger (1882, Pl. 10:2) as *Acrotreta* cf. *socialis* von Seebach.; Bjørkåsholmen Limestone; Krekling, Norway; PMO 1522; ×20. □F. Lateral view of E; ×20. □G–H. *Eurytreta* cf. *bisecta* (Matthew); Scandinavia. □G. Dorsal internal mould; figured by Moberg & Segerberg (1906, Pl. 1:25) as *Obolella sagittalis* Salter; Bjørkåsholmen Limestone; Fågelsång; LO 1774t; ×10. □H. Ventral internal mould; holotype by monotypy of *Acrotreta sagittalis lata*

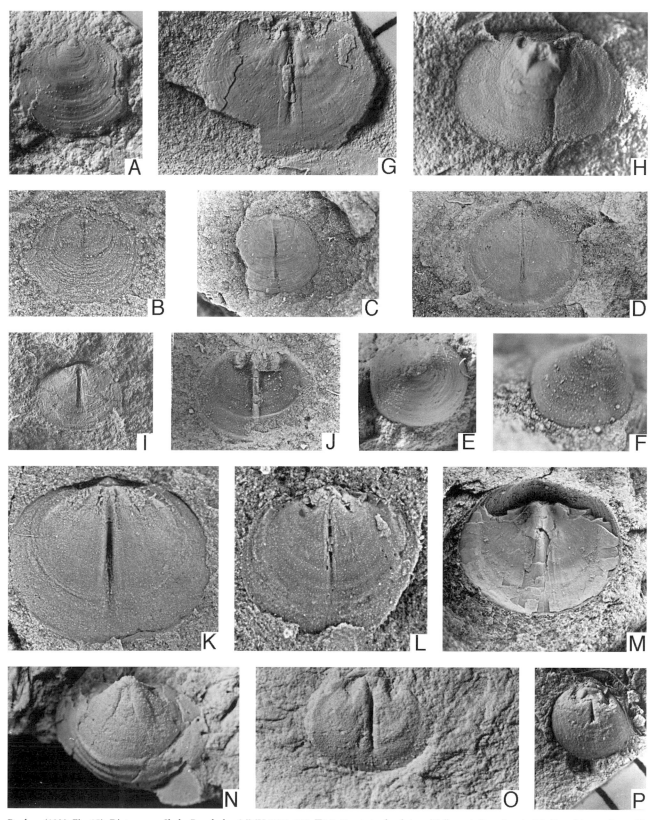

Poulsen (1922, Fig. 15); Dictyonema Shale; Bornholm; MMH 1892; ×15. □I–J. *Eurytreta* cf. *sabrinae* (Callaway); Scandinavia. □I. Dorsal internal mould; figured by Brøgger (1882, Pl. 10:7) as *Obolella sagittalis* Salter; Ceratopyge Shale; Vestfossen, Norway; PMO 1432; ×15. □J. Dorsal interior; Dictyonema Shale; Ottenby, Öland; SGU 8497; ×18. □K–M. *Acrotreta bisecta* (Matthew). □K. Dorsal internal mould; Shineton Shale; Shineton, Salop, Shropshire; BM B 47861; ×13. □L. Dorsal internal mould; Dictyonema Shale; Cape Breton, Nova Scotia; BM B 40377; ×18. □M. Dorsal interior; Shineton Shale; Bramton Bryan, Herefordshire; BM B 34934; ×13. □N–P. *Eurytreta sabrinae* (Callaway). □N. Ventral internal mould; Shineton Shale; Shineton, Salop, Shropshire; BM B 47867; ×8. □O. Dorsal internal mould; Shineton, Salop, Shropshire; BM B 47866; ×8. □P. Dorsal internal mould; *Clonograptus tenellus* Biozone; Dyfed, Carmarthen district, South Wales; NMW 77.1G.44; ×13.

*rytreta sabrinae* (Callaway) – Rushton & Bassett *in* Owens *et al.*, p. 26, Pl. 8a–h, j, k [synonymy].

*Lectotype* – Selected by Walcott (1912, p. 702); BU 1921; dorsal internal mould from the Shineton Shales; Shineton, Shropshire (topotypes figured here in Fig. 86N–O).

*Material.* – Figured from southern Kendyktas Range. Ventral valve: RM Br 136111. Dorsal valves: RM Br 136109; RM Br 136110; RM Br 136112; RM Br 136113; RM Br 136114. Total of 39 ventral and 58 dorsal valves.

Figured from Britain. Ventral valve: BM B47867. Dorsal valves: BM B 47866; NMW 77.1G.44.

*Measurements.* – See Table. 31.

*Diagnosis.* – Ventral valve conical with maximum height at umbo; ventral pseudointerarea apsacline. Dorsal valve transversely oval with pseudointerarea occupying about half of maximum valve width. Dorsal median septum extending about $\frac{4}{5}$ of valve length; dorsal cardinal muscle scars small, rounded, situated close to each other, extending about one-third of valve length and occupying somewhat less than half the valve width.

*Description of material from southern Kendyktas Range.* – The ventral valve is conical with the maximum height at the apex. The anterior slope of the valve is gently convex in lateral profile. The ventral pseudointerarea is slightly apsacline, with a weakly developed intertrough. The ventral interior has an apical process, forming an elongate, subtriangular platform anterior to the internal pedicle opening (Fig. 85C).

The dorsal valve is transversely oval, on average 78% as long as wide (Table 31), gently convex in lateral profile with the maximum height placed somewhat posterior to the mid-length. The dorsal pseudointerarea is orthocline to slightly apsacline, occupying on average 50% of the valve width (Table 31); it has a wide concave median groove. The dorsal interior has a low, triangular median septum, extending on average 80% of the valve length valve (Fig. 85A, E, F; Table 31). The dorsal cardinal muscle fields are small, subcircular in outline, and closely spaced; they occupy on average 48% of the total valve width and extend for 26% of the valve length (Fig. 85A, E, F; Table 31). The dorsal central muscle scars are elongate suboval, placed lateral to the median septum and somewhat anterior to the mid-length. The dorsal *vascula*

*media* are sometimes visible anterior to the median septum (Fig. 85A, E, F).

*Discussion.* – The specimens of *E. sabrinae* from the Kendyktas Formation of the southern Kendyktas Range, Kazakhstan, are closely comparable with the specimens from Wales and Shropshire illustrated by Rushton & Bassett (*in* Owens *et al.* 1982; see also Fig. 86N–P herein). As noted by Rushton & Bassett, *E. sabrinae* is closely similar (also in preservation) to *E. bisecta* (Matthew), originally described from the Tremadoc in Canada (see Fig. 86K–M); according to these authors, *E. sabrinae* differs from *E. bisecta* mainly in having (1) a smaller maximum size, (2) a flatter dorsal valve with a more rounded outline, (3) a shorter dorsal median septum, and (4) smaller and more closely spaced cardinal muscle fields.

As noted above, the type of preservation of acrotretid shells in clastic rocks makes it difficult to compare them with species described from isolated specimens from limestones.

*Occurrence.* – This species occurs in the Kendyktas Formation (Fig. 35; Appendix 1J).

## *Eurytreta* cf. *sabrinae* (Callaway, 1877)

Figs. 86I, J, 87, 88

*Synonymy.* – □cf. 1877 *Obolella sabrinae* sp. nov. – Callaway, p. 669, Pl. 24:12. □?1882 *Obolella sagittalis* Salter – Brøgger, p. 45, Pl. 10:6–8.

*Material.* – Figured from the South Urals. Ventral valves: RM Br 136121 (L 1.8, W 2.12, H 1.52). Dorsal valves: RM Br 136120 (L 1.32, W 1.46, LI 0.12, WI 0.64, WG 0.3, LM1 0.38, WM1 0.76, LS 0.82); RM Br 136122 (L 0.46, W 0.52, LI 0.06, WI 0.24, WG 0.12, LS 0.26); RM Br 136123. Total of 13 ventral and 33 dorsal valves.

Figured from Scandinavia. Ventral valves: LO 6542t; LO 6545t (L 1.28, W 1.54, H 0.84); LO 6573t. Dorsal valves: RM Br 129092; LO 6541t; LO 6543t (L 1.42, W 1.52, LI 0.18, WI 0.7, WG 0.4, LM1 0.34, WM1 0.7, LS 0.84); LO 6544t (L 1.9, W 2.1, LI 0.24, WI 0.82, WG 0.44, LM1 0.46, WM1 0.7, LS 1.18); PMO 1432; SGU 8497. Total of 25 ventral and 24 dorsal valves.

*Measurements.* – See Tables 32–33.

*Table 31. Eurytreta sabrinae,* average dimensions and ratios of dorsal valves.

| | L | W | LI | WI | ML | MG | LM1 | WM1 | LS | PHS | BS | $L/W$ | $LI/WI$ | $ML/L$ | $WI/W$ | $MG/WI$ | $LM1/L$ | $WM1/W$ | $LS/L$ | $PHS/L$ | $BS/L$ |
|---|---|---|---|---|---|---|---|---|---|---|---|---|---|---|---|---|---|---|---|---|---|
| Samples 551, 554, 558 | | | | | | | | | | | | | | | | | | | | | |
| N | 7 | 7 | 7 | 7 | 7 | 7 | 7 | 7 | 7 | 6 | 7 | 7 | 7 | 7 | 7 | 7 | 7 | 7 | 7 | 6 | 7 |
| X | 1.75 | 2.28 | 0.12 | 1.15 | 0.11 | 0.51 | 0.45 | 1.11 | 1.41 | 0.98 | 0.27 | 78% | 11% | 7% | 50% | 45% | 26% | 48% | 80% | 56% | 16% |
| S | 0.34 | 0.56 | 0.04 | 0.36 | 0.04 | 0.12 | 0.10 | 0.40 | 0.30 | 0.24 | 0.06 | 0.10 | 0.03 | 0.02 | 0.08 | 0.07 | 0.03 | 0.06 | 0.10 | 0.10 | 0.02 |
| MIN | 1.40 | 1.64 | 0.08 | 0.76 | 0.08 | 0.32 | 0.32 | 0.76 | 0.98 | 0.64 | 0.18 | 61% | 7% | 5% | 35% | 34% | 23% | 42% | 69% | 46% | 13% |
| MAX | 2.36 | 3.28 | 0.18 | 1.80 | 0.18 | 0.64 | 0.58 | 1.88 | 1.80 | 1.32 | 0.36 | 91% | 15% | 10% | 63% | 53% | 30% | 57% | 96% | 72% | 20% |

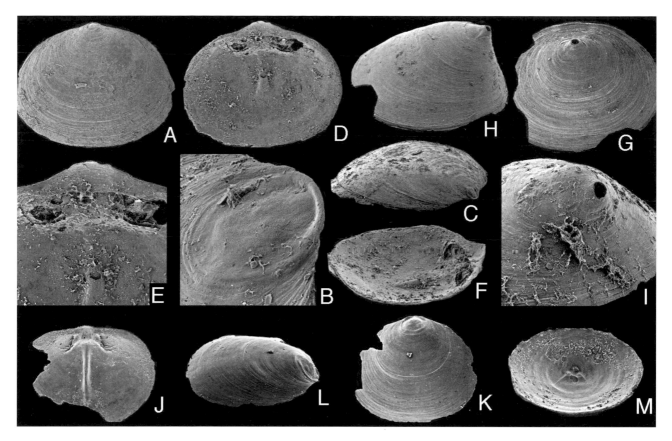

*Fig. 87. Eurytreta* cf. *sabrinae* (Callaway); Bjørkåsholmen Limestone; Sweden. □A. Dorsal exterior (sample Öl-1); Ottenby, Öland; LO 6543t; ×18. □B. Larval shell of A; ×195. □C. Lateral view of A; ×21. □D. Dorsal interior (sample Öl-1); Ottenby, Öland; LO 6544t; ×25. □E. Pseudointerarea of D; ×50. □F. Lateral view of D; ×30. □G. Ventral exterior (sample Öl-1); Ottenby, Öland; LO 6545t; ×27. □H. Lateral view of G; ×32. □I. Oblique posterior view of ventral exterior (sample Öl-1); Ottenby, Öland; LO 6573t; ×43. □J. Dorsal interior (sample Sk-1); Flagabro, Scania; RM Br 129092; ×18. □K. Dorsal exterior (sample Öl-1); Ottenby, Öland; LO 6541t; ×32. □L. Lateral view of dorsal exterior; ×40. □M. Ventral interior (sample Öl-1); Ottenby, Öland; LO 6542t; ×32.

*Description of Scandinavian and South Ural material.* – The ventral valve is conical, about 83–85% as long as wide and about 66–84% as high as long in two valves; the maximum height is at the apex. The anterior slope of the valve is evenly convex in lateral profile, and the lateral slopes are gently and evenly convex. The ventral pseudointerarea is cataclinc to slightly apsacline with a weakly defined intertrough (Figs. 87G–I, 88D–F). The pedicle foramen is small and rounded, placed within the larval shell (Fig. 87I). The ventral interior has a subtriangular apical prosess, placed anterior to the internal pedicle opening (Fig. 87M).

The dorsal valve is weakly sulcate, transversely oval, on average 88–90% as long as wide in two samples (Tables 32–33); in lateral profile it is gently convex with the maximum height somewhat posterior to the mid-length (Fig. 87C, L). The dorsal pseudointerarea is slightly anacline, occupying on average 44–46% of the total valve width and 21–27% of the length in two samples (Tables 32–33); the median groove is wide and subtriangular. The dorsal interior has a low median septum extending for, on average, 60–76% of the valve length in two samples (Tables 32–33). The median buttress is

transversely subrectangular. The dorsal cardinal muscle fields are subcircular and slightly thickened; they occupy on average 38–56% of the total valve width and 24–35% of the length in two samples (Tables 32–33). The central muscle scars are small and situated laterally to the median septum at about the centre of the valve (Figs. 87D–F, J, 88A–C). The larval shells of both valves are well defined, transversely suboval, around 0.2 mm wide and 0.15 mm long (Figs. 87B, 88H); the larval shell is ornamented by pits of two sizes, identical to those of *E. minor* (Fig. 88I).

*Discussion.* – The material of *E.* cf. *sabrinae* from the South Urals differs somewhat from the Scandinavian specimens in the morphology of the dorsal valve. The Scandinavian specimens are somewhat more convex and have more closely spaced and shorter cardinal muscle fields, in addition to a shorter median septum; however, they also have a larger maximum size than any of the South Ural specimens, and some of the differences might be related to size.

The material from both localities is somewhat similar to *E. sabrinae* in the convexity of the ventral valve and the internal

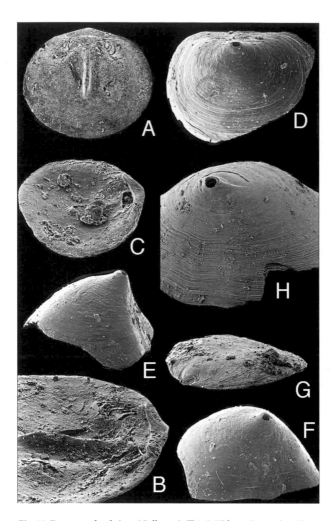

characters of both valves, such as the fairly closely spaced dorsal cardinal muscle fields.

The described material might represent one or two new species, but the material is too poor for a taxonomic decision.

*Occurrence.* – *E.* cf. *sabrinae* is known from the lower part of the Kidryas Formation at the Tyrmantau Ridge (Fig. 11; Appendix 1A) and from the Malaya and Bolshaya Kayala rivers (Appendix 1E). In Scandinavia, it was recorded from the Ceratopyge Shale and Bjørkåsholmen Limestone at Ottenby; Bjørkåsholmen Limestone at Flagabro; Bjørkåsholmen Limestone at Stora Backor; and from the Ceratopyge Shale and the Bjørkåsholmen Limestone in the Oslo region (Figs. 5–8; Appendix 1K).

## *Eurytreta* cf. *bisecta* (Matthew, 1901)
Fig. 86G–H

*Synonymy.* – □cf. 1901 *Acrotreta bisecta* sp. nov. – Matthew, p. 275, Pl. 5:5a–g. □? 1902 *Acrotreta seebachi* sp. nov. Walcott, p. 598. □?1912 *Acrotreta seebachi* Walcott – Walcott, p. 710, Pl. 77:3, 3a. □?1906 *Obolella* (*Acrotreta?*) *sagittalis* Salter – Moberg & Segerberg, p. 64, Pl. 1:25 [not 26]. □? 1922 *Acrotreta sagittalis* Salter var. *lata* var. nov. Poulsen, p. 16, Fig. 15.

*Material.* – Figured. Dorsal valve: LO 1774t. Ventral valve: MMH 1892.

*Fig. 88. Eurytreta* cf. *sabrinae* (Callaway). □A–I. Kidryas Formation; Tyrmantau Ridge (sample B-768-3), South Urals. □A. Dorsal interior; RM Br 136120; ×43. □B. Lateral view of A; ×45. □C. Lateral view of dorsal interior; RM Br 136122; ×33. □D. Ventral exterior; RM Br 136121; ×16. □E. Lateral view of D; ×18. □F. Posterior view of D; ×18. □G. Oblique lateral view of dorsal exterior; RM Br 136123; ×50. □H. Oblique posterior view of ventral exterior; Bjørkåsholmen Limestone (sample Öl-1); Ottenby, Öland; LO 6545t (same as Fig. 87G); ×90.

*Table 32. Eurytreta* cf. *sabrinae*, average dimensions and ratios of dorsal valves.

|  | L | W | LI | WI | WG | LM1 | WM1 | LS | $L/W$ | $LI/WI$ | $LI/L$ | $WI/W$ | $LM1/L$ | $WM1/W$ | $LS/L$ |
|---|---|---|---|---|---|---|---|---|---|---|---|---|---|---|---|
| Sample Öl-1 | | | | | | | | | | | | | | | |
| N | 4 | 4 | 4 | 4 | 4 | 4 | 4 | 4 | 4 | 4 | 4 | 4 | 4 | 4 | 4 |
| X | 1.70 | 1.90 | 0.23 | 0.84 | 0.45 | 0.41 | 0.71 | 1.01 | 90% | 27% | 13% | 44% | 24% | 38% | 60% |
| S | 0.21 | 0.26 | 0.04 | 0.11 | 0.05 | 0.05 | 0.02 | 0.14 | 0.04 | 0.03 | 0.01 | 0.04 | 0.01 | 0.06 | 0.03 |
| MIN | 1.42 | 1.52 | 0.18 | 0.70 | 0.40 | 0.34 | 0.70 | 0.84 | 84% | 24% | 12% | 39% | 23% | 33% | 56% |
| MAX | 1.90 | 2.10 | 0.28 | 0.98 | 0.52 | 0.46 | 0.74 | 1.18 | 93% | 29% | 15% | 49% | 24% | 46% | 62% |

*Table 33. Eurytreta* cf. *sabrinae*, average dimensions and ratios of dorsal valves.

|  | L | H | W | LI | WI | LM1 | WM1 | LS | PHS | BS | $H/L$ | $L/W$ | $LI/WI$ | $LI/L$ | $WI/W$ | $LM1/L$ | $WM1/W$ | $LS/L$ | $PHS/L$ | $BS/L$ |
|---|---|---|---|---|---|---|---|---|---|---|---|---|---|---|---|---|---|---|---|---|
| Samples B-768-1, B-768-3 | | | | | | | | | | | | | | | | | | | | |
| N | 17 | 11 | 16 | 17 | 17 | 16 | 16 | 17 | 15 | 17 | 11 | 16 | 17 | 17 | 16 | 16 | 15 | 17 | 15 | 17 |
| X | 0.92 | 0.25 | 1.02 | 0.10 | 0.48 | 0.31 | 0.58 | 0.70 | 0.52 | 0.19 | 25% | 88% | 21% | 11% | 46% | 35% | 56% | 76% | 56% | 20% |
| S | 0.28 | 0.10 | 0.33 | 0.03 | 0.17 | 0.11 | 0.22 | 0.22 | 0.14 | 0.05 | 0.05 | 0.03 | 0.04 | 0.02 | 0.08 | 0.07 | 0.05 | 0.09 | 0.06 | 0.03 |
| MIN | 0.60 | 0.13 | 0.63 | 0.06 | 0.27 | 0.19 | 0.39 | 0.41 | 0.33 | 0.10 | 18% | 83% | 16% | 9% | 34% | 25% | 48% | 61% | 48% | 16% |
| MAX | 1.37 | 0.44 | 1.66 | 0.14 | 0.89 | 0.54 | 0.97 | 1.14 | 0.74 | 0.27 | 32% | 95% | 33% | 15% | 60% | 49% | 63% | 93% | 66% | 25% |

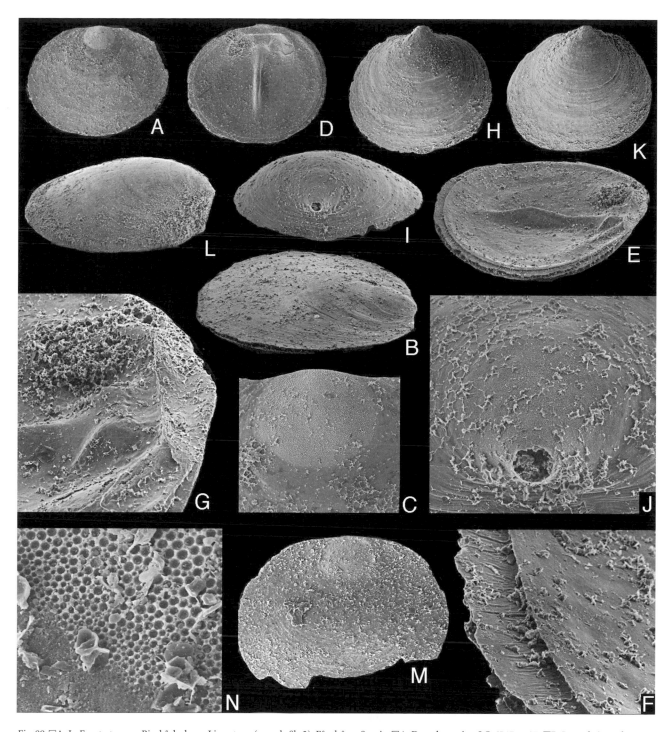

*Fig. 89.* □A–L. *Eurytreta* sp. a; Bjørkåsholmen Limestone (sample Sk-2); Fågelsång, Scania. □A. Dorsal exterior; LO 6547t; ×27. □B. Lateral view of A; ×55. □C. Larval shell of A; ×110. □D. Dorsal interior; LO 6548t; ×25. □E. Lateral view of D; ×45. □F. Detail of valve edge of E; ×292. □G. Pseudointerarea of E; ×135. □H. Ventral exterior; LO 6549t; ×40. □I. Posterior view of H; ×60. □J. Larval shell of H; ×225. □K. Ventral exterior; LO 6550t; ×25. □L. Lateral view of K; ×37. □M. Acrotretidae gen. et sp. nov a; dorsal exterior; Bjørkåsholmen Limestone (sample Sk-2); Fågelsång, Scania; LO 6551t; ×56. □N. Larval pitting of M; ×1200.

*Remarks.* – Considering the poor state of preservation and the limited number of specimens at hand, we only give a few remarks as to the morphology and affinity of this species. The complete distribution was also not investigated.

The outline of the shell is transversely oval, truncated posteriorly. The ventral valve appears to be procline to catacline; the apical process is wide and well developed, and in both valves there are well-developed baculate mantle canals with bifurcating *vascula lateralia*; the cardinal muscle scars of both valves are also distinctly raised and rather wide (Fig. 86G, H).

These *Eurytreta*-like acrotretids from the Dictyonema and Ceratopyge shales in Scandinavia appear to be somewhat different from those referred to *Eurytreta* cf. *sabrinae*; they are somewhat similar to *Eurytreta bisecta* described from Canada and Britain by Rushton & Bassett (*in* Owens *et al.* 1982).

*Acrotreta seebachi* Walcott, from the Bjørkåsholmen Limestone in the Oslo region (also recorded from the Ceratopyge Shale at Borgholm, island of Öland, by Walcott 1912), is also somewhat similar to *Eurytreta* cf. *bisecta* in the shape of the ventral valve and development of *vascula lateralia*, the cardinal muscle scars and apical process (Walcott 1912, Pl. 77:3). *Acrotreta sagittalis lata* of Poulsen (1922; Fig. 86H herein) from the Dictyonema Shale on the island of Bornholm, is also similar to *Eurytreta* cf. *bisecta* in these characters.

## *Eurytreta* sp. a
Fig. 89A–L

*Material.* – Figured. Ventral valves: LO 6549t; LO 6550t. Dorsal valves: LO 6547t; LO 6548t. Total of 2 ventral and 2 dorsal valves.

*Remarks.* – All valves of this species come from the Bjørkåsholmen Limestone in Scania (sample Sk-2), and it has not been found anywhere else. It would appear to be a species of *Eurytreta*, judging by the morphology of the dorsal valve; the ventral valve is extremely low and recurved as compared with any other species of *Eurytreta*. It probably represents a new species, but the definition of this is postponed until more material can be isolated.

## Genus *Galinella* gen. nov.

*Name.* – In honour of Galina T. Ushatinskaya.

*Type and only species.* – *Acrotreta retrorsa* Lermontova, 1951; Kujandy Formation.

*Diagnosis.* – Ventral valve widely conical, strongly geniculated dorsally; ventral pseudointerarea procline to catacline, undivided; pedicle foramen within larval shell. Ventral interior with ridge-like apical process, expanding posteriorly; ventral cardinal muscle scars and mantle canals strongly impressed; mantle canal system baculate with *vascula lateralia* slightly divergent anteriorly. Dorsal valve geniculated ventrally in adults. Dorsal median septum low, triangular, sometimes with bulbous projection; dorsal cardinal muscle fields small, subcircular, deeply impressed.

*Discussion.* – *Galinella* is similar to *Satpakella* Koneva, Popov & Ushatinskaya (*in* Koneva *et al.* 1990) in having geniculation in both valves, but it differs in possessing a ridge-like apical process expanding posteriorly and a simple, low triangular median septum in the dorsal valve.

## *Galinella retrorsa* (Lermontova, 1951)
Fig. 90

*Synonymy.* – □1951 *Acrotreta retrorsa* sp. nov. – Lermontova, p. 5, Pl. 1:2–6.

*Lectotype.* – Selected here (Lermontova 1951 selected two specimens as holotypes): CNIGR 2/7350; ventral valve; figured by Lermontova (1951, Pl. 1:2a) from the Kujandy Formation; core near Boshchekul.

*Material.* – Figured. Ventral valves: RM 136126; RM Br 136127 (L 1.16, W 1.44, H 0.96); RM Br 136129; RM Br 136130 (L 1.5, W 1.82, H 0.84); RM Br 136132; RM Br 136134; RM Br 136135 (Lmax 1.82, W 1.74, H 0.96); RM Br 136136; RM Br 136314. Dorsal valves: RM Br 136124 (L 1.7, W 1.92, LI 0.14, WI 1.06, WG 0.3, LM1 0.44, WM11.02, LS 0.9); RM Br 136125 (L 1.64, W 1.92, LI 0.12, WI 1.04, LM1 0.48, WM1 0.96, LS 1.16); RM Br 136128 (L 1.16, W 1.4, LI 0.14, WI 0.76, WG 0.26, LM1 0.4, WM1 0.72, LS 0.78); RM Br 136131 (L 1.16, W 1.92, LI 0.1, WI 1.04, LM1 0.32, WM1

---

*Fig. 90. Galinella retrorsa* (Lermontova); northeastern Central Kazakhstan. □A. Dorsal exterior; olistolith within Satpak Formation; Kujandy Section (sample 7825-12.5); RM Br 136124; ×18. □B. Lateral view of A; ×20. □C. Dorsal interior; olistolith within Satpak Formation; Kujandy Section (sample 7825-12.5); RM Br 136125; ×18. □D. Lateral view of C; ×22. □E. Ventral interior; olistolith within Satpak Formation; Kujandy Section (sample 7822-4); RM Br 136126; ×20. □F. Ventral interior; olistolith within Satpak Formation; Kujandy Section (sample 7825-12.5); RM Br 136132; ×33. □G. Ventral exterior; Kujandy Formation; Aksak–Kujandy Section (sample 7827); RM Br 136127; ×24. □H. Lateral view of ventral exterior; olistolith within Satpak Formation; Kujandy Section (sample 7822-5); RM Br 136130; ×25. □I. Ventral larval shell of H; ×160. □J. Pitted larval micro-ornament of H; ×750. □K. Dorsal interior; Kujandy Formation; Aksak–Kujandy Section (sample 7827); RM Br 136128;; ×24. □L. Lateral view of K; ×28. □M. Ventral interior; olistolith within Satpak Formation; Kujandy Section (sample 7822-5); RM Br 136129; ×33. □N. Dorsal interior; olistolith within Satpak Formation; Kujandy Section (sample 7825-12.5); RM Br 136131; ×15. □O. Lateral view of N; ×25. □P. Ventral exterior; olistolith within Satpak Formation; Kujandy Section (sample 7822-5); RM Br 136314; ×22. □Q. Dorsal exterior; olistolith within Satpak Formation; Kujandy Section (sample 7825-12.5); RM Br 136133; ×27. □R. Dorsal larval shell; olistolith within Satpak Formation; Kujandy Section (sample 7822-5); RM Br 136315; ×150. □S. Ventral exterior; olistolith within Satpak Formation; Kujandy Section (sample 7822-5); RM Br 136134; ×24. □T. Posterior view of Q; ×24. □U. Lateral view of Q; ×24. □V. Oblique lateral view of ventral exterior; olistolith within Satpak Formation; Kujandy Section (sample 7822-5); RM Br 136135; ×20. □W. Ventral interior; olistolith within Satpak Formation; Aksak–Kujandy Section (sample 8119); RM Br 136136; ×33. □X. Apical process and internal pedicle foramen of W; ×45. □Y. Dorsal interior; olistolith within Satpak Formation; Kujandy Section (sample 7822-5); RM Br 136137; ×33. □Z. Lateral view of Y; ×22.

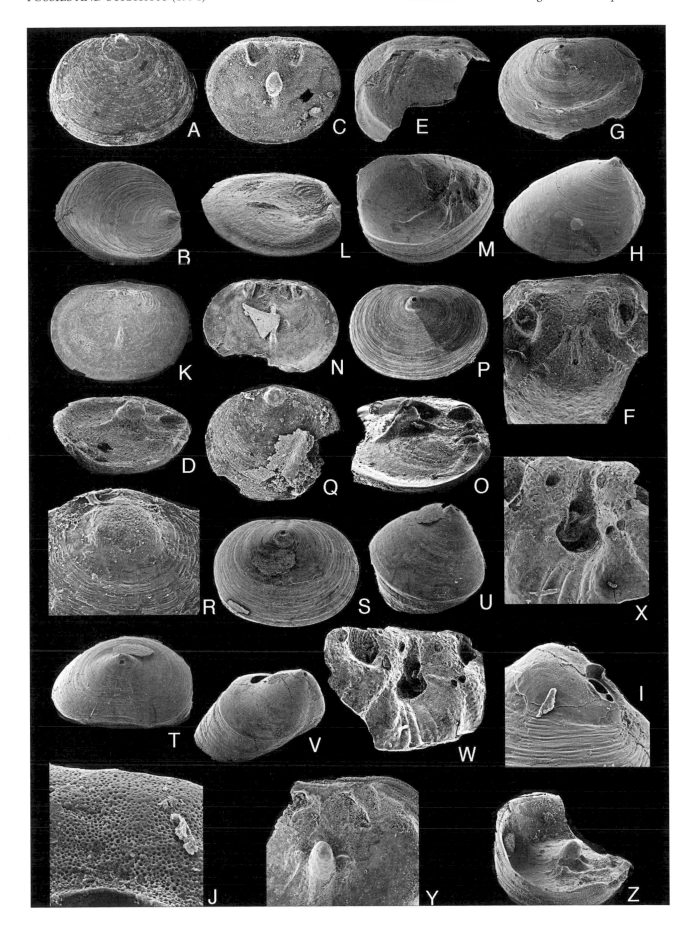

1.16, LS 0.78); RM Br 136133; RM Br 136137; RM Br 136315. Total of 142 ventral and 64 dorsal valves.

*Diagnosis.* – As for genus.

*Description.* – The valves are transversely oval. The ventral valve is widely conical, on average 82% as high as long (Table 34). In lateral profile, the sides of the valve are unevenly convex; adults usually have a strong dorsally directed geniculation (Fig. 90H, V). The ventral pseudointerarea is procline to cataline and undivided (Fig. 90T), gently convex in lateral profile (Fig. 90H, V). The pedicle foramen is small, circular and placed within the larval shell (Fig. 90I). The apical process is ridge-like and expands towards the internal pedicle opening (Fig. 90X). The ventral cardinal muscle scars are strongly thickened, elongate suboval. The ventral *vascula lateralia* are strongly impressed and widely spaced, slightly divergent anteriorly (Fig. 90E, F, M, W).

The dorsal valve is on average average 83% as long as wide (Table 35); in lateral view, it is flat to gently convex (Fig. 90D, L, O); adults usually have a strong, ventrally directed, geniculation near the anterior and lateral margins (Fig. 90Z). The dorsal pseudointerarea is slightly anacline; it occupies on average 52% of the maximum valve width (Table 35). The dorsal median septum is low and triangular; in adults it is sometimes strongly thickened and forms a bulbous projection. The septum extends on average 72% of the maximum valve length (Fig. 90C, D, K, L, N, O; Table 35). The median buttress is weakly developed. The dorsal cardinal muscle fields are subcircular in outline, strongly thickened peripherally, and deeply impressed medially; they occupy on average 54.3% of the maximum valve width and extend for 32% of the valve length (Fig. 90C, D, K, L, N, O; Table 35). The dorsal central muscle scars form a pair of low ridges extending to the centre of the valve (Fig. 90N, O).

The larval shells of both valves are well defined, close to circular, around 0.15–0.2 mm across (Fig. 90R); the larval pits are of somewhat varying sizes, up to 1.5 μm across (Fig. 90J). The outer rim of the larval shell has unusually well developed disturbances in the growth lines forming up to 12–15 sets of radial 'drapes' (Fig. 90R; see also p. 95).

*Occurrence.* – This species is present in the upper part of the Kujandy Formation at the Aksak–Kujandy Section (Fig. 27; Appendix 1H); it is also abundant in the olistoliths within the Satpak Formation at the Kujandy Section (Fig. 25; Appendix 1G).

## Genus *Longipegma* gen. nov.

*Name.* – Latin *longus*, long, and *pegma*, shelf; alluding to the extremely elongated dorsal pseudointerarea.

*Type species.* – *Longipegma gorjanskii* sp. nov.; Upper Cambrian Keti Stage (Aksai Series); Olenek River, northern Central Siberia.

*Diagnosis.* – Shell distinctly inequivalved. Ventral valve transversely suboval and low conical, less than half as high as long. Ventral pseudointerarea procline to cataline, divided by intertrough. Apical process high, ridge-like. Dorsal valve elongate subtriangular, with elongate subtriangular pseudointerarea, usually occupying more than half of valve width; median groove widely subtriangular. Dorsal cardinal muscle scars closely spaced; dorsal median ridge or septum starting directly anterior to pseudointerarea. Post-larval shell ornamented by evenly spaced rugellae.

*Species included.* – *Longipegma gorjanskii* sp. nov.; *Longipegma thulensis* sp. nov.

*Discussion.* – The species *Longipegma gorjanskii* is represented by very well preserved material, and although it originates from outside the study area, it is here selected as the type species of the new genus, which is also represented by much less well preserved material from Scandinavia. Adults of *Longipegma* are not closely similar to any other acrotretid genus, but as is the case with many other acrotretoids, the juveniles can be difficult to identify. *Physotreta* Rowell, 1966, also has an elongate dorsal pseudointerarea, but the ventral valve of this genus is highly conical with an apical process filling the entire apex.

*Table 34. Galinella retrorsa*, aeverage dimensions and ratios of ventral valves.

|  | L | W | H | Fp | $L/_W$ | $H/_L$ |
|---|---|---|---|---|---|---|
| Samples 7822-4, 7822-6 | | | | | | |
| N | 39 | 39 | 39 | 39 | 39 | 39 |
| X | 1.41 | 1.73 | 0.89 | 0.22 | 82% | 63% |
| S | 0.16 | 0.16 | 0.14 | 0.19 | 0.06 | 0.08 |
| MIN | 1.06 | 1.44 | 0.56 | 0.00 | 67% | 50% |
| MAX | 1.84 | 2.06 | 1.24 | 1.20 | 99% | 86% |

*Table 35. Galinella retrorsa*, aeverage dimensions and ratios of dorsal valves.

|  | L | W | LI | WI | ML | MG | LM1 | WM1 | LS | PHS | BS | $L/_W$ | $LI/_{WI}$ | $ML/_L$ | $MG/_W$ | $WI/_W$ | $LM1/_L$ | $WM1/_W$ | $LS/_L$ | $PHS/_L$ | $BS/_L$ |
|---|---|---|---|---|---|---|---|---|---|---|---|---|---|---|---|---|---|---|---|---|---|
| Samples 7822-4, 7822-6 | | | | | | | | | | | | | | | | | | | | | |
| N | 17 | 17 | 17 | 17 | 17 | 17 | 16 | 17 | 16 | 15 | 17 | 17 | 17 | 17 | 17 | 17 | 16 | 17 | 16 | 15 | 17 |
| X | 1.28 | 1.56 | 0.12 | 0.80 | 0.12 | 0.34 | 0.41 | 0.84 | 0.93 | 0.69 | 0.31 | 83% | 15% | 10% | 22% | 52% | 32% | 54% | 72% | 52% | 25% |
| S | 0.20 | 0.27 | 0.04 | 0.13 | 0.04 | 0.07 | 0.08 | 0.16 | 0.18 | 0.07 | 0.05 | 0.07 | 0.04 | 0.02 | 0.03 | 0.05 | 0.04 | 0.05 | 0.07 | 0.06 | 0.07 |
| MIN | 0.70 | 0.68 | 0.06 | 0.44 | 0.06 | 0.18 | 0.22 | 0.46 | 0.44 | 0.52 | 0.20 | 71% | 10% | 6% | 17% | 45% | 27% | 44% | 59% | 44% | 19% |
| MAX | 1.54 | 1.96 | 0.21 | 0.94 | 0.20 | 0.44 | 0.58 | 1.12 | 1.12 | 0.79 | 0.38 | 103% | 27% | 14% | 26% | 65% | 40% | 68% | 87% | 65% | 49% |

*Occurrence.* – Upper Cambrian – Lower Ordovician (Tremadoc); Siberia, Sweden.

## *Longipegma gorjanskii* sp. nov.
Fig. 91

*Name.* – In honour of Vladimir Yu. Gorjansky.

*Holotype.* – Fig. 91C–E; RM Br 136271 (L 2.2, W 2, LI 1, WI 1.8, WG 0.72, ML 0.68, LM1 1, WM1 0.92, LS 1.84, PHS 1.4) from the Upper Cambrian Keti Stage (Aksai Series); Olenek River, south of Sukhona village at the upper reaches of the Botorchuk Rivulet (NIIGA, St. Petersburg, sample number 5058; collected by F.F. Iljin 1957), northern Central Siberia.

*Material.* – Figured. Ventral valves: RM Br 136273 (L 1.12, W 1.28, H 0.48); RM Br 136274 (L 1.48, W 1.6, H 0.56); RM Br 136275 (L 1.6, W 2, H 0.72). Dorsal valves: RM Br 136270 (L 2.4, W 1.8); RM Br 136272 (L 2.04, W 1.72, LI 1, WI 1.68, WG 0.64, ML 0.8, LM1 1.04, WM1 0.8, LS 1.72, PHS 1.48); RM Br 136276. Total of 30 ventral and 15 dorsal valves.

*Diagnosis.* – Shell extremely inequivalved. Apical process forming high narrow ridge enclosing internal pedicle tube. Dorsal valve extremely elongate subtriangular, considerably longer than ventral valve. Dorsal pseudointerarea close to orthocline, extremely wide, occupying close to 9/10 of valve width and close to half of length. Dorsal median septum low, triangular, extending for 4/5 of total valve length; dorsal cardinal muscle fields closely spaced, with muscle track extending under the pseudointerarea.

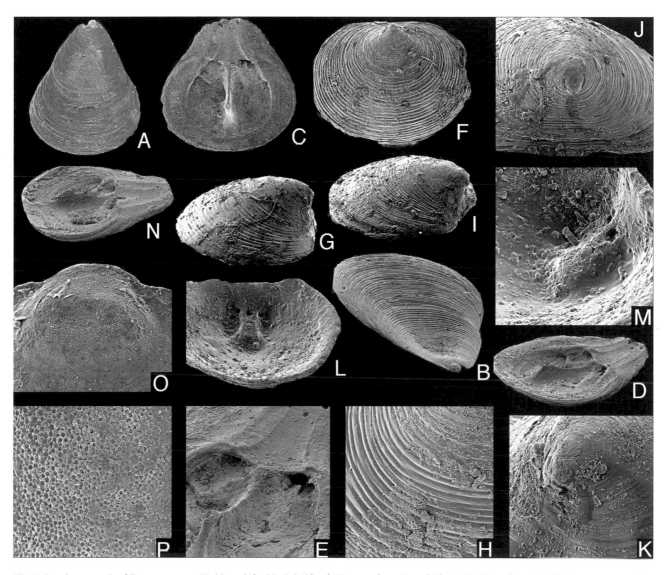

*Fig. 91. Longipegma gorjanskii* gen. et sp. nov.; Keti Stage (Aksai Series); Olenek River, northern Central Siberia. □A. Dorsal exterior; RM Br 136270; ×15. □B. Oblique posterior view of A; ×25. □C. Holotype; dorsal interior; RM Br 136271; ×17. □D. Lateral view of C; ×19. □E. Median septum and pseudointerarea of C; ×40. □F. Ventral exterior; RM Br 136273; ×30. □G. Lateral view of F; ×32. □H. Ornamentation of F; ×65. □I. Lateral view of ventral exterior; RM Br 136274; ×45. □J. Posterior view of I; ×45. □K. Larval shell of I; ×100. □L. Ventral interior; RM Br 136275; ×23. □M. Apical process of L; ×50. □N. Lateral view of dorsal interior; RM Br 136272; ×19. □O. Dorsal larval shell; RM Br 136276; ×275. □P. Larval pitting of O; ×1000.

*Description.* – The shell is extremely inequivalved. The ventral valve is transversely oval, on average 87% (OR 80–93%; *N* 5) as long as wide, and low conical, on average 41% (OR 37–45%; *N* 5) as high as long; the maximum height is anterior to the beak, at about the valve centre (Fig. 91G, I). The ventral pseudointerarea is procline to cataclone, divided by a broadly triangular intertrough, occupying most of the width of the pseudointerarea (Fig. 91J). The pedicle foramen is small, circular; it appears to be placed within the larval shell, but the details of the ventral larval shell are poorly visible on the available material (Fig. 91K). The apical process forms a high, narrow ridge enclosing the internal pedicle tube and extending anteriorly beyond the valve centre (Fig. 91L, M). The ventral muscle scars are not thickened, and the mantle canal system could not be observed.

The dorsal valve is extremely elongate subtriangular and much longer than the ventral valve; it is on average 121% (OR 110–133%; *N* 3) as long as wide; the maximum width is placed anterior to mid-length, at about 60% of the valve length; in lateral view, it is close to flat (Fig. 91A, B). The dorsal pseudointerarea is close to orthocline, elongate subtriangular, and extremely wide; it occupies 90–98% (*N* 2) of the maximum valve width, and close to half the valve length. The median groove is elongate subtriangular, extremely well developed and wide, occupying close to half the width of the pseudointerarea (Fig. 91C–E, N). The dorsal median septum starts directly anterior to the pseudointerarea and extends for 84% (*N* 2) of the total valve length; it is low and triangular with the maximum height somewhat anterior to mid-length. The median buttress is fairly small, triangular. The dorsal cardinal muscle fields are deeply impressed with muscle tracks extending under the pseudointerarea; they are closely spaced, subcircular in outline, occupying 90–98% (*N* 2) of the maximum valve width, and extend for 45–51% (*N* 2) of the valve length. The dorsal interior is surrounded by a limbus up to about 0.5 mm wide (Fig. 91C–E, N). The dorsal larval shell is transversely sub-oval, around 0.13 mm wide and 0.10 mm long; it has two distinct nodes at about the centre (Fig. 91); the larval pits are of one size, around 1 µm across. The postlarval shell is ornamented by closely spaced, well-developed rugellae, around 30 µm apart (Fig. 91H).

*Occurrence.* – At the type locality only, but according to G. Ushatinskaya (personal communication, 1991), it is found also at other localities in the same area.

## *Longipegma thulensis* sp. nov.
Fig. 92

*Synonymy.* – ☐1992 'Acrotretoid B' – Williams & Holmer, p. 659.

*Name.* – Latin *Thule*, farthest north.

*Holotype.* – Fig. 92J; LO 6553T; dorsal valve (L 0.9, W 0.96, LI 0.14, WI 0.5, WG 0.24, LM1 0.28, WM1 0.5, LS 0.82); Bjørkåsholmen Limestone (sample Öl-1), Ottenby, Öland.

*Paratypes.* – Figured. Ventral valves: LO 6554t (L 0.82, W 1, H 0.4); LO 6556t. Dorsal valves: LO 6552t; LO 6555t (L 1.08, W 1.14, LI 0.16, WI 0.52, LM1 0.3, WM1 0.4, LS 0.98). Total of 6 ventral and 28 dorsal valves.

*Measurements.* – See Table 36.

*Diagnosis.* – Shell moderately inequivalved. Dorsal valve moderately elongate subtriangular, longer than ventral valve. Dorsal valve slightly convex, with widely subtriangular pseudointerarea, occupying around half of valve width, and one-fifth of length. Dorsal median ridge extremely low.

*Description.* – The shell is moderately inequivalved. The ventral valve is transversely oval, 82% as long as wide, and 49% as high as long in one specimen; the maximum height is at the beak. The ventral pseudointerarea is procline to cataclone, divided by a poorly defined, triangular intertrough (Fig. 92D, E, K, L). The pedicle foramen is small, circular, forming a short tube that is placed within the larval shell. The ventral interior characters were poorly visible in the examined material, but the apical process appears to form a triangular ridge anterior to the internal pedicle opening.

The dorsal valve is elongate subtriangular and longer than the ventral valve; it is on average 196% as long as wide (Table 36); the maximum width is placed only somewhat anterior to mid-length (Fig. 92A); in lateral view, it is slightly convex. The dorsal pseudointerarea is flattened and elongate subtriangular, occupying on average 46% of the maximum valve width (Table 36) and close to 20% of the valve length. The median groove is subtriangular, wide (Fig. 92H, I). The dorsal median ridge is low, starting directly anterior to the pseudointerarea and extending for, on average, 91% of the total valve length (Table 36). The median buttress is poorly developed. The dorsal cardinal muscle fields are deeply impressed; they are closely spaced, subcircular in outline, occupying on average 43% of the maximum valve width, and extends for 28% of the valve length (Fig. 92G–J; Table 36).

The larval shells of both valves are well defined, surrounded by a raised rim, close to circular, around 0.25 mm across; the dorsal shell has two distinct nodes at about the centre and a smaller directly posterior (Fig. 92B); the pits are of two sizes, the larger about 3 µm across, surrounded by smaller, up to about 1 µm across. The outer rim of the larval shell has unusually well developed disturbances in the growth lines forming up to 9–10 sets of radial 'drapes' (Fig. 92B; see also p. 95). The postlarval shell is ornamented by closely spaced rugellae, about 20 µm apart.

*Discussion.* – The adults of *L. thulensis* are similar only to the type species, *L. gorjanskii*. They differ mainly in having (1) less inequivalved shells; (2) a less elongate subtriangular dorsal valve; (3) a less elongate subtriangular dorsal pseudointerarea; and (4) an extremely low dorsal median ridge. The

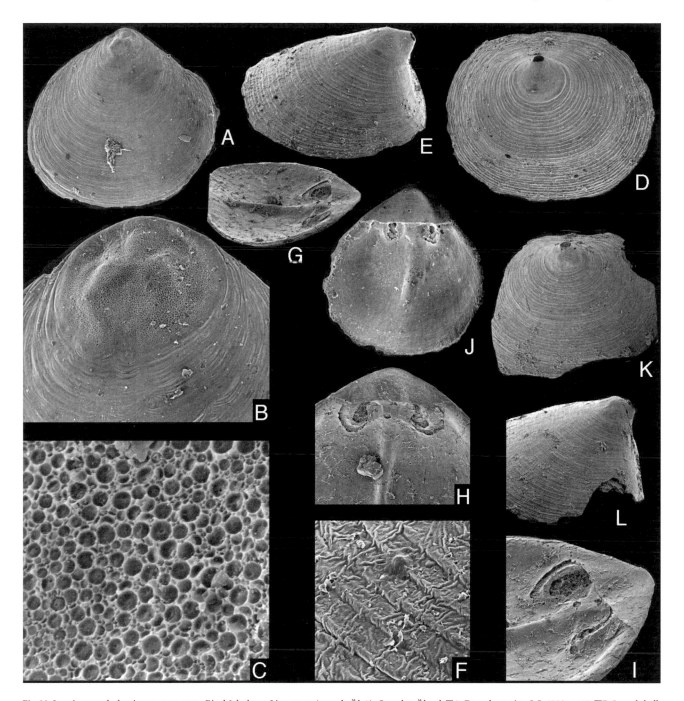

*Fig. 92. Longipegma thulensis* gen. et sp. nov.; Bjørkåsholmen Limestone (sample Öl-1); Ottenby, Öland. □A. Dorsal exterior; LO 6552t; ×45. □B. Larval shell of A; ×180. □C. Larval pitting of A; ×1800. □D. Ventral exterior; LO 6554t; ×56. □E. Lateral view of D; ×60. □F. Ornamentation of D; ×400. □G. Lateral view of dorsal interior; LO 6555t; ×45. □H. Pseudointerarea and cardinal muscle field of G; ×55. □I. Oblique lateral view of pseudointerarea and cardinal muscle field of G; ×80. □J. Holotype; dorsal interior; LO 6553T; ×45. □K. Ventral exterior; LO 6556t; ×37. □L. Lateral view of K; ×45.

*Table 36. Longipegma thulensis,* average dimensions and ratios of dorsal valves.

| | L | W | LI | WI | WG | LM1 | WM1 | LS | $L/_W$ | $LI/_{WI}$ | $MG/_W$ | $WI/_W$ | $LM1/_L$ | $WM1/_W$ | $LS/_L$ |
|---|---|---|---|---|---|---|---|---|---|---|---|---|---|---|---|
| Sample Öl-1 | | | | | | | | | | | | | | | |
| N | 3 | 3 | 3 | 3 | 3 | 3 | 3 | 3 | 3 | 3 | 3 | 3 | 3 | 3 | 3 |
| X | 1.01 | 1.06 | 0.15 | 0.49 | 0.17 | 0.28 | 0.45 | 0.92 | 96% | 32% | 17% | 46% | 28% | 43% | 91% |
| S | 0.10 | 0.09 | 0.01 | 0.04 | 0.15 | 0.02 | 0.05 | 0.09 | 0.02 | 0.04 | 0.15 | 0.06 | 0.03 | 0.09 | 0.00 |
| MIN | 0.90 | 0.96 | 0.14 | 0.44 | 0.00 | 0.26 | 0.40 | 0.82 | 94% | 28% | 0% | 41% | 25% | 35% | 91% |
| MAX | 1.08 | 1.14 | 0.16 | 0.52 | 0.28 | 0.30 | 0.50 | 0.98 | 98% | 36% | 26% | 52% | 31% | 52% | 91% |

Swedish species also has a much smaller maximum size (max L 1.08) as compared with the type species (max L 2.4). As noted above, the juvenile dorsal and ventral valves of *L. thulensis* can sometimes be difficult to distinguish from those of *Eurytreta* and *Ottenbyella*.

*Occurrence.* – The species is restricted to the Bjørkåsholmen Limestone in Sweden, where it was found at Ottenby and Stora Backor (Figs. 5–6; Appendix 1K).

## Genus *Ottenbyella* gen. nov.

*Name.* – After the occurrence at Ottenby, where the best specimens were found.

*Type and only species.* – *Acrotreta carinata* Moberg & Segerberg, 1906; Bjørkåsholmen Limestone; Fågelsång.

*Diagnosis.* – Ventral valve highly conical, about $^3/_4$ as high as long, with maximum height at beak; ventral pseudointerarea invariably procline, with distinct interridge; external pedicle tube very short. Apical process small and low; *vascula lateralia* well developed, straight. Dorsal valve flattened, broadly sulcate; dorsal pseudointerarea wide but short, broadly subtriangular. Dorsal median ridge very low; median buttress poorly developed; dorsal cardinal muscle fields small and short, transversely elongate suboval in outline.

*Discussion.* – The genus is somewhat similar to *Conotreta* in the morphology and shape and in the convexity of the ventral valve; a low, elongate apical process is also found in the type species, *C. rusti* Walcott. However, *Ottenbyella* differs mainly in having (1) a single pair of *vascula lateralia* in the ventral valve, never showing any bifurcation; (2) a flattened dorsal valve; (3) a low dorsal median ridge; and (4) small, short, transversely oval dorsal cardinal muscle scars.

The dorsal valves of *Ottenbyella* can possibly be confused with those of *Eurytreta* when they are preserved as flattened internal or external mould in shales (see, e.g., Fig. 86A–F); both genera have some kind of dorsal median ridge, and the details of the dorsal pseudointerarea are rarely visible in valves from shales. In fact, similarly preserved, distorted, and compressed ventral valves of *Ottenbyella* might also be mistaken for *Eurytreta*, both having some kind of apical process.

## *Ottenbyella carinata* (Moberg & Segerberg, 1906)

Figs. 86A–F, 93, 94A–I, 95, 96

*Synonymy.* – □1882 *Acrotreta* cf. *socialis* von Seebach – Brøgger, p. 46, Pl. 10:2–4. □?1902 *Acrotreta conula* sp. nov. – Walcott, p. 25. □1906 *Acrotreta carinata* sp. nov. – Moberg & Segerberg, p. 66, Pl. 3:5–6. □?1906 *Acrotreta circularis* sp. nov. – Moberg & Segerberg, p. 65, Pl. 3:4. □?1906 *Obolella* (*Acrotreta?*) *sagittalis* Salter – Moberg & Segerberg, p. 64, Pl.

1:26 [not 1:25]. □?1909 *Acrotreta oelandica* sp. nov. – Westergård, p. 76, Pl. 5:24a, b □?1912 *Acrotreta circularis* Moberg & Segerberg – Walcott, p. 680, Fig. 57A–C [synonymy]. □1912 *Acrotreta carinata* Moberg & Segerberg – Walcott, p. 679, Fig. 56A–D [synonymy]. □?1912 *Acrotreta conula* Walcott – Walcott, p. 681, Pl. 75:2, 2a–b. □?1952 *Acrotreta conula* Walcott – Waern, p. 235, Pl. 1:6–7 [synonymy]. □?1952 *Acrotreta* cf. *oelandica* Westergård – Waern, p. 235.

*Lectotype.* – Selected here: LO 1784T; ventral valve (L 2.7, W 2.9); figured by Moberg & Segerberg (1906, Pl. 3:5a–c); Bjørkåsholmen Limestone; Fågelsång.

*Material.* – Figured. Ventral valves: RM Br 133950 (L 2, W 2.6); RM Br 133953 (L 1.5, W 1.8, H 1.1); RM Br 133956 (L 1.2, W 1.6, H 0.9); RM Br 133958 (L 2.3, W 2.5, H 1.4); RM Br 135898; RM Br 133908; LO 1783T; LO 6560t; LO 6561t; PMO 1522. Dorsal valves: RM Br 21265; RM Br 133951 (L 1.4, W 1.6); RM Br 133957 (L 1.6, W 1.9); RM Br 135897; RM Br 135899; RM Br 135905; RM Br 135906; RM Br 136278; RM Br 136279; LO 1775t; LO 6557t; LO 6558t; LO 6559t; PMU Öl 110; PMO 19107. Total of 153 ventral and 109 dorsal valves.

*Measurements.* – See Tables 37–38.

*Diagnosis.* – As for genus.

*Description.* – The shell is somewhat transversely oval. The ventral valve is highly conical, on average 88% as long as wide and 76% as high as long (Table 37); maximum height is at the beak; in lateral profile the anterior slope can be slightly concave. The ventral pseudointerarea is invariably procline, somewhat flattened and broadly triangular, divided by a distinct interridge (Figs. 93E, F, K, L, 94D–F, 95A–F, 96A–H, J–Q). The pedicle foramen is small, rounded, placed within the larval shell and on a very short external pedicle tube (Figs. 93G, M, 94G). The ventral interior has distinct but rather small and low apical process at some distance anterior to the internal pedicle opening; a pair of distinct and straight *vascula lateralia* lie directly lateral to the apical process (Figs. 94I, 95D, 96E). The ventral cardinal muscle scars are thickened and rather small, situated high up on the posterior slope (Fig. 94I).

The dorsal valve is flattened, only slightly convex, on average 84% as long as wide (Table 38); there is invariably a

*Fig. 93.* *Ottenbyella carinata* (Moberg & Segerberg); Bjørkåsholmen Limestone (sample Ng-1); Bjørkåsholmen, Oslo region. □A. Dorsal exterior; RM Br 135905; ×29. □B. Dorsal interior; RM Br 135899; ×27. □C. Lateral view of B; ×32. □D. Pseudointerarea and cardinal muscle field of B; ×60. □E. Ventral exterior; RM Br 135898; ×37. □F. Posterior view of E; ×37. □G. Larval shell of E; ×165. □H. Lateral view of dorsal exterior; RM Br 135897; ×29. □I. Larval shell of H; ×82. □J. Larval pitting of H; ×825. □K. Ventral exterior; RM Br 133908; ×56. □L. Lateral view of K; ×60. □M. Larval shell of K; ×225. □N. Dorsal interior; RM Br 136279; ×37. □O. Lateral view of posterior of N; ×97. □P. Dorsal exterior; RM Br 135906; ×32. □Q. Dorsal exterior; RM Br 136278; ×56.

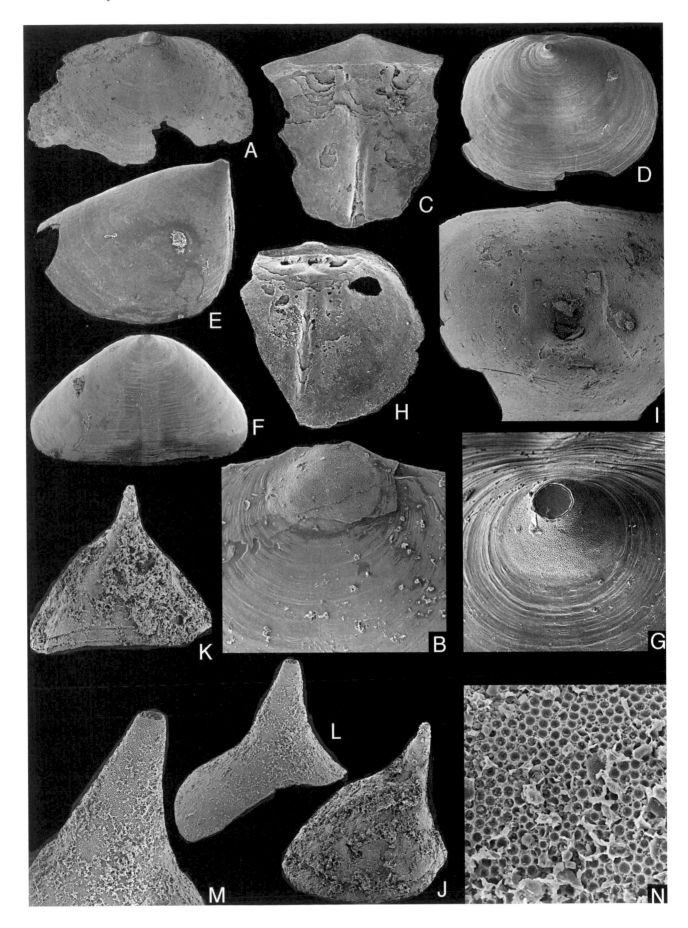

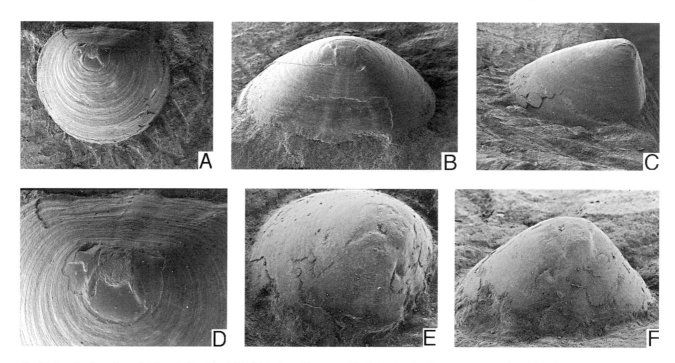

*Fig. 95. Ottenbyella carinata* (Moberg & Segerberg); Bjørkåsholmen Limestone; Fågelsång, Scania. □A. Lectotype; partly exfoliated ventral valve; figured by Moberg & Segerberg (1906, Pl. 3:5); LO 1784T; ×12. □B. Posterior view of A; ×15. □C. Lateral view of A; ×12. □D. Exfoliated apex of A, showing apical process and *vascula lateralia*; ×30. □E. Oblique lateral view of ventral internal mould; holotype by monotypy of *Acrotreta circularis* Moberg & Segerberg (1906, Pl. 3:4); LO 1783T; ×20. □F. Posterior view of E; ×18.

*Fig. 94* (opposite page). □A–I. *Ottenbyella carinata* (Moberg & Segerberg); Bjørkåsholmen Limestone (sample Öl-1); Ottenby, Öland. □A. Dorsal exterior; LO 6557t; ×28. □B. Larval shell of A; ×135. □C. Dorsal interior; LO 6558t; ×35. □D. Ventral exterior; LO 6561t; ×22. □E. Lateral view of D; ×27. □F. Posterior view of D; ×24. □G. Larval shell of D; ×150. □H. Dorsal interior; LO 6559t; ×32. □I. Ventral interior; LO 6560t; ×37. □J–N. Acrotretidae gen. et sp. nov. a.; Bjørkåsholmen Limestone. □J. Lateral view of ventral exterior; Mossebo (sample Vg-2), Västergötland; RM Br 136280; ×32. □K. Posterior view of J; ×32. □L. Lateral view of ventral exterior; Fågelsång (sample Sk-2), Scania; LO 6562t; ×60. □M. Pedicle tube of L; ×112. □N. Larval pitting of L; ×750.

*Table 37. Ottenbyella carinata*, average dimensions and ratios of ventral valves.

| | L | W | H | $L/W$ | $H/L$ |
|---|---|---|---|---|---|
| Sample Öl-1 | | | | | |
| N | 6 | 6 | 6 | 6 | 6 |
| X | 1.16 | 1.31 | 0.86 | 88% | 76% |
| S | 0.35 | 0.27 | 0.17 | 0.11 | 0.09 |
| MIN | 0.88 | 1.04 | 0.64 | 79% | 61% |
| MAX | 1.80 | 1.64 | 1.10 | 110% | 89% |

*Table 38. Ottenbyella carinata*, average dimensions and ratios of dorsal valves.

| | L | W | LI | WI | WG | LM1 | WM1 | LS | $L/W$ | $LI/WI$ | $MG/W$ | $WI/W$ | $LM1/L$ | $WM1/W$ | $LS/L$ |
|---|---|---|---|---|---|---|---|---|---|---|---|---|---|---|---|
| Sample Öl-1 | | | | | | | | | | | | | | | |
| N | 7 | 7 | 7 | 7 | 7 | 6 | 6 | 7 | 7 | 7 | 7 | 7 | 7 | 7 | 7 |
| X | 1.44 | 1.72 | 0.15 | 0.67 | 0.40 | 0.41 | 0.71 | 1.14 | 84% | 24% | 24% | 40% | 24% | 34% | 79% |
| S | 0.22 | 0.31 | 0.03 | 0.16 | 0.01 | 0.05 | 0.11 | 0.18 | 0.05 | 0.09 | 0.04 | 0.10 | 0.11 | 0.15 | 0.04 |
| MIN | 1.12 | 1.34 | 0.10 | 0.40 | 0.38 | 0.32 | 0.60 | 0.88 | 76% | 18% | 18% | 18% | 0% | 0% | 71% |
| MAX | 1.74 | 2.28 | 0.18 | 0.84 | 0.40 | 0.46 | 0.90 | 1.40 | 91% | 45% | 30% | 48% | 34% | 43% | 87% |

distinct and broadly shallow sulcus originating near the beak (Figs. 86A, 93H). The dorsal pseudointerarea is wide, broadly subtriangular, and slightly anacline, on average 24% as long as wide and occupying 40% of the valve width (Table 38). The median groove is wide and subtriangular (Figs. 93D, O, 94C). The dorsal interior has a very low median ridge that originates at some distance anterior to the pseudointerarea and extends on average 79% of the valve length (Figs. 86C,

93B, C, 94H, Table 38). The median buttress is poorly developed (Figs. 93D, O, 94C). The cardinal muscle fields are small and short, transversely elongate suboval, occupying on average 34% of the total valve width and extending for only 24% of the length (Figs. 93B–D, N, O, 94H; Table 38). The larval shells of both valves are well defined, close to circular, 0.18–0.22 mm across (Figs. 93G, I, 94B, G); the larval pits are of varying sizes, up to about 2.5 µm across (Fig. 93J).

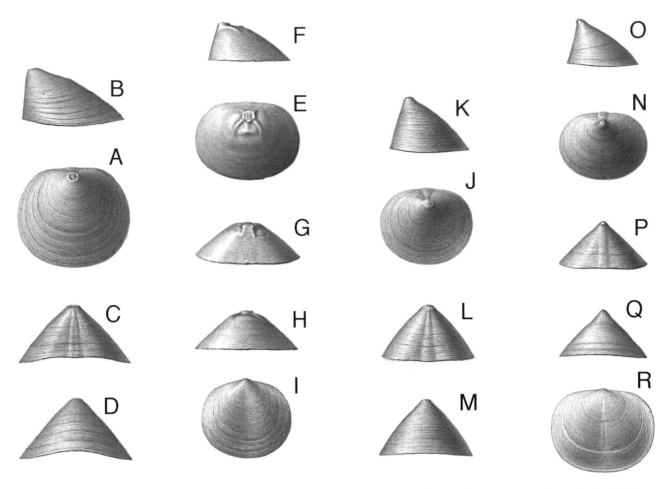

*Fig. 96. Ottenbyella carinata* (Moberg & Segerberg); Camera-lucida drawings prepared by G. Holm around 1900. □A. Ventral exterior; Bjørkåsholmen Limestone; Sörstrand, Norway (coll. G. Holm); RM Br 133958; ×11. □B. Lateral view of A; ×11. □C. Posterior view of A; ×11. □D. Anterior view of A; ×11. □E. Ventral internal mould, showing apical process and *vascula lateralia*; Bjørkåsholmen Limestone; Tyrifjord, Noway (coll. G. Holm); RM Br 133950; ×10. □F. Lateral view of E; ×10. □G. Posterior view of E; ×10. □H. Anterior view of E; ×10. □I. Dorsal exterior; Bjørkåsholmen Limestone; Bjørkåsholmen, Oslo region (coll. Holm); RM Br 133951; ×14. □J. Ventral exterior; Bjørkåsholmen Limestone; Bjørkåsholmen, Oslo region (coll. Holm); RM Br 133953; ×13. □K. Lateral view of J; ×13. □L. Posterior view of J; ×13. □M. Anterior view of J; ×13. □N. Ventral exterior; Ceratopyge Shale; Grönslunda, Öland (coll. G. Holm); RM Br 133956; ×14. □O. Lateral view of N; ×14. □P. Posterior view of N; ×14. □Q. Anterior view of N; ×14. □R. Dorsal exterior; Ceratopyge Shale; Grönslunda, Öland (coll. G. Holm); RM Br 133957; ×13.

*Discussion.* – The etched material of *O. carinata* described above is indistinguishable from the ventral valve (lectotype: LO 1784T; Fig. 95A–D herein) illustrated by Moberg & Segerberg (1906) from the Bjørkåsholmen Limestone at Fågelsång; the flattened pseudointerarea and interridge are well shown, as is the apical process and the single pair of *vascula lateralia*. The poorly preserved valve (holotype by monotypy: LO 1783T; Fig. 95E–F herein) described as *Acrotreta circularis* from the same locality by Moberg & Segerberg (1906) is in all probability a junior synonym of *O. carinata.*

Walcott (1902) described *Acrotreta conula* from an unnamed Upper Cambrian locality on Öland, but it clearly comes from the Ceratopyge Shale (see also p. 5). The extremely poorly preserved specimens would seem to be somewhat similar to *O. carinata* in the outline and convexity of the ventral valve, and *A. conula* might be a senior synonym, but

considering the problems concerning the type locality and preservation, it is equally well considered as a *nomen dubium*. Poorly preserved specimens from this level in the Bödahamn core were also referred to this species by Waern (1952; Fig. 88B herein).

Westergård (1909) described *Acrotreta oelandica* from the Ceratopyge Shale close to Ottenby; the single illustrated ventral valve is poorly preserved, and although it might be a junior synonym of *O. carinata*, it could also be related to the forms here referred to *Eurytreta* cf. *bisecta*; *A. oelandica* is here considered to be a *nomen dubium*.

*Occurrence.* – *O. carinata* occurs mainly in the Bjørkåsholmen Limestone at Ottenby, Flagabro, Stora Backor, Mossebo, and in the Oslo region (Figs. 5–8; Appendix 1K). It is possibly also present in the Ceratopyge Shale at numerous localitites throughout Scandinavia.

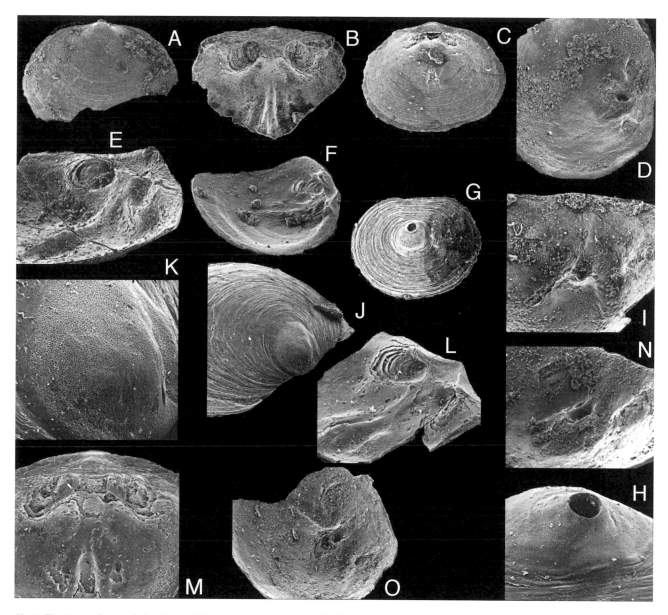

*Fig. 97.* □A–O. *Quadrisonia declivis* Koneva & Popov; northeastern Central Kazakhstan. □A. Dorsal exterior; Selety Limestone; Selety River (sample 325); RM Br 136323; ×20. □B. Dorsal interior; olistolith within Satpak Formation; Kujandy Section (sample 7822-5); RM Br 136321; ×33. □C. Dorsal exterior; olistolith within Satpak Formation; Kujandy Section (sample 7822-4); RM Br 136322; ×38. □D. Lateral view of ventral interior; Selety Limestone; Selety River (sample 325); RM Br 136316; ×18. □E. Oblique lateral view of dorsal interior; Selety Limestone; Selety River (sample 325); RM Br 136328; ×38. □F. Lateral view of dorsal interior; Kujandy Formation; Aksak–Kujandy Section (sample 7827-1); RM Br 136324; ×33. □G. Ventral exterior; olistolith within Satpak Formation; Kujandy Section (sample 7822-4); RM Br 136317; ×45. □H. Larval shell of G; ×165. □I. Ventral interior; Selety Limestone; Selety River (sample 325); RM Br 136318; ×38. □J. Oblique posterior view of dorsal exterior; Kujandy Formation; Aksak–Kujandy Section (sample 7827-1); RM Br 136327; ×60. □K. Dorsal larval shell of J; ×215. □L. Lateral view of dorsal interior; Kujandy Formation; Aksak–Kujandy Section (sample 7827-1); RM Br 136325; ×45. □M. Dorsal interior; Kujandy Formation; Aksak–Kujandy Section (sample 7827-1); RM Br 136326; ×50. □N. Ventral interior; Selety Limestone; Selety River (sample 325) RM Br 136319; ×40. □O. Ventral interior; Kujandy Formation; Aksak–Kujandy Section (sample 7827-1); RM Br 136320; ×38.

# Genus *Quadrisonia* Rowell & Henderson, 1978

*Type species.* – Original designation by Rowell & Henderson (1978, p. 6); *Quadrisonia minor* Rowell & Henderson, 1978; from the Upper Cambrian Orr Formation (*Taenicephalus* Biozone); Steamboat Pass, Utah.

*Diagnosis.* – Shell transversely oval, with relatively short, straight posterior margin. Ventral valve low subconical to conical; ventral pseudointerarea procline to cataline, undivided or with very shallow intertrough. Ventral interior with elongate subtriangular apical process, peforated posteriorly by internal pedicle tube. Dorsal valve slightly convex, with short, orthocline pseudointerarea, divided by wide median groove. Dorsal interior with variably developed median ridge; dorsal cardinal muscle fields small, subcircular in outline, closely spaced; dorsal central muscle scars well defined.

*Species included.* – *Quadrisonia minor* Rowell & Henderson, 1978; *Quadrisonia declivis* Koneva & Popov, 1988; *Quadrisonia suspensa* Koneva & Popov, 1988; *Quadrisonia simplex* Koneva, Popov & Ushatinskaya, 1990; *Quadrisonia* sp. nov. Henderson (*in* Henderson et al.), 1992; *Quadrisonia*? sp. Puura & Holmer, 1993.

*Occurrence.* – Upper Cambrian; Australia, North America, Kazakhstan, Antarctica, ?Sweden.

## *Quadrisonia declivis* Koneva & Popov, 1988
Fig. 97

*Synonymy.* – □1988 *Quadrisonia declivis* sp. nov. – Koneva & Popov, p. 54, Pl. 1:1–8.

*Holotype.* – CNIGR 4/12034; dorsal valve; figured by Koneva & Popov (1988, Pl. 1:5); Upper Cambrian (*Trisulcagnostus trisulcus* Biozone); Batyrbay Section, Malyj Karatau Range, southern Kazakhstan.

*Material.* – Figured. Ventral valves: RM Br 136316 (L 2, W 2.04, H 0.56); RM Br 136317 (L 0.62, W 0.66, H 0.29); RM Br 136318; RM Br 136319; RM Br 136320. Dorsal valves: RM Br 136321; RM Br 136322 (L 0.72, W 0.92, LI 0.08, WI 0.44, ML 0.2, LM1 0.22, WM1 0.48, LS 0.44); RM Br 136323; RM Br 136324; RM Br 136325; RM Br 136326; RM Br 136327; RM Br 136328. Total of 86 ventral and 55 dorsal valves.

*Measurements.* – See Table 39.

*Diagnosis.* – Ventral valve somewhat less than half as high as long; ventral pseudonterarea procline, flattened, with weakly defined and shallow intertrough. Apical process elongate, subtriangular extending along most of anterior valve surface.

*Description.* – The shell is ventri-biconvex and transversely oval. The ventral valve is on average 82% as long as wide and 44% as high as long (Table 39); the maximum height is near the beak, on average 27% of the valve length from the

*Table 39. Quadrisonis declivis, average dimensions and ratios of ventral valves.*

|  | L | W | H | M | L/W | H/L | M/L |
|---|---|---|---|---|---|---|---|
| Samples 7822-4, 7822-6, 7825-13.5 | | | | | | | |
| N | 12 | 12 | 12 | 12 | 12 | 12 | 12 |
| X | 1.04 | 1.26 | 0.46 | 0.28 | 82% | 44% | 27% |
| S | 0.26 | 0.30 | 0.13 | 0.07 | 0.05 | 0.07 | 0.04 |
| MIN | 0.70 | 0.88 | 0.28 | 0.20 | 72% | 34% | 20% |
| MAX | 1.44 | 1.64 | 0.62 | 0.40 | 91% | 56% | 34% |

posterior margin (Table 39). The ventral pseudointerarea is procline and flattened, with a weakly defined and shallow intertrough (Fig. 97G). In lateral view, the anterior slope of the valve is gently and evenly convex. The pedicle foramen forms a very short tube within the larval shell. The ventral interior has an elongate, subtriangular apical process, extending along most of the anterior slope of the valve; it is perforated posteriorly by the internal pedicle opening. The ventral *vascula lateralia* are baculate and widely divergent (Fig. 97D, I, N, O).

The dorsal valve is gently convex and 78% (N1) as long as wide. The dorsal pseudointerarea is orthocline occupying about 40% (N1) of the valve width; the median groove is broadly triangular. The dorsal cardinal muscle fields are suboval and closely spaced, occupying 49% of the valve width and about 32% of the length in one valve. The median ridge is low and triangular; it originates directly anterior to the median buttress, and extends for about 26% (N1) of the valve length. The dorsal scars of the central and anterior lateral muscles are closely spaced and form a small field directly lateral to the median ridge (Fig. 97B, E, F, L, M).

*Remarks.* – The material described here is less well preserved than that from the type locality, but there are no major differences in morphology. Koneva & Popov (1988, Fig. 1) published a reconstruction of the dorsal interior characters of this species.

*Occurrence.* – This species occurs in the Kujandy Formation at the following localities: Aksak–Kujandy Section (Fig. 27; Appendix 1H); Selety River (Appendix 1H); and in olistoliths within the Satpak Formation at the Kujandy Section (Fig. 25; Appendix 1G) and at the Aksak–Kujandy Section. In southern Kazakhstan, it is known only from the type locality.

## *Quadrisonia simplex* Koneva, Popov & Ushatinskaya, 1990
Fig. 98

*Synonymy.* – □1990 *Quadrisonia simplex* sp. nov. – Koneva, Popov & Ushatinskaya *in* Koneva et al., p. 159, Pl. 28:1–8.

*Holotype.* – PIN 4321/1; ventral valve (L 1.05, W 1.15, H 0.4); Kujandy Formation; Aksak–Kujandy.

*Material.* – Figured. Ventral valves: RM Br 136117 (L 1.1, W 1.38, H 0.36); RM Br 136119 (L 1.04, W 1.16, H 0.34); RM Br 136331 (L 0.94, W 1.3, H 0.34). Dorsal valves: RM Br 136115 (L 1.04, W 1.22, LI 0.08, WI 0.72, WG 0.24, LM1 0.32, WM10.76, LS 0.48); RM Br 136116 (L 1.06, W 1.16, LI 0.1, WI 0.68, WG 0.26, LM1 0.36, WM10.78, LS 0.72); RM Br 136118; RM Br 136329; RM Br 136330. Total of 21 ventral and 39 dorsal valves.

*Measurements.* – See Table 40.

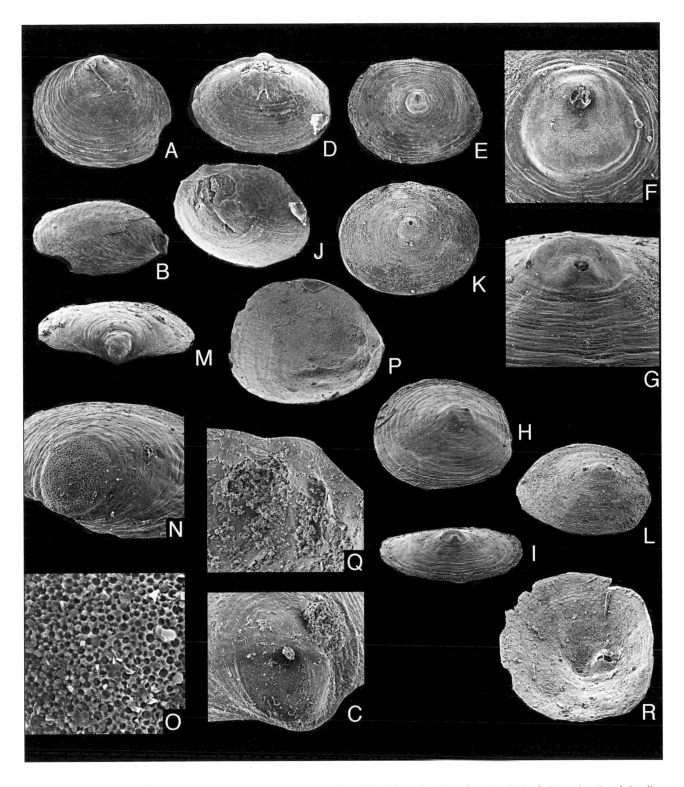

*Fig. 98. Quadrisonia simplex* Koneva, Popov & Ushatinskaya; northeastern Central Kazakhstan. □A. Dorsal exterior; Kujandy Formation; Satpak Syncline (sample 7843); RM Br 136115; ×28. □B. Lateral view of A; ×33. □C. Larval shell of A; ×120. □D. Dorsal interior; Kujandy Formation; Satpak Syncline (sample 7843); RM Br 136116, ×25. □E. Ventral exterior; Kujandy Formation; Satpak Syncline (sample 7843); RM Br 136117; ×25. □F. Ventral larval shell of E; ×120. □G. Posterior view of E; ×100. □H. Lateral view of E; ×28. □I. Posterior view of E; ×28. □J. Lateral view of dorsal interior; Kujandy Formation; Satpak Syncline (sample 7843); RM Br 136118; ×30. □K. Ventral exterior; olistolith within Satpak Formation; Aksak–Kujandy Section (sample 7844-2); RM Br 136119; ×27. □L. Lateral view of K; ×33. □M. Posterior view of dorsal exterior; Kujandy Formation; Satpak Syncline (sample 7843); RM Br 136330; ×33. □N. Larval shell of M; ×110. □O. Pitted microornament of larval shell of M; ×900. □P. Lateral view of dorsal interior; olistolith within Satpak Formation; Aksak–Kujandy Section (sample 7844-2); RM Br 136329; ×40. □Q. Dorsal pseudointerarea of P; ×80. □R. Ventral interior; olistolith within Satpak Formation; Aksak–Kujandy Section (sample 7844-2); RM Br 136331; ×40.

*Diagnosis.* – Shell subequally biconvex. Ventral valve forms low, wide cone; both valves less than one-third as long as high. Ventral pseudointerarea strongly procline with narrow interridge. Apical process short and low. Dorsal median ridge low.

*Description.* – The shell is subequally biconvex. The ventral valve forms a very low and wide cone with the maximum height situated on average 32% from the posterior margin (Fig. 98H, L; Table 40); it is transverse suboval, on average 79% as long as wide and 29% as high as long (Table 40). Ventral pseudointerarea strongly procline, with a narrow interridge (Fig. 98I). The pedicle foramen forms a very short tube that is enclosed within the larval shell (Fig. 98F, G). The ventral interior has a short, low and small apical process that is perforated by the pedicle opening; the dorsal cardinal muscle fields are poorly defined (Fig. 98R).

The dorsal valve is moderately and evenly convex in lateral profile (Fig. 98B) and transversely oval, on average 72% as long as wide and 29% as high as long (Table 41). The dorsal pseudointerarea is low and orthocline, occupying on average 45% of the valve length (Table 41); the median groove is broad, concave, and lens-shaped (Fig. 98Q). The dorsal cardinal muscle fields are closely spaced and transversely suboval in outline, occupying on average 45% of the valve width and 29% of the length (Fig. 98D, P; Table 41); the median buttress is low and elongated (Fig. 98P). The dorsal median ridge is low, extending on average 53% of the valve length (Fig. 98P; Table 41). The larval shells of both valves are well defined, surrounded by a raised rim, close to circular, around 0.27 mm across (Fig. 98C, N, F), covered by pits of somewhat varying sizes, up to about 2 μm across (Fig. 98O).

*Discussion.* – *Q. simplex* differs from all other species of the genus by having an almost equibiconvex shell, with both valves less than one-third as high as long. It is somewhat similar to *Q.* sp. nov. described by Henderson (*in* Henderson *et al.* 1992, Pl. 2:1–4) from the Upper Cambrian of West Antarctica, but the ventral valve of this species is proportionally higher, and has a more marginally placed apex.

*Occurrence.* – This is a relatively rare species that occurs in the Kujandy Formation at the Kujandy Section (Fig. 25; Appendix 1G) and at the Satpak Syncline (Fig. 29; Appendix 1I). It also occurs in olistoliths within the Satpak Formation at the Aksak–Kujandy Section (Fig. 27; Appendix 1H).

# Genus *Semitreta* Biernat, 1973

*Type species.* – Original designation by Biernat (1973, p. 75); *Semitreta maior* Biernat, 1973; Tremadoc chalcedonites; Holy Cross Mountains, Poland.

*Diagnosis.* – Shell subcircular, with short convex posterior margin. Ventral valve highly conical; ventral pseudointerarea poorly defined, procline to apsacline, with poorly defined, narrow interridge. Apical process poorly developed. Dorsal valve flattened; dorsal pseudointerarea raised, with triangular median groove and anacline propareas. Dorsal interior with elongate, cardinal muscle fields, and well-developed median ridge.

*Species included.* – *Semitreta maior* Biernat, 1973; *Torynelasma? magnum* Gorjansky, 1969.

*Discussion.* – In addition to the species listed above, Rushton & Bassett (*in* Owens *et al.* 1982, Pl. 6m–p) illustrated the ventral valve of an unnamed species of '*Torynelasma?*' that is similar to *Semitreta*.

The genus is similar to *Hansotreta* Krause & Rowell, 1975, but according to these authors, *Semitreta* differs mainly in the lack of an incurved dorsal apex.

*Occurrence.* – Lower Ordovician (Tremadoc–?Arenig); Poland, ?Estonia, ?Britain.

*Table 40. Quadrisonis simplex, average dimensions and ratios of ventral valves.*

| | L | W | H | M | L/W | H/L | M/L |
|---|---|---|---|---|---|---|---|
| Samples 7823-6, 7843 | | | | | | | |
| N | 8 | 8 | 8 | 5 | 8 | 8 | 5 |
| X | 0.98 | 1.25 | 0.29 | 0.31 | 79% | 29% | 32% |
| S | 0.15 | 0.21 | 0.09 | 0.11 | 0.06 | 0.07 | 0.06 |
| MIN | 1.10 | 1.40 | 0.30 | 0.33 | 79% | 27% | 30% |
| MAX | 1.11 | 1.50 | 0.40 | 0.46 | 90% | 36% | 41% |

*Table 41. Quadrisonis simplex, average dimensions and ratios of dorsal valves.*

| | L | W | H | LI | WI | MG | LM1 | WM1 | LS | BS | L/W | LI/WI | H/L | WI/W | LM1/L | WM1/W | LS/L | BS/L |
|---|---|---|---|---|---|---|---|---|---|---|---|---|---|---|---|---|---|---|
| Samples 7823-6, 7843 | | | | | | | | | | | | | | | | | | |
| N | 9 | 10 | 8 | 9 | 9 | 9 | 7 | 6 | 8 | 6 | 10 | 9 | 7 | 9 | 6 | 9 | 7 | 6 |
| X | 0.90 | 1.13 | 0.35 | 0.12 | 0.55 | 0.28 | 0.29 | 0.62 | 0.56 | 0.22 | 72% | 61% | 29% | 45% | 29% | 45% | 53% | 22% |
| S | 0.25 | 0.35 | 0.33 | 0.03 | 0.20 | 0.05 | 0.07 | 0.12 | 0.08 | 0.06 | 0.26 | 1.27 | 0.05 | 0.17 | 0.01 | 0.17 | 0.04 | 0.07 |
| MIN | 0.46 | 0.56 | 0.16 | 0.08 | 0.04 | 0.19 | 0.14 | 0.47 | 0.47 | 0.14 | 0% | 11% | 22% | 7% | 28% | 7% | 46% | 13% |
| MAX | 1.10 | 1.71 | 1.16 | 0.16 | 0.72 | 0.33 | 0.36 | 0.78 | 0.72 | 0.30 | 89% | 400% | 37% | 64% | 31% | 64% | 58% | 34% |

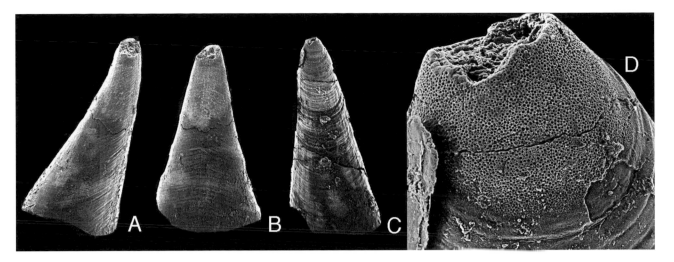

*Fig. 99. Semitreta?* aff. *magna* (Gorjansky); Akbulaksai Formation; Alimbet River, South Urals (sample B-780-1). □A. Lateral view of ventral exterior; RM Br 136150; ×33. □B. Posterior view of A; ×27. □C. Lateral view of ventral exterior; RM Br 136151 ; ×36. □D. Larval shell of C; ×215.

## *Semitreta?* aff. *magna* (Gorjansky, 1969)

Figs 99

*Material.* – Figured. Ventral valves: RM Br 136150; RM Br 136151. Total of 25 ventral valves.

*Description.* – The ventral valve forms a high conus that is about as high as long. The pseudointerarea is slightly flattened and apsacline; the intertrough forms a narrow and shallow furrow (Fig. 99B). The pedicle foramen is within the larval shell (Fig. 99D). In lateral view the anterior slope of the valve is steep and gently concave (Fig. 99A). The ventral interior is not known. The larval shell is circular, about 0.3 mm across, and ornamented by fine pits up to about 2 µm across (Fig. 99D). The postlarval shell is covered by closely spaced fila.

*Discussion.* – In size and external characters of the ventral valve, these specimens are closely comparable with *S.? magna* from the lower part of the Leetse beds in northern Estonia. The internal characters and dorsal valves are still unknown, both from Estonia and the South Urals, and their generic attribution is very uncertain.

*Occurrence.* – This rare species occurs in the upper part of the Alimbet Formation at the Alimbet Farm (Appendix 1B); it was also found in the Akbalaksai Formation at Alimbet River (Appendix 1C), and at Koagasli (Appendix 1F).

## Genus *Treptotreta* Henderson & MacKinnon, 1981

*Type species.* – Original designation by Henderson & MacKinnon (1981, p. 293); *Treptotreta jucunda*; Middle Cambrian (*Goniagnostus nathorsti* Biozone) Mailchange Limestone; Georgina Basin, Western Queensland.

*Diagnosis.* – Shell transversely oval; posterior margin moderately wide and straight to slightly convex. Ventral valve conical; ventral pseudointerarea procline, with intertrough poorly developed or absent; pedicle foramen within larval shell. Ventral interior with apical process widening anteriorly and occluding apex, perforated posteriorly by internal pedicle tube. Dorsal valve slightly convex to flattened; dorsal pseudointerarea wide, orthocline, divided by median groove; dorsal median septum high, triangular.

*Species included.* – *Treptotreta jucunda* Henderson & MacKinnon, 1981; *Treptotreta bella* Koneva, Popov & Ushatinskaya, 1990; *Treptotreta* sp. nov. Henderson (*in* Henderson *et al.*), 1992; *Treptotreta?* sp. Puura & Holmer, 1993.

*Discussion.* – *T. conversa* and *T. mutabilis*, described by Mei (1993) from the Middle Cambrian of China (Hebei), differ from all other species by having a pedicle foramen that is not enclosed within the larval shell, and their systematic relationship is uncertain.

*Occurrence.* – Middle–Upper Cambrian; Australia, New Zealand, Antarctica, Kazakhstan, ?Sweden.

## *Treptotreta bella* Koneva, Popov & Ushatinskaya, 1990

Figs. 39A–D, 100

*Synonymy.* – □1990 *Treptotreta bella* sp. nov. – Koneva, Popov & Ushatinskaya *in* Koneva *et al.*, p. 160, Pls. 25:6–9; 26:2–8.

*Holotype.* – PIN 4321/40; ventral valve; from an olistolith within the Erzhan Formation; Aksak–Kujandy Section.

*Material.* – Figured. Ventral valves: RM Br 136155 (L 1.44, W 1.8, H 0.84); RM Br 136154; RM Br 136160; RM Br 136161.

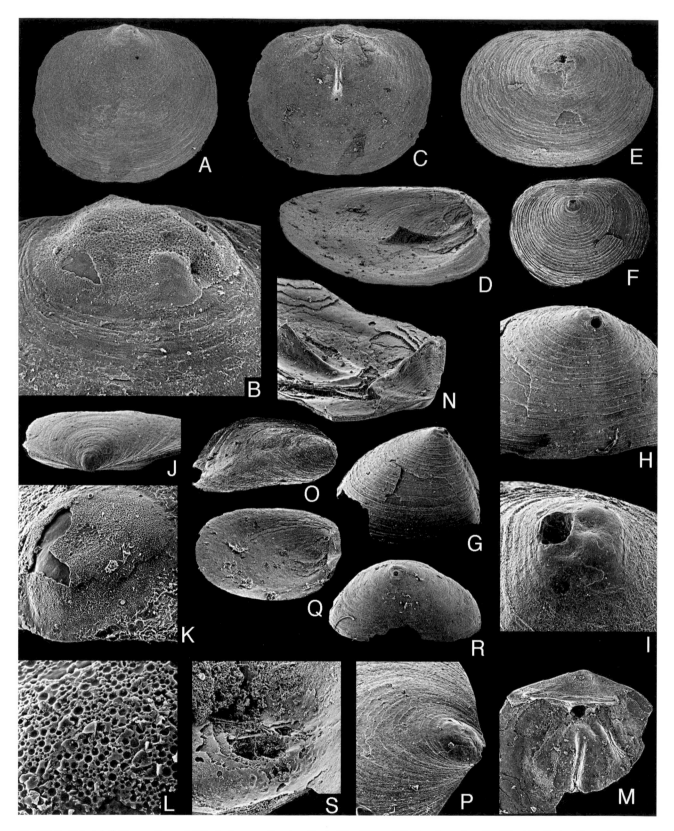

*Fig. 100. Treptotreta bella* Koneva, Popov & Ushatinskaya; northeastern Central Kazakhstan. □A. Dorsal exterior; olistolith within Erzhan Formation; Aksak–Kujandy Section (sample 79101); RM Br 136152; ×27. □B. Larval shell of A; ×225. □C. Dorsal interior; olistolith within Erzhan Formation; Aksak–Kujandy Section (sample 79101); RM Br 136153; ×27. □D. Lateral view of C; ×30. □E. Ventral exterior; olistolith within Erzhan Formation; Aksak–Kujandy Section (sample 79101); RM Br 136154; ×23. □F. Ventral exterior; olistolith within Olenty Formation (sample 79108); Aksak–Kujandy Section; RM Br 136155; ×22. □G. Ventral exterior, lateral view of F; ×25. □H. Posterior view of F; ×40. □I. Larval shell of F; ×110. □J. Posterior view of dorsal exterior; olistolith within Olenty Formation (sample 79108); Aksak–Kujandy Section; RM Br 136157; ×40. □K. Larval shell of J; ×180. □L. Pitted micro-ornament of larval shell of J;

Dorsal valves: RM Br 136152 (L 1.52, W 1.64, LI 0.14, WI 0.74, WG 0.32, LM1 0.56, WM10.84, LS 0.84); RM Br 136153 (L 1.74, W 1.86, LI 0.14, WI 0.92, WG 0.34, LM1 0.46, WM11.08, LS 0.92); RM Br 136156; RM Br 136157; RM Br 136158; RM Br 136159. Total of 135 ventral and 211 dorsal valves.

*Diagnosis.* – Ventral valve close to half as high as long; ventral pseudointerarea procline with poorly defined intertrough. Dorsal median septum low, extending for slightly less than $\frac{2}{5}$ of valve length.

*Description.* – The shell is subcircular. The ventral valve is broadly conical, on average about 90% as long as wide and close to half as high as long; the ventral pseudointerarea is procline and gently convex in lateral profile (Fig. 100G). The intertrough is poorly defined (Fig. 100H, R). The pedicle foramen is small and rounded, situated within the larval shell (Fig. 100I). The ventral interior has an apical process forming a wide ridge along the anterior and posterior slopes of the valve; the apical process is perforated by internal pedicle tube near its posterior end (Fig. 100S). The ventral cardinal muscle fields are slightly thickened, situated posterolaterally on the posterior slope of the valve. The ventral *vascula lateralia* are baculate and widely divergent.

The dorsal valve is gently convex in lateral profile, about 90% as long as wide. The dorsal pseudointerarea is orthocline and occupies more than half the valve width; the median groove is concave (Fig. 100M, N). The dorsal cardinal muscle fields are large and elongate suboval, occupying slightly more than 60% of the valve width, and extending anteriorly for about 40% of the length. The dorsal median septum is low and triangular, extending for slightly less than 40% of the valve length; the maximum height is at about the valve centre (Fig. 100C, D, Q). The larval shells of both valves are well defined, close to circular, around 0.3 mm across; they are covered by pits of several sizes, up to about 2 μm across.

*Discussion.* – This species differs from the type species in having (1) an undivided ventral pseudointerarea, (2) relatively low and short median septum, and (3) deeply impressed cardinal muscle fields in both valves.

*Occurrence.* – *T. bella* is common in the Kujandy Formation at the Kujandy and Aksak–Kujandy sections and the Satpak Syncline; it is also present in olistoliths within the Satpak, Olenty, and Erzhan formations (Figs. 25, 27, 29; Appendix 1H–J).

---

×900. □M. Dorsal interior; olistolith within Olenty Formation (sample 79108); Aksak–Kujandy Section; RM Br 136156; ×30. □N. Lateral view of M; ×45. □O. Lateral view of dorsal exterior; olistolith within Erzhan Formation; Aksak–Kujandy Section (sample 79101); RM Br 136158; ×30. □P. Umbonal area of O; ×60. □Q. Lateral view of dorsal interior; olistolith within Erzhan Formation; Aksak–Kujandy Section (sample 79101); RM Br 136159; ×14. □R. Posterior view of ventral exterior; olistolith within Erzhan Formation; Aksak–Kujandy Section (sample 79101); RM Br 136161; ×25. □S. Lateral view of ventral interior; olistolith within Olenty Formation (sample 79108); RM Br 136160; ×55.

## Family Acrotretidae gen. et sp. nov. a

Figs 89M–N, 94J–N

*Material.* – Figured. Ventral valves: RM Br 136280; LO 6562t. Dorsal valve: LO 6551t. Total of 1 dorsal and 3 ventral valves.

*Remarks.* – The ventral valve is transversely oval and widely conical, almost as high as long, with an extremely drawn out external pedicle tube (Fig. 94J–M). The ventral interior is invariably filled with matrix. The dorsal valve is only known from a somewhat questionable specimen, which is flattened in lateral view and has an outline that seems to fit to the ventral valve (Fig. 89M); it is filled with matrix, and the interior characters are not seen. The larval shells of both valves are pitted with deep pits of one size (Figs. 89N, 94N).

This extremely highly conical acrotretid probably represents a new genus. However, since it is represented by extremely poorly preserved material, it is treated under open nomenclature.

Similar, undescribed forms are known from the Arenig in Baltoscandia (Holmer, unpublished).

*Occurrence.* – It is only known from the Bjørkåsholmen Limestone at Fågelsång and Mossebo (Fig. 7; Appendix 1K).

## Family Ephippelasmatidae Rowell, 1965

*nomen translatum* herein (*ex* Ephippelasmatinae Rowell, 1965, p. H279), including Myotretinae Biernat, 1973, p. 80

*Diagnosis.* – Shell with narrow, straight posterior margin; usually with fine rugellae. Ventral valve recurved conical; pseudointerarea catacline to strongly apsacline with intertrough; foramen usually at end of tube within larval shell. Internal pedicle tube usually present along posterior slope; apical process reduced, usually forming low ridge near apex. Dorsal valve slightly convex to concave; dorsal pseudointerarea divided by short lens-like median groove. Dorsal median septum variable developed; median buttress commonly absent. Larval shell with unequally distributed pits of varying size.

*Genera included.* – *Pomeraniotreta* Bednarczyk, 1986 [=?*Anatreta* Mei, 1993]; *Akmolina* gen. nov.; *Mamatia* gen. nov; in addition to those listed by Holmer (1986, p. 112; 1989, p. 112).

*Occurrence.* – Upper Cambrian – Upper Ordovician.

## Genus *Akmolina* gen. nov.

*Name.* – After the Akmolinsk – the former name of Tselinograd.

*Type species.* – *Akmolina olentensis* sp. nov.; Kujandy Formation.

*Diagnosis.* – Ventral valve widely conical; ventral pseudo-interarea usually cataclne, more rarely somewhat procline or apsacline; intertrough wide and shallow with median plication; pedicle foramen at the end of short pedicle tube, enclosed within larval shell. Ventral interior with short internal pedicle tube fused with the posterior valve slope; apical process forming low elevation anterior and lateral to pedicle tube, becoming slightly ridged anteriorly. Dorsal pseudo-interarea low, divided by median groove. Median buttress and dorsal median ridge weakly developed or absent.

*Species assigned.* – *Akmolina olentensis* sp. nov.; *Eurytreta?* *exigua* Popov (*in* Koneva & Popov), 1988.

*Discussion.* – *Akmolina* is most similar to *Pomeraniotreta*, but differs in having (1) a wider conical ventral valve with a divided pseudointerarea, (2) a smaller apical process, and (3) a wider dorsal pseudointerarea with a well-defined median groove. In some characters of the ventral valve it comparable with *Lurgiticoma* Popov, 1980, but it differs from this genus in having a weakly developed or absent dorsal median ridge.

## *Akmolina olentensis* sp. nov.

Fig. 101

*Name.* – After the Olenty River, close to the type locality.

*Holotype.* – Fig. 101U–V; RM Br 136169; complete shell (L 0.48, W 0.62, H 0.34); Kujandy Formation (sample 7827); Aksak–Kujandy Section.

*Table 42. Akmolina olentensis*, average dimensions and ratios of ventral valves.

| | L | W | H | $L/W$ | $H/L$ |
|---|---|---|---|---|---|
| Sample 7827, 7827-1 | | | | | |
| N | 6 | 6 | 6 | 6 | 6 |
| X | 0.50 | 0.64 | 0.36 | 78% | 71% |
| S | 0.07 | 0.11 | 0.14 | 0.03 | 0.16 |
| MIN | 0.40 | 0.48 | 0.20 | 75% | 50% |
| MAX | 0.62 | 0.82 | 0.62 | 83% | 100% |

*Paratypes.* – Figured. Ventral valves: RM Br 136164 (L 0.62, W 0.82, H 0.62); RM Br 136332 (L 0.52, W 0.68, H 0.36); RM Br 136171. Dorsal valves: RM Br 136162 (L 0.96, W 1.02, LI 0.2, WI 0.62, WG 0.3); RM Br 136163 (L 0.88, W 0.94, LI 0.12, WI 0.48, WG 0.26, LM1 0.34, WM1 0.56); RM Br 136165 (L 0.9, W 1.04, LI 0.16, WI 0.6, ML 0.32, LM1 0.38, WM1 0.82); RM Br 136166; RM Br 136167 ( L 0.94, W 1.08, LI 0.14, WI 0.41, WG 0.28); RM Br 136168 (L 0.42, 0.44 , LI 0.04, WI 0.22, WG 0.12). Total of 2 complete shells, 13 ventral valves, and 85 dorsal valves.

*Mesurements.* – See Tables 42–43.

*Diagnosis.* – Ventral valve widely conical with catacline pseudointerarea and wide intertrough. Dorsal cardinal muscle scars large, elongate oval, weakly impressed; dorsal median ridge lacking.

*Description.* – The ventral valve is widely conical, on average 78% as long as wide and 71% as high as long (Table 42), with the maximum height placed at the beak (Fig. 101L). The pedicle foramen forms a short pedicle tube within the larval shell (Fig. 101J). The ventral pseudointerarea is usually catacline or more rarely slightly procline; the intertrough is wide and very shallow, with a weak median plication. In lateral profile the outline of the anterior valve slope is moderately and evenly convex. The internal pedicle tube is short, placed along the posterior slope, and surrounded by a small apical process that becomes slightly ridged anteriorly (Fig. 101W).

The dorsal valve is flat, slightly sulcate, with a swollen beak (Fig. 101S); it is on average 89% as long as wide (Table 43). The dorsal pseudointerarea is apsacline and on average 27% as long as wide, occupying 55% of the valve width (Table 43). The median groove is wide and subtriangular (Fig. 101H, P). The dorsal interior has a subrectangular median buttress but lacks a median septum or ridge. The dorsal cardinal muscle fields are large, weakly impressed, elongate oval, and occupy on average 55% of the maximum valve width, extending for 43% of the valve length (Fig. 101F–H, O, P; Table 43). The larval shells of both valves are well defined, transversely suboval, around 0.25 mm wide and 0.2 mm long, surrounded by a raised rim (Fig. 101B, J, V); the larval pits are of varying sizes, up to about 3.5 μm across (Fig. 101C, D).

*Discussion.* – This species differs from *A. exigua* in having (1) a moderately convex lateral profile of the ventral anterior

*Table 43. Akmolina olentensis*, average dimensions and ratios of dorsal valves.

| | L | W | LI | WI | MG | LM1 | WM1 | $L/W$ | $LI/WI$ | $MG/WI$ | $WI/W$ | $LM1/L$ | $WM1/W$ |
|---|---|---|---|---|---|---|---|---|---|---|---|---|---|
| Sample 7827, 7827-1 | | | | | | | | | | | | | |
| N | 16 | 16 | 16 | 16 | 16 | 4 | 4 | 16 | 16 | 16 | 16 | 4 | 4 |
| X | 0.79 | 0.89 | 0.13 | 0.49 | 0.31 | 0.36 | 0.50 | 89% | 27% | 62% | 55% | 43% | 55% |
| S | 0.14 | 0.15 | 0.03 | 0.10 | 0.08 | 0.06 | 0.08 | 0.05 | 0.04 | 0.10 | 0.07 | 0.04 | 0.05 |
| MIN | 0.59 | 0.64 | 0.07 | 0.30 | 0.16 | 0.27 | 0.43 | 75% | 19% | 40% | 45% | 39% | 48% |
| MAX | 0.99 | 1.14 | 0.17 | 0.69 | 0.43 | 0.40 | 0.61 | 97% | 33% | 80% | 69% | 47% | 61% |

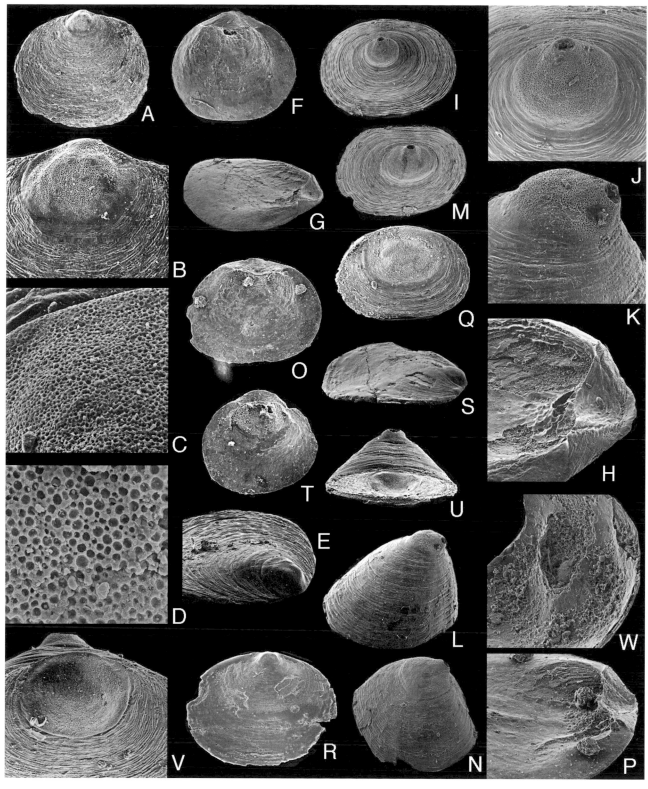

*Fig. 101. Akmolina olentensis* sp. nov.; Kujandy Formation; Aksak–Kujandy Section, northeastern Central Kazakhstan. □A. Dorsal exterior (sample 7827-1); RM Br 136162; ×33. □B. Larval shell of A; ×120. □C. Margin of larval shell of A; ×500. □D. Larval pitting of A; ×800. □E. Oblique posterior view of A; ×50. □F. Dorsal interior (sample 7827); RM Br 136163; ×33. □G. Lateral view of F; ×45. □H. Dorsal pseudointerarea of F; ×110. □I. Ventral exterior (sample 7827); RM Br 136164; ×40. □J. Larval shell of I; ×120. □K. Lateral view of larval shell of I; ×150. □L. Lateral view of I; ×50. □M. Ventral exterior (sample 7827); RM Br 136332; ×45. □N. Lateral view of M; ×55. □O. Dorsal interior (sample 7827-1); RM Br 136165; ×30. □P. Lateral view of O; ×60. □Q. Dorsal exterior (sample 7827); RM Br 136166; ×60. □R. Dorsal exterior (sample 7827); RM Br 136167; ×33. □S. Lateral view of R; ×60. □T. Dorsal interior (sample 7827-1); RM Br 136168; ×90. □U. Holotype; posterior view of complete shell (sample 7827); RM Br 136169; ×55. □V. Dorsal larval shell of U; ×110. □W. Ventral interior, showing internal pedicle tube and apical process (sample 7827); RM Br 136171; ×100.

slope, (2) a catacline ventral pseudointerarea with an inter-trough that is ridged medially, and (3) large, elongate oval, weakly impressed dorsal cardinal muscle fields, and (4) in lacking a dorsal median ridge.

*Occurrence.* – The species occurs only in the upper part of the Kujandy Formation at the Aksak–Kujandy section (Fig. 27; Appendix 1H).

## Genus *Mamatia* gen. nov.

*Name.* – After the Mamat Mountain in the Chingiz Range, close to the type locality.

*Type and only species.* – *Paratreta retracta* Popov (*in* Nazarov & Popov), 1980; Lower Ordovician (upper Tremadoc) Mamat Formation; Chingiz Range, Kazakhstan.

*Diagnosis.* – Ventral valve conical; ventral pseudointerarea undivided, catacline to apsacline; pedicle foramen forming short external pedicle tube within larval shell. Apical process completely occluding apex, perforated by short internal pedicle tube; ventral *vascula lateralia* baculate. Dorsal valve evenly convex, with wide, narrow pseudointerarea, and well-developed propareas. Median buttress well developed; dorsal median septum triangular, low to moderately high, with a single septal rod; dorsal cardinal muscle fields strongly impressed.

*Discussion.* – *Mamatia* is comparable with *Pomeraniotreta* and *Lurgiticoma* in the shape of the ventral valve and apical process, but differs in having a triangular dorsal median septum with a single rod. The type (and only) species of *Mamatia* was assigned previously to *Paratreta* Biernat, 1973; however, although the internal morphology of *Paratreta* is still virtually unknown, it seems to be more closely comparable to some early torynelasmatids (such as *Cristicoma*).

## *Mamatia retracta* (Popov, 1980)

Fig. 102

*Synonymy.* – □1980 *Paratreta retracta* sp. nov. – Popov *in* Nazarov & Popov, p. 95, Pl. 25:1–6.

*Holotype.* – CNIGR 110/11352; ventral valve; Lower Ordovician Mamat Formation; Chingiz Range, eastern Kasakhstan.

*Material.* – Figured. Ventral valves: RM Br 136175 (L 0.6, W 0.74, H 0.44); RM Br 136176; RM Br 136178 (L 0.74, W 0.82, H 0.62); RM Br 136181; RM Br 136183; RM Br 136185 (L 0.6, W 0.74, H 0.44); RM Br 136187; RM Br 136190; RM Br 136333. Dorsal valves: RM Br 136173; RM Br 136174; RM Br 136177; RM Br 136179; RM Br 136180; RM Br 136182; RM Br 136184 (L 0.52, W 0.72, LI 0.04, WI 0.24, WG 0.18); RM Br 136186. Total of 57 ventral and 105 dorsal valves.

*Measurements.* – See Tables 44–45.

*Diagnosis.* – As for genus.

*Description of material from South Urals and northeastern Central Kazakhstan.* – The shell is dorsi-biconvex and transversely suboval. The posterior margin is straight, occupying about 75% of the maximum valve width. The ventral valve is broadly conical, on average 75% as long as wide and 75% as high as long (Table 44) with the maximum height at the umbo; the anterior slope of the valve is gently and evenly convex in lateral profile (Fig. 102G, O, V). The ventral pseudointerarea is procline to slightly catacline and gently convex in lateral profile; it is bisected by a poorly defined intertrough (Fig. 102P, R). The pedicle foramen forms a short external pedicle tube, enclosed within the larval shell (Fig. 102Q, S). The apical process occludes the apical part of the valve and is slightly ridged anteriorly (Fig. 102Y).

The dorsal valve is moderately and evenly convex, with the maximum height in the posterior third of the valve length (Fig. 102C), on average 76% as long as wide (Table 45). The dorsal pseudointerarea is low, orthocline, and occupies on average 57% of the valve width. The dorsal median septum is triangular and has a single upper rod, extending for on average 74% of the valve length (Table 45). The median buttress is subtriangular. The dorsal cardinal muscle fields are elongate suboval, slightly divergent anteriorly; they occupy on average 61% of the valve width and extend for 42% of the length (Fig. 102K–M, T, U, X, Z; Table 45). The larval shells of both valves are well defined, transversely suboval, around 0.23×0.17 mm, surrounded by a raised rim (Fig. 102B, J, S); the larval pits are of varying sizes, up to about 2 μm across (Fig. 102N).

*Discussion.* – The specimens from the Olenty Formation agree closely with the type material described by Popov (*in* Nazarov & Popov 1980, p. 94; Fig. 102M, N) from the Lower Ordovician (presumably lower Arenig) Mamat Formation

*Fig. 102. Mamatia retracta* (Popov). □A. Dorsal exterior; Olenty Formation; Sasyksor Lake, northeastern Central Kazakhstan (sample 601); RM Br 136173; ×40. □B. Larval shell of A; ×150. □C. Lateral view of A; ×50. □D. Dorsal exterior; Koagash Formation; Karabutak River, South Urals (sample 1163); RM Br 136174; ×40. □E. Ventral exterior; Koagash Formation; Karabutak River, South Urals (sample 1163); RM Br 136175; ×38. □F. Ventral exterior; Olenty Formation; Sasyksor Lake, northeastern Central Kazakhstan (sample 601); RM Br 136176; ×38. □G. Lateral view of F; ×45. □H. Ventral exterior; Olenty Formation; Sasyksor Lake, northeastern Central Kazakhstan (sample 601); RM Br 136178; ×50. □I. Dorsal exterior; Koagash Formation; Karabutak River, South Urals (sample 1163); RM Br 136179; ×40. □J. Larval shell of I; ×150. □K. Dorsal interior; Koagash Formation; Karabutak River, South Urals (sample 1163); RM Br 136177; ×50. □L. Lateral view of K; ×40. □M. Dorsal interior; topotype; Olenty Formation; Chingiz Range, Kazakhstan; RM Br 136180; ×50. □N. Pitted microornament of larval shell of L; ×910. □O. Lateral view of ventral exterior; Olenty Formation; Sasyksor Lake, northeastern Central Kazakhstan (sample 601); RM Br 136181; ×50. □P. Posterior view of O; ×45. □Q. Lateral view of larval shell of O. □R. Ventral exterior, posterior view;

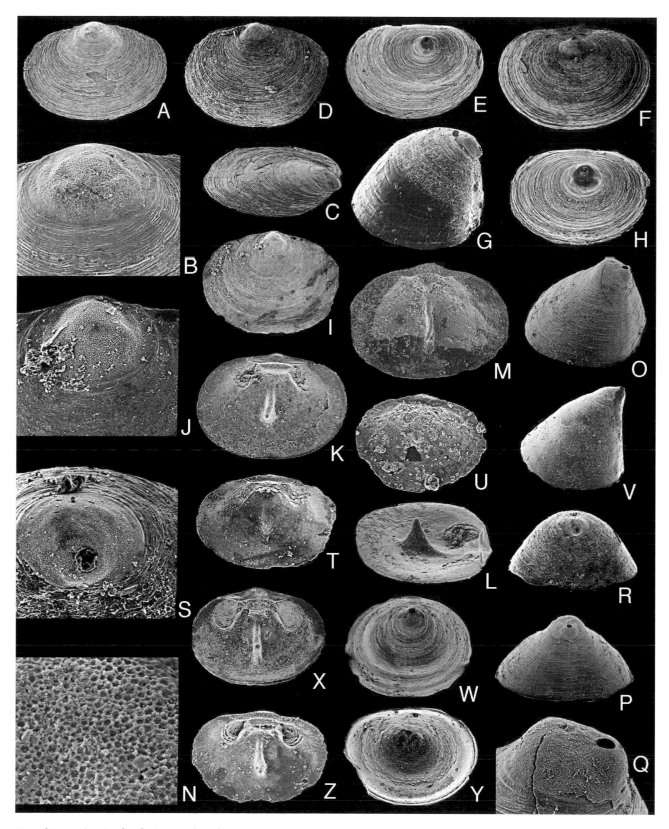

Koagash Formation; Karabutak River, South Urals (sample 1163); RM Br 136190; ×38. □S. Larval shell of R; ×150. □T. Dorsal interior; Koagash Formation; Karabutak River, South Urals (sample 1163); RM Br 136184; ×45. □U. Dorsal interior; Olenty Formation; Sasyksor Lake, northeastern Central Kazakhstan (sample 601); RM Br 136182; ×50. □V. Lateral view of ventral exterior; Koagash Formation; Karabutak River, South Urals (sample 1163); RM Br 136333; ×45. □W. Plane view of V; ×90. □X. Dorsal interior; Olenty Formation; Sasyksor Lake, northeastern Central Kazakhstan (sample 601); RM Br 136185; ×30. □Y. Ventral interior; Koagash Formation; Karabutak River, South Urals (sample 1163); RM Br 136187; ×40. □Z. Dorsal interior of juvenile specimen; Olenty Formation; Sasyksor Lake, northeastern Central Kazakhstan (sample 601); RM Br 136186; ×40.

in the Chingiz Range. The specimens from the South Urals are also closely comparable with the type material.

*Occurrence.* – Outside the type locality, *M. retracta* occurs in the Olenty Formation at Sasyksor Lake (Appendix 1H) and at the Karabutak and Koagash rivers (Appendix 1F–G). This species is possibly also present in the upper part of the Agalatas Formation (Fig. 35, Appendix 1J).

*Table 44. Mammatia retracta*, average dimensions and ratios of ventral valves.

|  | L | W | H | $L/W$ | $H/L$ |
|---|---|---|---|---|---|
| **Sample 601-a** | | | | | |
| N | 24 | 24 | 24 | 24 | 24 |
| X | 0.63 | 0.85 | 0.48 | 75% | 75% |
| S | 0.08 | 0.12 | 0.11 | 0.05 | 0.09 |
| MIN | 0.50 | 0.64 | 0.30 | 68% | 60% |
| MAX | 0.82 | 1.10 | 0.74 | 88% | 90% |

*Table 45. Mammatia retrac ta*, average dimensions and ratios of dorsal valves.

| | L | W | LI | WI | ML | MG | LM1 | WM1 | LS | PHS | BS | H | $L/W$ | $LI/WI$ | $MG/WI$ | $WI/W$ | $LM1/L$ | $WM1/W$ | $LS/L$ | $PHS/L$ | $BS/L$ | $H/L$ |
|---|---|---|---|---|---|---|---|---|---|---|---|---|---|---|---|---|---|---|---|---|---|---|
| **Sample 601** | | | | | | | | | | | | | | | | | | | | | | |
| N | 30 | 30 | 30 | 30 | 30 | 30 | 29 | 30 | 30 | 30 | 30 | 30 | 30 | 30 | 30 | 30 | 29 | 30 | 30 | 30 | 30 | 30 |
| X | 0.63 | 0.84 | 0.09 | 0.48 | 0.10 | 0.18 | 0.27 | 0.51 | 0.47 | 0.43 | 0.19 | 0.18 | 76% | 19% | 39% | 57% | 42% | 61% | 75% | 68% | 30% | 29% |
| S | 0.08 | 0.12 | 0.02 | 0.09 | 0.07 | 0.03 | 0.05 | 0.09 | 0.07 | 0.06 | 0.04 | 0.04 | 0.05 | 0.04 | 0.06 | 0.07 | 0.05 | 0.06 | 0.05 | 0.05 | 0.05 | 0.05 |
| MIN | 0.48 | 0.58 | 0.06 | 0.32 | 0.06 | 0.12 | 0.18 | 0.34 | 0.34 | 0.32 | 0.14 | 0.12 | 65% | 11% | 28% | 48% | 36% | 49% | 69% | 61% | 17% | 20% |
| MAX | 0.83 | 1.10 | 0.14 | 0.72 | 0.46 | 0.26 | 0.36 | 0.72 | 0.68 | 0.62 | 0.30 | 0.32 | 88% | 32% | 52% | 82% | 58% | 75% | 91% | 81% | 42% | 43% |

## Genus *Pomeraniotreta* Bednarczyk, 1986

*Synonymy.* – ☐1986 *Pomeraniotreta* gen. nov. – Bednarczyk, p. 415. ☐?1993 *Anatreta* gen. nov. Mei, p. 405 [in part].

*Type species.* – *Pomeraniotreta biernatae* Bednarczyk, 1986; Lower Ordovician (?Arenig); northern Poland.

*Diagnosis.* – Ventral valve highly conical with undivided ventral pseudointerarea, strongly apsacline and concave in lateral profile; ventral interior with well-developed apical process occluding apex; dorsal valve flattened with short, undivided pseudointerarea. dorsal interior with median buttress and poorly defined median ridge.

*Species assigned.* – *Pomeraniotreta biernatae* Bednarczyk, 1986; *Pomeraniotreta* sp. Puura & Holmer, 1993; ?*Anatreta transversa* Mei, 1993.

*Discussion.* – According to Bednarczyk (1986), *Pomeraniotreta* is to be referred to the torynelasmatids. This view is not supported here (see also Holmer 1986, p. 112); the genus is closely related to *Myotreta* and *Numericoma*. The torynelasmatids are characterized by a wide, straight posterior margin and a well-defined, flattened ventral pseudointerarea (Holmer 1989).

The new genus and species *Anatreta transversa*, described by Mei (1993) from the Upper Cambrian of China (Hebei), appears to be closely related to *Pomeraniotreta*. However, the ventral valve of this taxon is poorly known; it appears to be lower than that of *Pomeraniotreta*. The second species described by Mei, *Anatreta cava*, probably does not belong to the same genus; the morphology of the dorsal valve suggests that it possibly belongs to *Quadrisonia*.

## *Pomeraniotreta biernatae* Bednarczyk, 1986

Figs. 103–105

*Synonymy.* – ☐1971 'Acrotretacean' – Poulsen, Pls. 1:1–2; 2:1. ☐1986 *Pomeraniotreta biernati* [*sic*] sp. nov. – Bednarczyk, p. 415, Pls. 2:1–3; 3:1–3, Fig. 2.

*Holotype.* – ING PAN 9–78; dorsal valve; figured by Bednarczyk (1986, Pl. 2:2–3); Lower Ordovician (?*Paroistodus proteus* Biozone); Bialogóra core near Łeba, depth 2701.4–2700.0 m, northern Poland.

*Material.* – Figured. Ventral valves: RM Br 129088 (L 0.81, W 0.82, H 1.56); RM Br 129089 (L 0.72, W 0.77, H 1.4); RM Br 129099; RM Br 133955 (L 0.65, W 0.7, H 1); RM Br 136282; RM Br 136286. Dorsal valves: RM Br 129080 (L 0.51, W 0.45, LI 0.07, WI 0.22); RM Br 129085 (L 0.85, W 0.8); RM Br 129086 (L 0.9, W 0.77, LI 0.5, WI 0.17); RM Br 129098 (L 0.9, W 0.73, LI 0.22, WI 0.51); RM Br 136283 (L 0.57, W 0.47, LI

*Fig. 103. Pomeraniotreta biernatae* Bednarczyk; Bjørkåsholmen Limestone. ☐A. Dorsal exterior; Flagabro (sample Sk-1), Scania; RM Br 129085; ×60. ☐B. Lateral view of A; ×60. ☐C. Larval shell of A; ×202. ☐D. Larval pitting of A; ×900. ☐E. Dorsal interior; Flagabro (sample Sk-1), Scania; RM Br 129086; ×60. ☐F. Pseudointerarea of E; ×165. ☐G. Ventral exterior; Flagabro (sample Sk-1), Scania; RM Br 129089; ×60. ☐H. Posterior view of G; ×112. ☐I. Lateral view of dorsal interior; Ottenby (sample Öl-1), Öland; LO 6563t; ×135. ☐J. Dorsal interior; Flagabro (sample Sk-1), Scania; RM Br 129098; ×49. ☐K. Posterior view of ventral exterior; Flagabro (sample Sk-1), Scania; RM Br 129088; ×35. ☐L. Lateral view of K; ×32. ☐M. Dorsal interior; Flagabro (sample Sk-1), Scania; RM Br 129080; ×97. ☐N. Lateral view of ventral exterior; Ottenby, Öland (coll. S. Bengtson); RM Br 136282; ×80. ☐O. Larval shell of N; ×200.

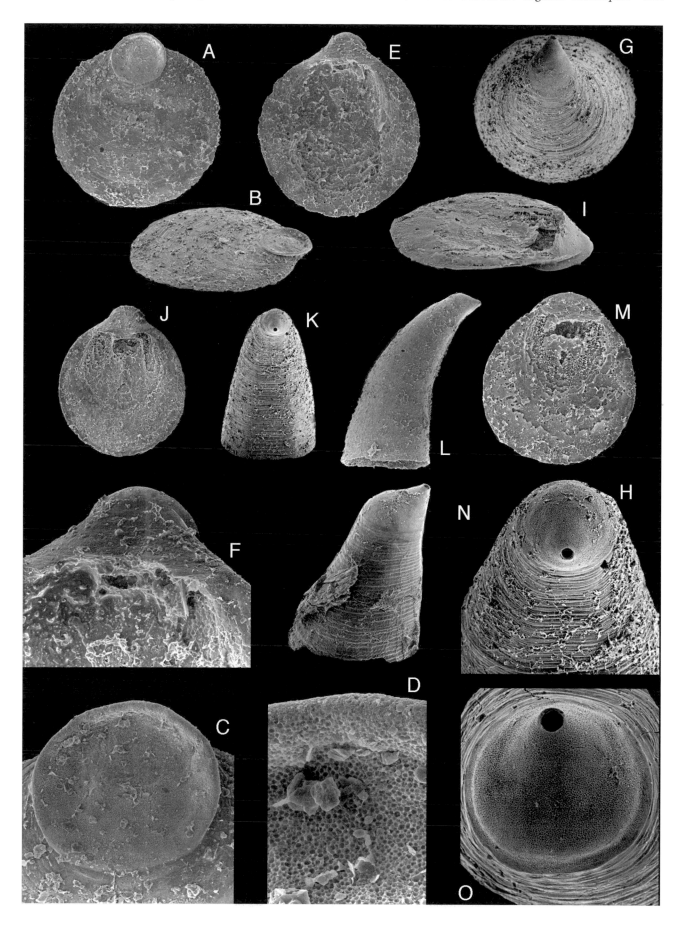

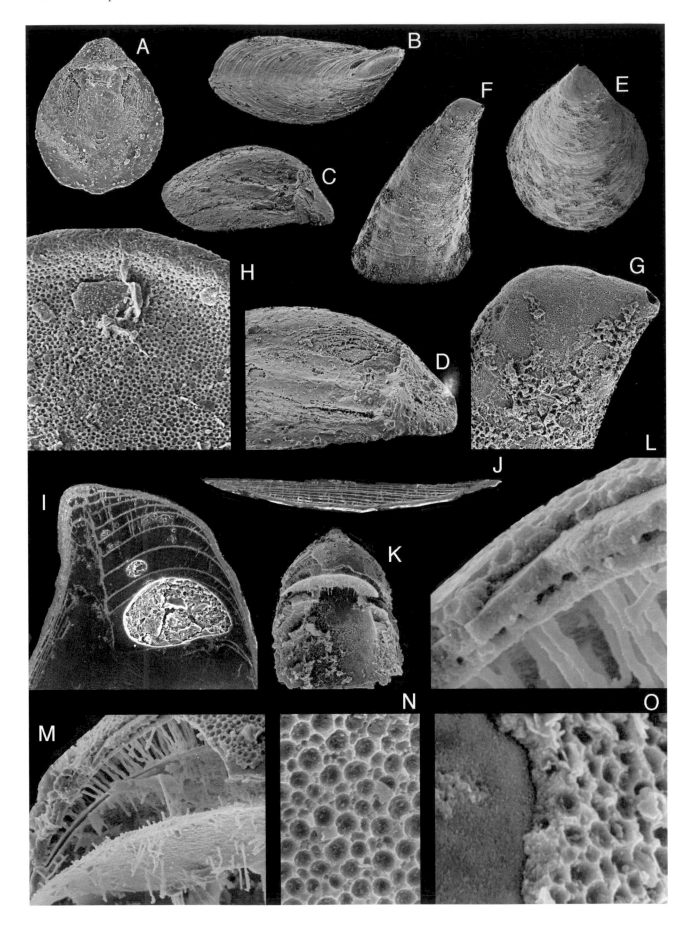

0.12, WI 0.27); RM Br 136284; RM Br 136285; RM Br 136287; LO 6563t. Total of 736 ventral and 594 dorsal valves.

*Mesurements.* – See Tables 46–47.

*Diagnosis.* – As for genus.

*Description.* – The ventral valve is highly conical, on average 95% as long as wide and 156% as high as long (Table 46). The ventral pseudointerarea is flattened, undivided and strongly apsacline (Figs. 103K, 105C); in lateral view it is strongly concave, and the ventral apex is about 0.11 mm posterior to the margin (Figs. 103L, 104F, 105B). The pedicle foramen forms a short pedicle tube within the larval shell (Figs. 103N, 104G). The apical process fills the entire apex, occupying about one-third of the total height, perforated posteriorly by the internal pedicle tube, which is up to 60 µm wide (Fig. 104I, K).

*Fig. 105. Pomeraniotreta biernatae* Bednarczyk; Camera lucida drawings prepared by G. Holm around 1900; Bjørkåsholmen Limestone; Ottenby, Öland (coll. G. Holm). □A. Ventral exterior; RM Br 133955; ×22. □B. Lateral view of A; ×22. □C. Posterior view of A; ×22. □D. Anterior view of A; ×22.

*Table 46. Pomeraniotreta biernatae*, average dimensions and ratios of ventral valves.

|        | Lmax | W    | H    | L    | H/L  | L/W  |
|--------|------|------|------|------|------|------|
| **Sample Öl-1** |      |      |      |      |      |      |
| N      | 30   | 30   | 30   | 30   | 30   | 30   |
| X      | 0.65 | 0.57 | 0.84 | 0.54 | 156% | 95%  |
| S      | 0.10 | 0.06 | 0.15 | 0.10 | 0.18 | 0.11 |
| MIN    | 0.44 | 0.44 | 0.56 | 0.40 | 127% | 85%  |
| MAX    | 0.92 | 0.66 | 1.12 | 0.88 | 204% | 138% |

The dorsal valve is flattened (Fig. 103B) to concave (Fig. 104B–D) and subcircular (Fig. 103A) to elongate suboval (Fig. 104A), on average 108% as long as wide (Table 47). The dorsal pseudointerarea is reduced to a small, concave, and undivided plate, on average 53% as wide as the valve (Figs. 103E, I, J, 104A, C, D; Table 47). The median buttress is wide and subrectangular. The dorsal cardinal muscle fields are poorly defined, occupying on average 57% of the valve width and extending for 44% of the length (Figs. 103E, I, J, 104A, C, D). The dorsal median ridge is poorly developed or absent.

*Table 47. Pomeraniotreta biernatae*, average dimensions and ratios of dorsal valves.

|        | L    | W    | LI   | WI   | LM1  | WM1  | LS   | L/W  | LI/L | LI/WI | WI/W | LM1/L | WM1/W | LS/L |
|--------|------|------|------|------|------|------|------|------|------|-------|------|-------|-------|------|
| **Sample Öl-1** |      |      |      |      |      |      |      |      |      |       |      |       |       |      |
| N      | 30   | 30   | 30   | 30   | 30   | 30   | 25   | 30   | 30   | 30    | 30   | 30    | 30    | 25   |
| X      | 0.56 | 0.52 | 0.12 | 0.27 | 0.25 | 0.30 | 0.33 | 109% | 22%  | 46%   | 53%  | 44%   | 57%   | 58%  |
| S      | 0.08 | 0.07 | 0.03 | 0.03 | 0.03 | 0.05 | 0.06 | 0.05 | 0.04 | 0.07  | 0.05 | 0.05  | 0.06  | 0.06 |
| MIN    | 0.41 | 0.40 | 0.08 | 0.22 | 0.18 | 0.20 | 0.22 | 100% | 17%  | 33%   | 44%  | 37%   | 46%   | 48%  |
| MAX    | 0.68 | 0.62 | 0.19 | 0.36 | 0.30 | 0.38 | 0.44 | 121% | 32%  | 68%   | 65%  | 64%   | 73%   | 76%  |

*Fig. 104. Pomeraniotreta biernatae* Bednarczyk; Bjørkåsholmen Limestone. □A. Dorsal interior; Bjørkåsholmen (sample Ng-1), Oslo region; RM Br 136283; ×75. □B. Lateral view of dorsal exterior; Bjørkåsholmen (sample Ng-1), Oslo region; RM Br 136284; ×82. □C. Lateral view of dorsal interior; Bjørkåsholmen (sample Ng-1), Oslo region; RM Br 136285; ×82. □D. Pseudointerarea of C; ×150. □E. Ventral exterior; Mossebo (Vg-2), Västergötland; RM Br 136286; ×60. □F. Lateral view of E; ×49. □G. Lateral view of larval shell of E; ×225. □H. Larval pitting of dorsal exterior; Mossebo (Vg-2), Västergötland; RM Br 136287; ×900. □I. Polished section through ventral apex, showing apical process and pedicle tube; Flagabro (sample Sk-1), Scania; ×255. □J. Polished section through dorsal valve; Flagabro (sample Sk-1), Scania; ×225. □K. Anterior view of broken ventral valve, showing section through apical process; Flagabro (sample Sk-1), Scania; RM Br 129099; ×112. □L. Detail of K, showing larval shell and underlying columnar secondary shell; ×2500. □M. Detail of K, showing alternation of columnar layers in apical process; ×600. □N. Detail of K, showing larval pitting; ×2400. □O. Detail of K, showing partly exfoliated larval shell; ×2500.

The larval shells of both valves are well defined, subcircular, about 0.26 mm across; the dorsal larval shell is surrounded by a strongly elevated rim 25 µm wide (Figs. 104C, D, O, 104H); the larval pits are of two sizes, the larger up to 2.5 µm across, surrounded by smaller, up to 1.5 µm across (Fig. 104N, O). The columnar shell structure in the secondary layer is well developed in sections through the apical process, where successive columnar laminae, up to 0.3 mm thick, have columns that are 1.6 µm thick connecting the upper and lower lamellae, up to 2 µm thick (Fig. 104I, K–O). The dorsal valve consists of a secondary layer with subparallel columnar laminae (Fig. 104J).

*Discussion.* – As the species was named in honour of Gertruda Biernat, the spelling of the name is here corrected. In all known characters, the Swedish specimens are identical to those described by Bednarczyk (1986); however the internal morphology of the ventral valve of the Polish form is still unknown.

*Occurrence. – P. biernatae* is the most common species in the Bjørkåsholmen Limestone at almost all investigated localities (Fig. 5–8; Appendix 1K). According to Bednarczyk (1986) it occurs in the Arenig of northern Poland.

## Family Torynelasmatidae Rowell, 1965

*nomen translatum,* herein (*ex* Torynelasmatinae Rowell, 1965, p. H279)

*Diagnosis. –* Shell with wide straight posterior margin. Ventral valve subpyramidal with flat, well-defined, undivided pseudointerarea, procline to apsacline; foramen within larval shell. Ventral interior with low, ridge-like apical process, supporting pedicle tube. Dorsal pseudointerarea poorly divided. Dorsal valve with high, triangular median septum, frequently with variably developed surmounting platform; broad median buttress elevated above valve floor.

*Genera included. – Cristicoma* Popov, 1980; *Sasyksoria* gen. nov.; in addition to those listed by Holmer (1989, p. 106).

*Occurrence. –* Lower Ordovician – Silurian.

## Genus *Cristicoma* Popov, 1980

*Type species. –* Original designation by Popov (1980, p. 89); *Cristicoma sincera* Popov, 1980; Llanvirn Kopalin Stage; Central Kazakhstan.

*Diagnosis. –* Apical process filling entire apex and forming septum, supported anteriorly by long pedicle tube. Dorsal median septum high, triangular, with numerous anterior spines and bundles of spines at top.

*Species included. – Cristicoma sincera* Popov, 1980; *Cristicoma? keskentassica* sp. nov.

*Occurrence. –* Ordovician (Tremadoc–Llanvirn); Kazakhstan.

## *Cristicoma? keskentassica* sp. nov.
Fig.106,

*Name. –* After Keskentas Ridge.

*Holotype. –* Fig. 106A–D; RM Br 136249; dorsal valve (L 2.48, W 2.8, WI 1.4, WG 0.72, LS 1.8, LM1 0.72, WM1 1.52); Agalatas Formation (sample 564); Keskentas Ridge.

*Paratypes. –* Figured. Ventral valve: RM Br 136250. Total of 1 dorsal and 5 ventral valves.

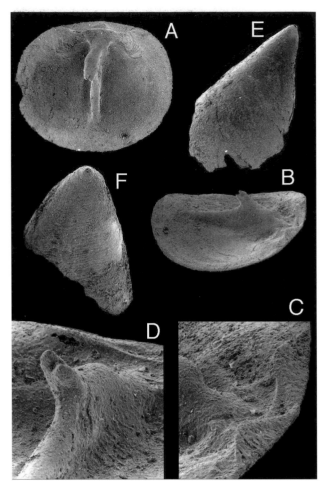

*Fig. 106. Cristicoma? keskentassica* sp. nov.; Agalatas Formation; southern Kendyktas Range, Kazakhstan (sample 564). □A. Holotype; dorsal interior; RM Br 136249; ×12. □B. Lateral view of A; ×13. □C. Dorsal pseudointerarea and cardinal muscle fields of A; ×50. □D. Median septum of A; ×50. □E. Lateral view of ventral exterior; RM Br 136250; ×13. □F. posterior view; ×13.

*Diagnosis. –* Ventral valve about as high as long; ventral pseudointerarea procline to slightly apsacline. Apical process filling entire apex. Dorsal valve gently sulcate; dorsal pseudointerarea anacline. Dorsal cardinal muscle fields not extending to mid-length; dorsal median septum triangular, moderately high with single upper rod bifurcating near top.

*Description. –* The shell is transversely suboval to subcircular. The ventral valve is highly conical, about as high as long. The ventral pseudointerarea is procline to slightly apsacline with indistinct intertrough (Fig. 106E, F). The pedicle foramen is placed within the larval shell. The apical process fills the entire apex.

The dorsal valve is gently convex, with a wide, shallow sulcus originating somewhat anterior to the umbo; the valve is about 89–99% (*N* 2) as long as wide. The dorsal pseudointerarea is anacline, occupying about 50% (*N* 1) of the valve width; the median groove is deeply concave (Fig. 106C). The dorsal median septum is moderately high and triangular and has a single upper rod that bifurcates near the top of the septum. The septum originates 16% of the valve length from the posterior margin and occupies 73% (*N* 1) of the valve length; the median buttress is subtriangular (Fig. 106A–D). The dorsal cardinal muscle fields are thickened anteriorly and transversely suboval in outline (Fig. 106A); they occupy 54% of the valve width and extend for 29% (*N* 1) of the length.

*Discussion.* – This rare species differs from the type species in having (1) a shorter dorsal pseudointerarea with a well-defined median groove, (2) shorter dorsal cardinal muscle fields that do not extend to the mid-length, and (3) a moderately high dorsal median septum with a single upper bifurcating rod.

*Occurrence.* – At the type locality only (Fig. 35; Appendix 1J).

## Genus *Sasyksoria* gen. nov.

*Name.* – After Sasyksor Lake.

*Type and only species.* – *Sasyksoria rugosa* sp. nov.; Olenty Formation.

*Diagnosis.* – Ventral valve low conical; ventral pseudointerarea flattened, procline to catacline, divided by wide interridge. Apical process low, ridge-like, penetrated posteriorly by internal pedicle foramen. Dorsal pseudointerarea wide; median groove wide, poorly defined. Dorsal median septum simple, triangular; cardinal buttress high and wide. Postlarval shell covered by strong, evenly spaced lamellae.

*Discussion.* – *Sasyksoria* differs from all other torynelasmatids in lacking a dorsal median septum with a surmounting platform or septal rods. It also differs in having a lamellose ornamentation and a divided ventral pseudointerarea.

## *Sasyksoria rugosa* sp. nov.
Fig. 107

*Name.* – Latin *rugosus*, wrinkled; alluding to the lamellose ornamentation.

*Holotype.* – Fig. 107E–G; RM Br 136252; ventral valve (L 0.72, W 18, H 0.42); Olenty Formation (sample 601); Sasyksor Lake.

*Paratypes.* – Figured. Dorsal valve: RM Br 136251 (L 1.24, LI 0.16, LS 2.24). Total of 4 ventral and 1 dorsal valve.

*Diagnosis.* – As for genus.

*Description.* – The valves are transversely oval, on average 71% (OR 74–61%; *N* 5) as long as wide; the posterior margin is wide and straight (Fig. 107A, C). The ventral valve is

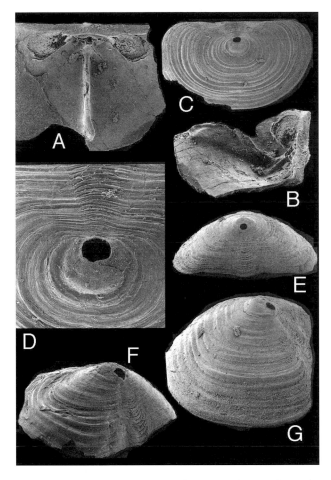

*Fig. 107.* *Sasyksoria rugosa* sp. nov.; Olenty Formation; Sasyksor Lake (sample 601); □A. Dorsal interior; RM Br 136251; ×28. □B. Lateral view of A; ×25. □C. Holotype; ventral exterior; RM Br 136252; ×25. □D. Larval shell of A; ×100. □E. Posterior view of C; ×25. □F. Oblique lateral view of C; ×50. □G. Lateral view of C; ×40.

broadly conical, on average 57% (OR 65–53%; *N* 5) as high as long (Fig. 107). The ventral pseudointerarea is procline and has an interridge (Fig. 107D, G). The pedicle foramen is situated within the larval shell but close to the posterior margin (Fig. 107D). In lateral profile, the anterior and posterior surfaces are slightly convex (Fig. 107G). The apical process forms a low, thickened area anterior and lateral to the internal pedicle opening. The ventral cardinal muscle fields are large and subtriangular in outline, situated on the posterolateral parts of the valve.

The dorsal valve is gently and evenly convex. The dorsal pseudointerarea is low and wide, slightly anacline; it occupies about 14% of the valve length; the median groove is shallow and poorly defined (Fig. 107A, B). The dorsal median septum is high and long, occupying up to 92% of the valve length (Fig. 107A, B). The postlarval shell is covered by strong, evenly spaced lamellae, about 40 µm apart (Fig. 107C).

*Occurrence.* – At the type locality only (Appendix 1H).

## Family Biernatidae Holmer, 1989

*nomen translatum*, herein (*ex* Biernatinae Holmer, 1989, p. 131).

*Diagnosis.* – Shell with narrow, convex posterior margin. Ventral valve narrowly cone-shaped; ventral pseudointerarea strongly apsacline, poorly defined laterally; intertrough weak to absent. Apical process absent. Dorsal pseudointerarea small, divided, with tendency to reduction. Dorsal interior with high triangular median septum, bearing convex surmounting plate or rod on posterior slope; dorsal scars of anterior lateral and central muscles absent.

*Occurrence.* – Lower Ordovician – Devonian.

## Genus *Biernatia* Holmer, 1989

*Type species.* – Original designation by Holmer (1989, p. 133); *Torynelasma minor rossicum* Gorjansky, 1969; Lower Ordovician (lower Llanvirn) Kunda Stage; Pechory core, Pskov district, Russia.

*Diagnosis.* – Larval shell with regularly spaced pits of two sizes. Dorsal median septum with surmounting plate.

*Species included.* – *Biernatia circularis* sp. nov.; in addition to the species listed by Holmer (1989, p. 133).

*Occurrence.* – Ordovician (Tremadoc–Ashgill); North America, Baltoscandia, Kazakhstan.

## *Biernatia circularis* sp. nov.

Fig.108

*Synonymy.* – ☐1976 *Torynelasma rossicum* Gorjansky – Gjessing (unpublished manuscript), p. 50, Pl. 3D–F, Fig. 16.

*Name.* – Latin *circularis*, round; alluding to the outline.

*Holotype.* – Fig. 108C; PMO 97262B; dorsal valve (W 0.4, LI 0.03, WI 0.17); Bjørkåsholmen Limestone in the Teigen–Stablum cores [the specimens were collected by Gjessing 1976, who apparently did not distinguish between material from the two cores]; Eiker–Sandsvaer district, Norway.

*Paratypes.* – Figured. Ventral valves: PMO 97263A; PMO 97263B. Dorsal valves: PMO 97262A; PMO 97262C; PMO 97262D. Total of 91 ventral valves and 80 dorsal valves.

*Diagnosis.* – Shell small for genus; subcircular. Ventral valve apsacline with somewhat concave posterior slope. Dorsal pseudointerarea minute, occupied mainly by median groove. Dorsal median septum extending for most of valve length, with narrow, only slightly convex surmounting plate.

*Description.* – The shell is small for the genus, with a recorded maximum width of 0.4 mm; Ventral valve is highly conical (Fig. 108H). The ventral pseudointerarea is strongly apsacline and undivided, in lateral profile somewhat concave,

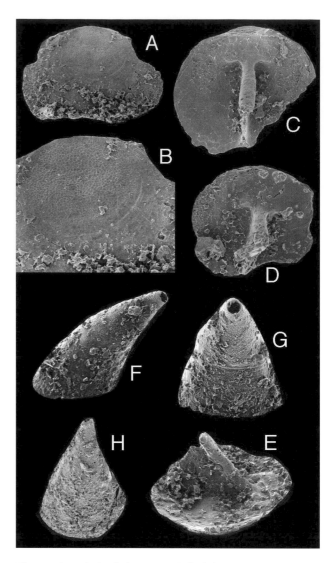

*Fig. 108. Biernatia circularis* sp. nov.; Bjørkåsholmen Limestone; Teigen–Stablum core, Norway. ☐A. Dorsal exterior; PMO 97262A; ×80. ☐B. Larval shell of A; ×150. ☐C. Holotype; dorsal interior; PMO 97262B; ×100. ☐D. Dorsal interior; PMO 97262C; ×120. ☐E. Lateral view of dorsal interior; PMO 97262D; ×100. ☐F. Lateral view of ventral exterior; PMO 97263A; ×75. ☐G. Posterior view of F; ×90. ☐H. Ventral exterior; PMO 97263B; ×55.

poorly defined laterally. The pedical foramen is small, apical, and placed within the larval shell. In lateral profile the ventral anterior slope is gently and moderately convex (Fig. 108F, G). The ventral interior is invariably filled with matrix.

The dorsal valve is gently and evenly convex, subcircular; the dorsal pseudointerarea is minute, and mainly occupied by a shallow median groove (Fig. 108C). The dorsal interior has a high triangular median septum that extends for close to about 90% of the valve length (Fig. 108C, D, E). The surmounting plate is narrow and convex in cross-section; the anterior slope of the median septum is undercut near the top. The dorsal cardinal muscle fields are bounded anteriorly by elevated ridges that continue to the base of the surmounting plate (Fig. 108C). The dorsal larval shell is around 0.22×0.17 mm (Fig. 108B).

*Discussion.* – This species is very similar to the type species; it differs mainly in having (1) an even smaller maximum size, (2) a less apsaconical ventral valve, (3) a longer dorsal median septum, and (4) a narrower and less convex surmounting plate.

*Occurrence.* – At the type locality only (see Gjessing 1976 for details).

# Family Scaphelasmatidae Rowell, 1965

*nomen translatum* by Koneva, Popov & Ushatinskaya 1990, p. 163 (*ex* Scaphelasmatinae Rowell, 1965, p. H278).

*Diagnosis.* – Shell biconvex to concavo-convex, with straight or slightly concave posterior margin. Ventral valve with wide triangular intertrough dividing pseudointerarea; large, elongate oval pedicle foramen posterior to apex, usually not within larval shell. Apical process forming low thickening anterior to foramen or absent. Dorsal pseudointerarea with short median groove and wide propareas. Dorsal median septum reduced, near anterior margin. Larval shell with pits of two sizes. Post-larval shell ornamented by coarse regularly spaced rugae or rugellae, usually lamellose peripherally.

*Genera included.* – *Scaphelasma* Cooper, 1956; *Artiotreta* Ireland, 1961; *Rhysotreta* Cooper, 1956; *Eoscaphelasma* Koneva, Popov & Ushatinskaya 1990; *Tobejalotreta* Koneva, Popov & Ushatinskaya 1990.

*Occurrence.* – Middle Cambrian – Silurian.

# Genus *Eoscaphelasma* Koneva, Popov & Ushatinskaya, 1990

*Type and only species.* – Original designation by Koneva, Popov & Ushatinskaya (*in* Koneva *et al.* 1990, p. 164); *Eoscaphelasma satpakensis* Koneva, Popov & Ushatinskaya, 1990; Kujandy Formation.

*Diagnosis.* – Shell with widely conical ventral valve; ventral pseudointerarea procline, poorly defined laterally, with well-developed intertrough; pedicle foramen elongate oval, not enclosed within larval shell. Short internal pedicle tube may be present. Dorsal valve gently convex; dorsal pseudointerarea low, orthocline, with median groove. Larval shell covered by pits of two sizes.

*Discussion.* – Externally *Eoscaphelasma* is closely comparable with *Scaphelasma*, but it lacks a dorsal median septum. The genus is also somewhat similar to the Middle Cambrian *Kotylotreta* Koneva, 1990; however, *Eoscaphelasma* differs in having (1) a rugellose ornamentation, (2) a pedicle foramen that is not placed within the larval shell, and (3) a short internal pedicle tube.

# *Eoscaphelasma satpakensis* Koneva, Popov & Ushatinskaya, 1990

Fig. 109

*Synonymy.* – □1990 *Eoscaphelasma satpakensis* sp. nov. – Koneva, Popov & Ushatinskaya *in* Koneva *et al.*, p. 165, Pl. 30:1–7.

*Holotype.* – PIN 4321/143; ventral valve (L 0.7, W 0.9, H 0.5); figured by Koneva, Popov & Ushatinskaya (*in* Koneva *et al.*), Pl. 30:1; Kujandy Formation; Satpak Syncline.

*Material.* – Figured. Complete shell: RM Br 136198. Ventral valves: RM Br 136193; RM Br 136194 (L 0.8, W 0.92, H 0.32); RM Br 136196; RM Br 136197. Dorsal valves: RM Br 136192; RM Br 136195 (L 1.06, W 1.24, LI 0.08, WI 0.62, WG 0.28); RM Br 136199; RM Br 136200; RM Br 136201; RM Br 136202; RM Br 136334. Total of 1 complete shell, 105 ventral valves, and 1061 dorsal valves.

*Measurements.* – See Tables 48–49.

*Diagnosis.* – As for genus.

*Description.* – The shell is ventri-biconvex and transversely oval. The ventral valve is widely conical, on average 77% as long as wide and 35% as high as long (Table 48); the ventral apex is situated close to mid-length, on average 42% of the total valve length from the posterior margin (Table 48). The pedicle foramen is large and elongate oval, up to 0.1×0.15 mm long, not enclosed within the larval shell, and in some valves a small plate covers the anterior part of the foramen (Fig. 109I, N). The ventral pseudointerarea is poorly defined and procline with a wide intertrough (Fig. 109K–M). The ventral cardinal muscle fields are poorly defined. The internal pedicle tube is usually enclosed within a poorly developed weak apical process (Fig. 109G).

The dorsal valve is gently convex with the maximum height placed somewhat anterior to the umbo (Fig. 109B). The dorsal pseudointerarea is low and orthocline, occupying on average 48% of the valve width (Table 49); the median groove is well defined and wide, occupying half of the width of the pseudointerarea (Fig. 109F). The dorsal cardinal muscle fields are large and elongate oval in outline; they occupy on average 71% of the valve width and extend for 46% of the length (Fig. 109D, J, U; Table 49).

The larval shells of both valves are well defined, transversely suboval, around 0.2×0.15 mm (Fig. 109C, I, N), ornamented by pits of two sizes, the larger are around 3 μm across, surrounded by smaller, up to about 1 μm across (Fig. 109O). The postlarval shell is covered with evenly spaced fila superposed on concentric rugellae, about 20 μm apart.

*Occurrence.* – *E. satpakensis* is common in the Kujandy Formation at the Satpak Syncline (Fig. 29; Appendix 1I), and the Kujandy section (Fig. 25; Appendix 1G); it also occurs in olistoliths within the Satpak, Olenty, and Erzhan formations in the Aksak–Kujandy Section (Fig. 27; Appendix 1H) and at the Satpak Syncline (Fig. 29, Appendix 1I).

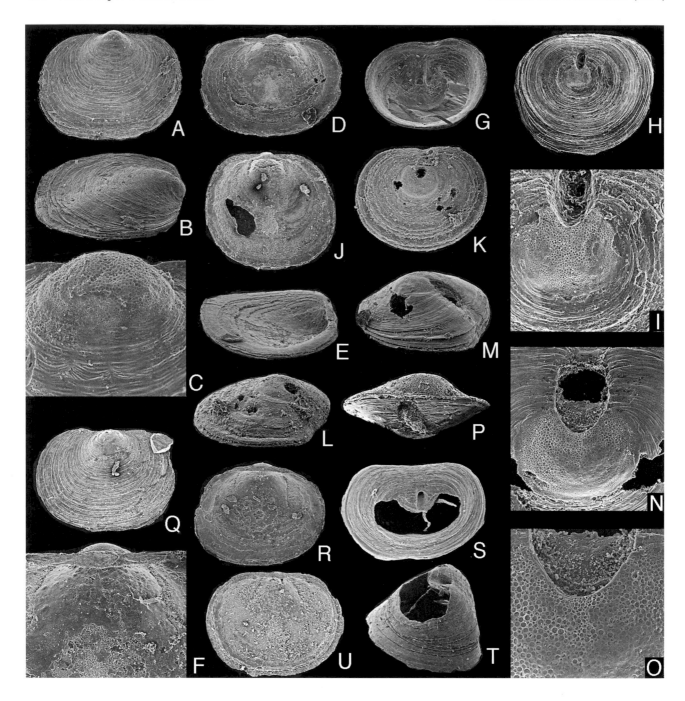

*Fig. 109. Eoscaphelasma satpakensis* Koneva, Popov & Ushatinskaya; Kujandy Formation; northeastern Central Kazakhstan. □A. Dorsal exterior; Satpak Syncline (sample 7836b); RM Br 136199; ×38. □B. Lateral view of A; ×45. □C. Larval shell of A; ×165. □D. Dorsal interior; Satpak Syncline (sample 7836b); RM Br 136192; ×30. □E. Lateral view of D; ×40. □F. Pseudointerarea and visceral area of D; ×75. □G. Ventral interor; Satpak Syncline (sample 7836b); RM Br 136193; ×38. □H. Ventral exterior; Satpak Syncline (sample 7836a); RM Br 136194; ×40. □I. Larval shell of H; ×150. □J. Dorsal interior; Satpak Syncline (sample 7836a); RM Br 136195; ×28. □K. Ventral exterior; Chingiz Range (coll. D.S. Avrov); RM Br 136196; ×50. □L. Lateral view of K; ×60. □M. Lateral view of ventral exterior; Satpak Syncline (sample 7836b); RM Br 136197; ×80. □N. Larval shell of M; ×150. □O. Larval pitting of M; ×300. □P. Posterior view of complete shell; Satpak Syncline (sample 7836a); RM Br 136198; ×80. □Q. Dorsal exterior; olistolith within Erzhan Formation; Aksak–Kujandy Section (sample 79101); RM Br 136200; ×30. □R. Dorsal interior; Satpak Syncline (sample 7836b); RM Br 136201; ×55. □S. Ventral exterior; Satpak Syncline (sample 7836b); RM Br 136334; × 38. □T. Lateral view of S; ×45. □U. Dorsal interior; olistolith within Erzhan Formation; Aksak–Kujandy Section (sample 79101); RM Br 136202; ×30.

*Table 48. Eoscaphelasma satpakensis,* average dimensions and ratios of ventral valves.

|  | L | W | H | Fa | Fp | M | $L/W$ | $H/W$ | Lped | $M/L$ |
|---|---|---|---|---|---|---|---|---|---|---|
| Sample 7836 | | | | | | | | | | |
| N | 32 | 32 | 32 | 32 | 32 | 32 | 32 | 32 | 32 | 32 |
| X | 0.63 | 0.81 | 0.29 | 0.22 | 0.15 | 0.26 | 77% | 35% | 7% | 42% |
| S | 0.11 | 0.12 | 0.07 | 0.12 | 0.04 | 0.06 | 0.07 | 0.06 | 0.11 | 0.09 |
| MIN | 0.43 | 0.57 | 0.17 | 0.13 | 0.06 | 0.13 | 64% | 24% | 1% | 18% |
| MAX | 0.87 | 1.04 | 0.43 | 0.81 | 0.24 | 0.36 | 94% | 47% | 64% | 64% |

*Table 49. Eoscaphelasma satpakensis,* average dimensions and ratios of dorsal valves.

|  | L | W | T | LI | WI | MG | LS | LM1 | WM1 | $L/W$ | $MG/WI$ | $WI/W$ | $LS/L$ | $LM1/L$ | $WM1/W$ |
|---|---|---|---|---|---|---|---|---|---|---|---|---|---|---|---|
| Sample 7836 | | | | | | | | | | | | | | | |
| N | 31 | 31 | 31 | 31 | 31 | 31 | 9 | 6 | 6 | 31 | 31 | 31 | 9 | 6 | 6 |
| X | 0.99 | 0.88 | 0.17 | 0.07 | 0.42 | 0.19 | 0.56 | 0.35 | 0.67 | 117% | 46% | 48% | 69% | 46% | 71% |
| S | 1.60 | 0.13 | 0.03 | 0.02 | 0.07 | 0.04 | 0.09 | 0.06 | 0.07 | 2.13 | 0.07 | 0.05 | 0.12 | 0.07 | 0.02 |
| MIN | 0.44 | 0.57 | 0.11 | 0.04 | 0.27 | 0.13 | 0.50 | 0.30 | 0.53 | 69% | 30% | 40% | 56% | 39% | 68% |
| MAX | 0.44 | 0.57 | 0.11 | 0.04 | 0.27 | 0.13 | 0.50 | 0.30 | 0.53 | 69% | 30% | 40% | 56% | 39% | 68% |

# Family Eoconulidae Cooper, 1956

*Diagnosis.* – Ventral valve truncated conical, attached by cementation. Pedicle opening may become closed. Ventral pseudointerarea usually not developed. Ventral interior with boss-like apical process anterior to foramen. Dorsal pseudo-interarea variably developed but frequently absent; larval shell with regularly spaced pits of two sizes.

*Genera included.* – *Eoconulus* Cooper, 1956; *Undiferina* Cooper, 1956; *Otariella* gen. nov.

*Occurrence.* – Ordovician (Tremadoc–Ashgill).

# Genus *Eoconulus* Cooper, 1956

*Type species.* – Original designation by Cooper (1956, p. 282); *Eoconulus rectangulatus* Cooper, 1956; Middle Ordovician Pratt Ferry beds, Alabama.

*Diagnosis.* – Shell smooth or finely rugellose. Ventral valve low, truncated conical. Dorsal valve conical with apex usually subcentral. Dorsal and ventral pseudointerareas absent. Ventral interior with variably developed apical process; ventral cardinal muscle fields forming elevated elongate platforms. Dorsal interior with thickened cardinal muscle fields.

*Species included.* – *Eoconulus primus* sp. nov.; in addition to the species listed by Holmer (1989, p. 147).

*Occurrence.* – Ordovician (Tremadoc–Ashgill); North America, Europe, Kazakhstan.

# *Eoconulus primus* sp. nov.

Fig. 110

*Name.* – Latin *primus*, the first; alluding to the fact that it is the oldest known species of *Eoconulus*

*Holotype.* – Fig. 110D–F; RM Br 136205 (L 0.71, W 0.92); ventral valve; Olenty Formation (sample 601); Sasyksor Lake.

*Paratypes.* – Figured. Ventral valves: RM Br 136207 (L 0.7, W 0.92); RM Br 136208 (L 0.6, W 0.7). Dorsal valves: RM Br 136203 (L 0.6, W 0.9); RM Br 136204 (L 0.53, W 0.73); RM Br 136206. Total of 4 ventral and 59 dorsal valves.

*Measurements.* – See Table 50.

*Diagnosis.* – Shell ventri-biconvex, asymmetrical, transversely suboval. Ventral valve subrectangular in lateral profile; pedicle foramen placed at about one-third of valve length from posterior margin. Dorsal valve moderately convex with apex placed posterior to mid-length.

*Description.* – The shell is ventri-biconvex and strongly asymmetrical, transverse suboval. The ventral valve is sub-cylindrical and strongly geniculated dorsally 110E); the attachment scar is large, irregular and flattened (Fig. 110D, E, J). The pedicle foramen is placed on average 35% (N 3; OR 30–39%) of the valve length from the posterior margin at about the centre of a poorly defined subcircular larval shell, 0.4–0.5 mm across (Fig. 110G). The ventral interior has a minute apical process, about 0.1 mm wide, directly anterior to pedicle opening (Fig. 110L).

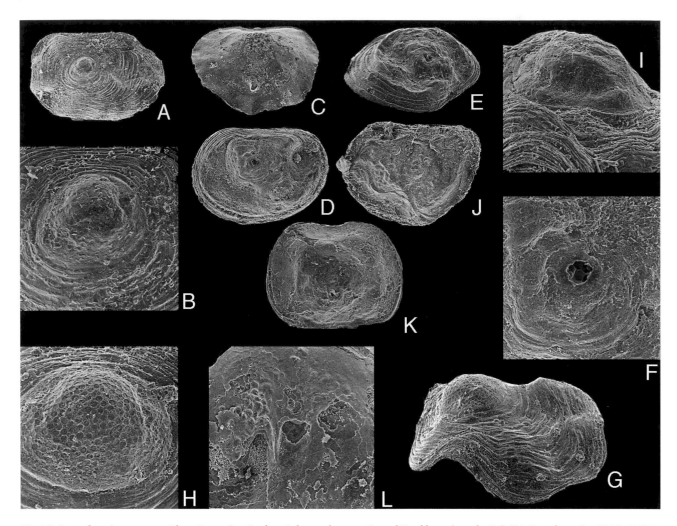

*Fig. 110.* *Eoconulus primus* sp. nov.; Olenty Formation; Sasyksor Lake, northeastern Central Kazakhstan (sample 601). □A. Dorsal exterior; RM Br 136203; ×40. □B. Larval shell of A; ×150. □C. Dorsal interior; RM Br 136204; ×45. □D. Holotype; ventral exterior; RM Br 136205; ×38. □E. Lateral view of D; ×50. □F. Pedicle foramen and larval shell of D; ×180. □G. Dorsal exterior; RM Br 136206; ×65. □H. Larval shell of G; ×180. □I. Lateral view of larval shell of G; ×165, □J. Ventral exterior; RM Br 136207; ×40. □K. Ventral interior; RM Br 136208; ×50. □L. Apical process of K; ×150.

The dorsal valve is moderately convex and strongly asymmetrical, on average 41% as long as wide and 41% as high as long (Fig. 110G; Table 50). In lateral profile, the anterior and lateral slopes of the valves are somewhat convex (Fig. 110G). The apex is situated on average 31% of the valve length from the posterior margin (Table 50). The dorsal cardinal muscle fields are thickened and placed on the inner side of the posterolateral slopes of the valve; the median ridge is vestigial, forming a minute thickening close to the anterior margin (Fig. 110C). The dorsal larval shell is well defined, conical, close to circular, 0.18–0.23 mm wide, and has a bulbous apex; it is covered by pits of two sizes: the larger are up to 10 μm across and surrounded by smaller that are less than 3 μm across (Fig. 110B, C, I). The ventral larval shell lacks pitted micro-ornament (Fig. 110F). The postlarval shell is covered by fine, closely spaced rugellae, about 7 μm apart.

*Discussion. –* *E. primus* is characterized by a relatively low conical dorsal valve, with convex anterior and lateral slopes, while most other species of *Eoconulus*, such as *E. cryptomyus* Gorjansky, 1969, *E. clivosus* Popov, 1975, and *E. robustus* Holmer, 1989, have highly conical dorsal valves with steep

*Table 50. Eoconulus primus*, average dimensions and ratios of dorsal valves.

| | L | W | H | M | $L/_W$ | $H/_L$ | $M/_L$ |
|---|---|---|---|---|---|---|---|
| Sample B-523 | | | | | | | |
| N | 18 | 18 | 7 | 7 | 18 | 7 | 7 |
| X | 0.69 | 0.78 | 0.29 | 0.20 | 89% | 41% | 31% |
| S | 0.08 | 0.10 | 0.10 | 0.05 | 0.04 | 0.11 | 0.05 |
| MIN | 0.51 | 0.56 | 0.19 | 0.16 | 77% | 26% | 25% |
| MAX | 0.83 | 0.90 | 0.46 | 0.31 | 94% | 55% | 37% |

and straight anterior and lateral slopes, in addition to a subcentrally placed apex.

The ventral valve of *E. primus* is most similar to the ventral valve of *E. clivosus* but differs in having a rectangular lateral profile; moreover, the apical process of *E. primus* is much smaller compared with all other species.

*Occurrence.* – This species is present in the Koagash Formation (sample B-523; Appendix 1F) at Koagash River, South Urals; it is also known from the Olenty Formation at Sasyksor Lake (Appendix 1H).

## Genus *Otariella* gen. nov.

*Name.* – After the town of Otar.

*Type species.* – *Otariella prisca* sp. nov.; Satpak Formation.

*Diagnosis.* – Shell unequally biconvex, usually more or less asymmetrical. Pedicle foramen partly outside larval shell. Apical process anterior to internal foramen; ventral cardinal muscle fields large, thickened, placed posterolaterally. Dorsal valve convex with marginal beak; dorsal pseudointerarea well defined. Dorsal cardinal muscle fields large, elongate oval; dorsal median ridge usually present.

*Species assigned.* – *Otariella prisca* sp. nov.; *Otariella intermedia* sp. nov; *Econulus*? spp. Puura & Holmer, 1993 [in part].

*Discussion.* – *Otariella* is similar to *Eoconulus* but differs mainly in having (1) a marginal dorsal beak, and (2) a well-defined dorsal pseudointerarea with (4) a median groove. At the same time, *Otariella* has features suggesting that it is related closely with the Family Scaphelasmatidae – in particular with *Eoscaphelasma*. The dorsal valve of the earliest known species of *Otariella* is similar to *Eoscaphelasma* in most exterior and interior characters, such as (1) a submarginal beak, (2) a dorsal pseudointerarea with well-defined median groove, (3) and a small dorsal median ridge. The ventral larval shell of *Otariella* is also very similar to that of the Scaphelasmatidae, but smaller. The pitted larval micro-ornament, with circular pits of two sizes, is a further

character in common with the Scaphelasmatidae. It seems that *Otariella* repesents a stock that is a morphological intermediate in the transition between the earliest Scaphelasmatidae and the first Eoconulidae.

## *Otariella prisca* sp. nov.
Figs. 111A–M

*Name.* – Latin *priscus*, old.

*Holotype.* – Fig. 111E–G; RM Br 136211; ventral valve (L 0.92, W 1.04, H 0.56); Satpak Formation (sample 79109); Aksak–Kujandy Section.

*Paratype.* – Figured from northeastern Central Kazakhstan. Ventral valve: RM Br 136212. Dorsal valves: RM Br 136210 (L 0.78, W 0.88, LI 0.12, WI 0.58, WG 0.32, LM1 0.36, WM 10.7, LS 0.62); RM Br 136213 (L 0.72. W 0.96, LI 0.12, WI 0.42, WG 0.24, LM1 0.28, WM10.64, LS 0.62); Br136214 (0.78, W 0.96, LI 0.06, WI 0.48, WG 0.16, LM1 0.4, WM 10.7, LS 0.52); 136215 (L 0.82, W 1, LI 0.14, WI 0.58, WG 0.32, LM1 0.36, WM1 0.78, LS 0.72). Total of 3 ventral and 25 dorsal valves.

*Measurements.* – See Table 51.

*Diagnosis.* – Shell slightly asymmetrical, transversely suboval. Dorsal valve gently to moderately convex; dorsal pseudointerarea widely subtriangular with well-defined median groove. Dorsal cardinal muscle fields large, elongate oval, slightly thickened, extending to mid-length; median ridge low, thickened anteriorly.

*Description.* – The shell is ventri-biconvex, usually asymmetrical, transversely suboval. The ventral valve is gently to moderately convex and geniculated (Fig. 111G). The pedicle foramen is small, circular, placed at the centre of an elongate oval pedicle track, about 50×100 μm long, half of which is not within the larval shell (Fig. 111I). The apicle process is anterior to the internal pedicle opening. The ventral cardinal muscle fields are large, thickened and elongate oval in outline, extending to the mid-length.

*Table 51. Otariella prisca*, average dimensions and ratios of dorsal valves.

|  | L | W | H | ML | MG | WI | $^L/_W$ | $^H/_L$ | $^{ML}/_L$ | $^{MG}/_{WI}$ | $^{WI}/_W$ |
|---|---|---|---|---|---|---|---|---|---|---|---|
| **Sample 8119** | | | | | | | | | | | |
| *N* | 18 | 18 | 18 | 18 | 18 | 18 | 18 | 18 | 18 | 18 | 18 |
| *X* | 0.73 | 0.95 | 0.24 | 0.09 | 0.26 | 0.52 | 77% | 32% | 12% | 27% | 55% |
| *S* | 0.05 | 0.09 | 0.03 | 0.02 | 0.04 | 0.10 | 0.04 | 0.04 | 0.03 | 0.04 | 0.09 |
| MIN | 0.64 | 0.80 | 0.18 | 0.04 | 0.14 | 0.32 | 70% | 25% | 5% | 17% | 39% |
| MAX | 0.82 | 1.12 | 0.30 | 0.14 | 0.32 | 0.72 | 90% | 38% | 19% | 36% | 73% |

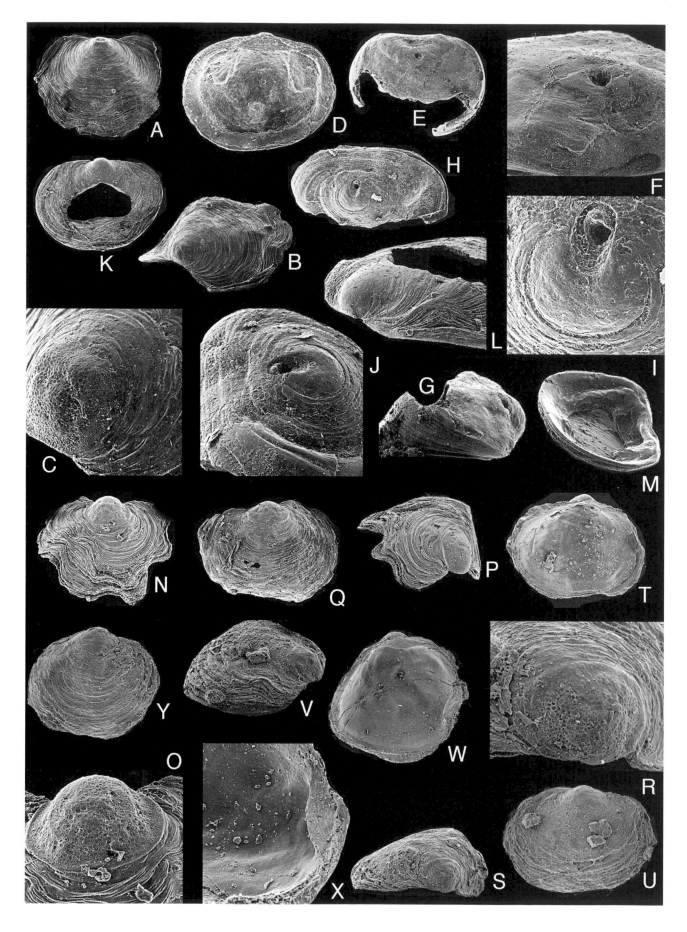

The dorsal valve is gently to moderately convex in lateral profile, with the maximum height in the posterior third of the valve length (Fig. 111B, L), on average 77% as long as wide and 32% as high as long (Table 51). The dorsal umbo is marginal. The dorsal pseudointerarea is well developed and orthocline; it occupies on average 55% of the valve width; the median groove is well developed (Fig. 111D, M). The dorsal median ridge is poorly developed, extending somewhat anterior to mid-length; the cardinal muscle fields are elongate oval in outline, extending somewhat anterior to mid-length (Fig. 111D).

The larval shells of both valves are well defined, somewhat transversely suboval, around 0.17 mm wide and 0.14 mm long; the larval pits are of two sizes, the larger pits up to 5 µm across, surrounded by smaller ones up to 1 µm across (Fig. 111C, I). The postlarval shell is covered by fine, closely spaced rugellae, around 5 µm apart (Fig. 111L).

*Occurrence.* – *O. prisca* was found in the Satpak Formation, and in an olistolith within the Olenty Formation at the Aksak–Kujandy section (Fig. 27; Appendix 1H).

## *Otariella intermedia* sp. nov.

Fig. 111N–Y

*Name.* – Latin *intermedius*, that is between.

*Holotype.* – Fig. 111W–X; RM Br 136218; dorsal valve (L 0.82, W 0.8, LI 0.12, WI 0.34, WG 0.24); upper part of the Agalatas Formation (sample 564).

*Paratypes.* – Figured. Dorsal valves: RM Br 136209; Br136216 (L 0.48, W 0.56, LI 0.06, WI 0.32, WG 0.12); RM Br136217 (L 0.6, W 0.7, LI 0.06, WI 0.32, WG 0.12); RM Br136219 (L 0.66, W 0.88, LI 0.06, WI 0.32, WG 0.18); RM Br136220 (L 0.44, W 0.66, LI 0.08, WI 0.34, WG 0.24). Total of 13 dorsal valves.

*Fig. 111.* □A–M. *Otariella prisca* sp. nov.; Satpak Formation; Aksak–Kujandy Section, northeastern Central Kazakhstan. □A. Dorsal exterior; olistolith within Olenty Formation (sample 8119); RM Br 136214; ×38. □B. Lateral view of A; ×40. □C. Larval shell of A; ×180. □D. Dorsal interior; olistolith within Olenty Formation (sample 8119); RM Br 136210; ×45. □E. Holotype; ventral exterior, lateral view (sample 79109); RM Br 136211; ×38. □F. Pedicle foramen and larval shell of E; ×75. □G. Lateral view of E; ×38. □H. Ventral exterior; olistolith within Olenty Formation (sample 8119); RM Br 136212; ×45. □I. Pedicle foramen and larval shell of H; ×180. □J. Lateral view of H; ×100. □K. Dorsal exterior (sample 79109); RM Br 136213; ×38. □L. Lateral view of K; ×75. □M. Lateral view of dorsal interior (sample 79106); RM Br 136215; ×38. □N–Y. *Otariella intermedia* sp. nov. □N. Dorsal exterior; Koagash Formation; Karabutak River (sample 1163); RM Br 136216; ×55. □O. Larval shell of N; ×155. □P. Lateral view of N; ×60. □Q. Dorsal exterior; Koagash Formation; Karabutak River (sample 1163); RM Br 136217; ×40. □R. Larval shell of Q; ×150. □S. Posterior view of Q; ×55. □T. Dorsal interior; Koagash Formation; Karabutak River (sample 1163); RM Br 136219; ×40. □U. Dorsal exterior; Agalatas Formation; Kendyktas Range (sample 564); RM Br 136220; 38. □V. Lateral view of U; ×50. □W. Holotype; dorsal exterior; Agalatas Formation; Kendyktas Range (sample 564); RM Br 136218; ×38. □X. Pseudointerarea of W; ×110. □Y. Dorsal exterior; Agalatas Formation; Kendyktas Range (sample 564); RM Br 136209; ×50.

*Measurements.* – See Table 52.

*Diagnosis.* – Dorsal valve asymmetrical, with transversely suboval to subcircular outline, moderately convex. Dorsal pseudointerarea small, with wide median groove and reduced proparea.

*Description.* – The dorsal valve is transversely oval to subcircular and moderately convex, on average 84% as long as wide and 12% as high as long (Table 52), with maximum height posterior to the mid-length; the beak is marginal and strongly swollen (Fig. 111P, V). The pseudointerarea is low, occupying on average 28% of the valve width (Table 52); the median groove is well developed (Fig. 111W, X). The dorsal interior has large but poorly defined cardinal muscle fields that extend to the mid-length. The dorsal larval shell is well defined, close to circular, around 0.2 mm across, with pits identical to those of *O. prisca* (Fig. 111O, R). The postlarval shell is covered by rugellae, identical to those of *O. prisca*, superposed on closely spaced growth lamellae, about 50 µm apart (Fig. 111N–P).

*Discussion.* – This species is closely comparable with *O. prisca* in having a marginal dorsal beak and dorsal pseudointerarea; however, it differs in having (1) a moderately convex lateral profile with the maximum height placed posterior to mid-length, (2) a narrow dorsal pseudointerarea with reduced propareas, and (3) no dorsal median ridge.

*O. prisca* is also similar in ornamentation, and in the morphology of the dorsal pseudointerarea, to *Eoconulus?* spp. described by Puura & Holmer (1993, p. 229, Fig. 4E–I) from the 'Obolus' beds in the Siljan district, and the latter form is here referred to *Otariella*.

*Occurrence.* – This species occurs in the upper part of the Agalatas Formation in the southern Kendyktas Range (Fig. 35; Appendix 1J). In the South Urals, *O. intermedia* was recorded from the Karabutak River (Appendix 1E) and the Koagash Formation at the Koagash Section (Appendix 1F).

## *Otariella* sp.

Fig. 112

*Material.* – Figured from Sweden. Dorsal valves: RM Br 136287 (L 0.7, W 1.03); RM Br 136288 (L 0.89, W 1.05, LI

*Table 52. Otariella intermedia,* average dimensions and ratios of dorsal valves.

| | L | W | H | LI | WI | $L/_W$ | $H/_L$ | $LI/_{WI}$ | $WI/_W$ |
|---|---|---|---|---|---|---|---|---|---|
| Sample 8119 | | | | | | | | | |
| N | 7 | 7 | 7 | 7 | 7 | 7 | 7 | 7 | 7 |
| X | 0.60 | 0.71 | 0.07 | 0.31 | 0.19 | 84% | 12% | 164% | 28% |
| S | 0.13 | 0.11 | 0.02 | 0.05 | 0.04 | 0.11 | 0.03 | 0.49 | 0.07 |
| MIN | 0.44 | 0.56 | 0.06 | 0.22 | 0.12 | 67% | 9% | 120% | 17% |
| MAX | 0.82 | 0.88 | 0.12 | 0.36 | 0.24 | 103% | 18% | 267% | 36% |

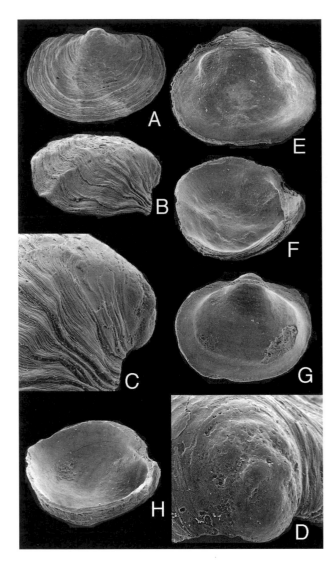

*Fig. 112.* Otariella sp.; Bjørkåsholmen Limestone (sample Vg-1); Stora backor, Västergötland. □A. Dorsal exterior; RM Br 136287; ×37. □B. Lateral view of A; ×50. □C. Umbo of A; ×120. □D. Larval shell of A; ×165. □E. Dorsal interior; RM Br 136288; ×37. □F. Lateral view of E; ×40. □G. Dorsal interior; RM Br 136289; ×45. □H. Lateral view of G; ×55.

0.16, WI 0.54, WG 0.4); RM Br 136289 (L 0.66, W 0.82, LI 0.06, WI 0.33). Total of 3 dorsal valves.

*Description.* – The dorsal valve is moderately convex and asymmetrical, transversely oval, 68–85% as long as wide (*N* 3), with maximum height posterior to mid-length; the beak is marginal and strongly swollen (Fig. 112A–C). The pseudo-interarea is low and wide, occupying 40–51% of the valve width (*N* 2); the median groove is well developed and extremely wide in one of the specimens (Fig. 112E). The dorsal interior has large, but poorly defined cardinal muscle fields that do not extend to the mid-length (Fig. 112E–H). The dorsal larval shell is well defined, close to circular, around 0.2 mm across, with pits identical to those of *O. prisca* (Fig.

112D). The postlarval shell is ornamented by closely spaced rugellae, identical to those of *O. prisca*, superposed on poorly developed growth lamellae, up to about 0.1 mm apart (Fig. 112C).

*Remarks.* – The scant material is similar to *O. intermedia* in the lack of a dorsal median ridge; however, the ornamentation is less lamellose, and the dorsal pseudointerarea is more similar to *O. prisca*.

*Occurrence.* – The speciess has only been found in the Bjørkåsholmen Limestone at Stora Backor (Fig. 6; Appendix 1K).

## Superfamily Botsfordioidea Schindewolf, 1955

*nomen correctum*, herein (*pro* Botsfordiacea Schindewolf, 1955, p. 545)

*Diagnosis.* – Shell with short convex posterior margin. Ventral valve convex or low conical; ventral pseudointerarea vestigial or lacking; pedicle usually within listrium-like track. Dorsal pseudointerarea vestigial or lacking. Mantle system baculate. Larval shell with two ventral apical spines and four dorsal spines. Post-larval shell ornamented by fine, evenly distributed pustules.

*Occurrence.* – Lower Cambrian (Botom) – Lower Ordovician (Arenig).

## Family Acrothelidae Walcott & Schuchert, 1908

*Diagnosis.* – Shell ventri-biconvex. Ventral valve low conical to subconical, lacking pseudointerarea or with poorly defined pseudointerarea; pedicle foramen circular, immediately posterior to apex, not enclosed within larval shell. Dorsal pseudointerarea forming crescent-shaped rim.

## Subfamily Acrothelinae Walcott & Schuchert, 1908

*Diagnosis.* – Dorsal valve flat to gently convex, with marginal beak, internally with low median ridge extending forward from beak, never developed as septum.

## Genus *Orbithele* Sdzuy, 1955

*Type species.* – Original designation by Sdzuy (1955, p. 9); *Discina contraria* Barrande, 1868; Leimitz Shale [of reputed Tremadoc age]; Bavaria.

*Diagnosis.* – Shell with marginal spines and internal pedicle tube supported anteriorly by short septum.

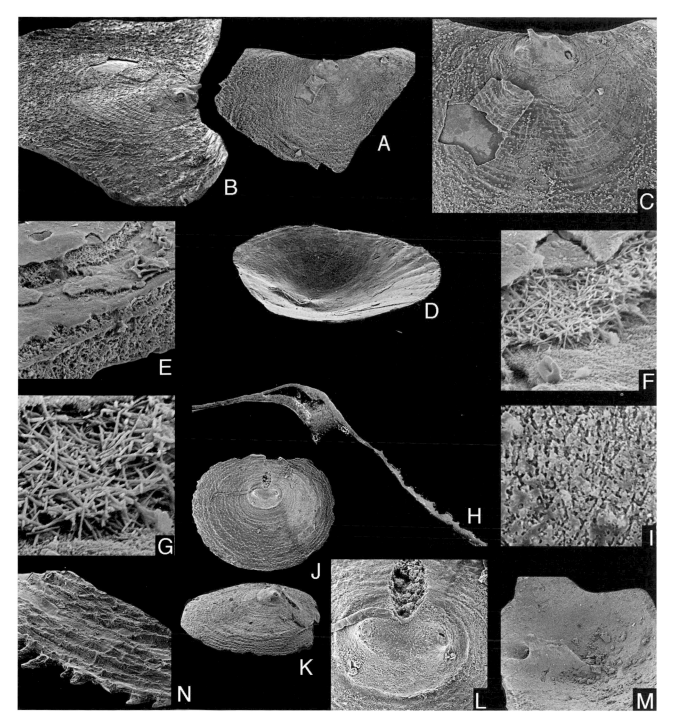

*Fig. 113. Orbithele ceratopygarum* (Brøgger). □A. Dorsal exterior; Bjørkåsholmen Limestone (sample Ng-1); Bjørkåsholmen, Oslo region; RM Br 136290; ×12. □B. Lateral view of A; ×27. □C. Larval and juvenile shell of A; ×27. □D. Lateral view of ventral interior; Bjørkåsholmen Limestone; Ottenby, Öland; RM Br 20793; ×19. □E. Detail of broken ventral valve; Bjørkåsholmen Limestone (sample Öl-1); Ottenby, Öland; LO 6566t; ×500. □F. Detail of E; ×1800. □G. Detail of E, showing baculate shell structure; ×3000. □H. Polished and etched section through ventral valve; Bjørkåsholmen Limestone (sample Öl-1); Ottenby, Öland; LO 6567t; ×44. □I. Detail of H, showing cast of baculae in plastic resin; ×1800. □J. Ventral exterior; Kidryas Formation (sample B-768-1); Tyrmantau Ridge, South Urals; RM Br 136221; ×21. □K. Lateral view of J; ×25. □L. Larval shell of J; ×80. □M. Ventral interior, showing pedicle tube and median septum; Kidryas Formation (sample B-768-1); Tyrmantau Ridge, South Urals; RM Br 136222; ×27. □N. Detail of fragment showing ornamentation with marginal spines; Kidryas Formation (sample B-768-1); Tyrmantau Ridge, South Urals; RM Br 136291; ×50.

*Table 53.* Orbithele ceratopygarum, average dimensions and ratios of dorsal valves.

| | L | W | ML | LI | WI | LM1 | WM1 | $L/W$ | $WI/W$ | $LM1/L$ | $WM1/W$ |
|---|---|---|---|---|---|---|---|---|---|---|---|
| **Sample Öl-1** | | | | | | | | | | | |
| N | 5 | 5 | 1 | 2 | 3 | 3 | 3 | 5 | 3 | 3 | 3 |
| X | 4.06 | 5.00 | 0.20 | 0.19 | 2.17 | 0.87 | 1.40 | 81% | 54% | 27% | 35% |
| S | 1.13 | 1.42 | | 0.16 | 0.38 | 0.12 | 0.00 | 0.04 | 0.04 | 0.00 | 0.04 |
| MIN | 3.00 | 3.50 | 0.20 | 0.08 | 1.90 | 0.80 | 1.40 | 75% | 50% | 26% | 31% |
| MAX | 5.30 | 6.60 | 0.20 | 0.30 | 2.60 | 1.00 | 1.40 | 86% | 58% | 27% | 40% |

*Species included.* – *Discina contraria* Barrande, 1868; *Discina secedens* Barrande, 1879 [=?*Orbithele discontinua* Mergl, 1981; ?*Orbithele maior* Mergl, 1981]; *Discina undulosa* Barrande, 1879 [=?*Orbithele rimosa* Mergl, 1981]; *Discina ceratopygarum* Brøgger, 1882; *Acrothele sougyi* Poulsen, 1960 [=?*Acrothele spinulosa* Poulsen, 1960]; *Orbithele vana* Mergl, 1981.

*Discussion.* – The type species is extremely poorly known, and it is diffcult to compare it in detail with any other species; however, the possession of an internal pedicle tube and septum immediately distinguishes the genus from *Acrothele* Linnarsson. The two species of *Acrothele* from Mauritania described by Poulsen (1960) are here considered to be species of *Orbithele*; this was also noted by Mergl (1981, 1986), who redescribed and discussed the Lower Ordovician *Orbithele* from Bohemia and Morocco. Some of the species discussed by him might possibly be synonymous with the approximately coeval type species from Bavaria. Henderson (1974) described the earliest ontogeny of an unnamed Upper Cambrian species from Australia. The genus has also been recorded recently by Rowell (1980) from Nevada, and by Zell & Rowell (1988), from the Middle Cambrian of Greenland.

*Occurrence.* – Upper Middle Cambrian – Lower Ordovician (Tremadoc–Arenig); Europe, North Africa, South Urals, ?Greenland, Australia.

## *Orbithele ceratopygarum* Brøgger, 1882

Figs.113–115.

*Synonymy.* – □1882 *Discina* (*Acrotreta?*) *ceratopygarum* sp. nov. – Brøgger, p. 47, Pl. 10:1–1b. □1906 *Acrothele barbata* sp. nov. – Moberg & Segerberg, p. 67, Pl. 3:7–10. □1908 *Acrothele borgholmensis* sp. nov. – Walcott, p. 84, Pl. 8:12. □1911 *Acrothele borgholmensis* n.sp. – Walcott, p. 639, Pl. 43:2, 2a–b. □1912 *Acrothele ceratopygarum* (Brøgger) – Walcott, p. 640, Pl. 43:1a–c. □1969 *Acrothele*(?) *barbata* Moberg & Segerberg – Gorjansky, p. 74, Pl. 16:7. □?1969 *Orbithele* sp. – Gorjansky, p. 79, Pl. 13:8, 9. □?1973 *Orbithele bicornis* sp. nov. – Biernat, p. 101, Pl. 27:1–3, Fig. 27.

*Holotype.* – By monotypy; exfoliated ventral valve (L 5.2, W 6, H 1.5–2; according to Brøgger 1882, p. 47; specimen lost) illustrated by Brøgger (1882, Pl. 10:1a, b); Bjørkåsholmen Limestone; Vestfossen (Walcott 1912; locality 323f).

*Material.* – Figured from Scandinavia. Dorsal valves: RM Br 20733 (L 3, W 3.5, LI 0.3, WI 1.9, LM1 0.8, WM1 1.4, LM2 1.4, WM2 0.6, LM3 1.6, WM3 0.2); RM Br 20790 (L 3.8, W 4.5); RM Br 20792 (L 5.2, W 6.4); RM Br 20791b (L 5.3, W 6.6); RM Br 136290; RM Br 136294; LO 6564t. Ventral valves: RM Br 14967; RM Br 20791a (L 4.7, W 5.7, H 1.4); RM Br 20793 (L 2.5, W 3.4); RM Br 20707 (L 3.6, W 4.2); RM Br 136292; RM Br 136293; LO 1786T; LO 1787t; LO 6565t; LO 6566t; LO 6567t; LO 6571t (L 4.2, W 5.1); USNM 51974; USNM 51977a. Total of 21 ventral and 7 dorsal valves.

Figured from the South Urals. Ventral valves: RM Br 136221 (L 1.5, W 1.8); RM Br 136222; RM Br 136292; RM Br 136293. Dorsal valve: RM Br 136294. Fragment: RM Br136291. Total of 8 ventral and 2 dorsal valves.

*Measurements.* – See Table 53.

*Diagnosis.* – Ventral valve about one-third as high as long with apex about one-third from posterior margin; ventral pseudointerarea comparatively poorly defined, with intertrough; internal pedicle tube extending almost along entire posterior slope, forming knob-like median septum; dorsal pseudointerarea undivided, occupying more than half of valve width; cardinal buttress small, continuing as small very low ridge; juvenile shell, around 1 mm wide, with ornamentation of fine costellae and rugellae; adult shell with irregularly spaced knob-shaped pustules forming high, irregular, wavy rows.

*Description.* – Shell transversely oval. The ventral valve is moderately convex, about 73–90% as long as wide and 29% (*N* 1) as high as long; in lateral view, the posterior slope is flattened and the anterior is only slightly convex (Fig. 115G).

*Fig. 114.* Orbithele ceratopygarum (Brøgger). □A. Partly exfoliated ventral exterior; the holotype of *Acrothele borgholmensis* Walcott (1912, Pl. 63:2, 2a); Ceratopyge Shale; Borgholm (Walcott 1912; his locality 310d), Öland; USNM 51974; ×6. □B. Ventral part internal mould; figured by Moberg & Segerberg (1906, Pl. 3:9) as *Acrothele barbata* (counterpart to Fig. 116H); Bjørkåsholmen Limestone; Ottenby, Öland; LO 1787t; ×8. □C. Ventral interior; Ceratopyge Shale; Borgholm, Öland; RM Br 14967; ×10. □D. Dorsal interior; Bjørkåsholmen Limestone; Ottenby, Öland; RM Br 20790; ×10. □E. Dorsal interior; Kidryas Formation (sample B-768-1); Tyrmantau Ridge, South Urals; RM Br 136294; ×12. □F. Posterior view of ventral internal mould; figured by Walcott (1912, Pl. 63:1b, c); Bjørkåsholmen

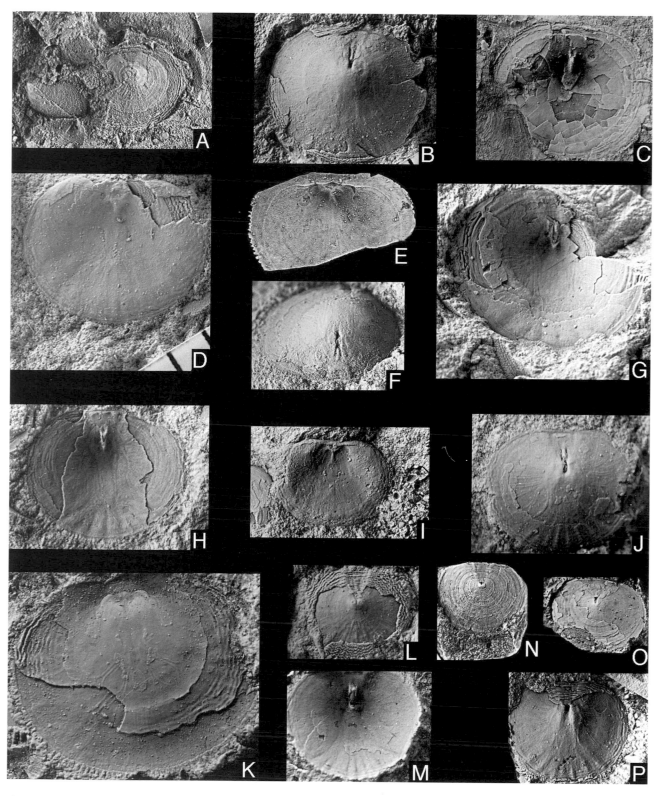

Limestone; Vestfossen (Walcott 1912; his locality 323h), Norway; USNM 51977a; ×6. □G. Ventral interior; Bjørkåsholmen Limestone; Ottenby, Öland; specimen lost; ×10. □H. Ventral interior; figured by Moberg & Segerberg (1906, Pl. 3:9) as *Acrothele barbata* (part to Fig. 116B); Bjørkåsholmen Limestone; Ottenby, Öland; LO 1787t; ×8. □I. Dorsal interior; Bjørkåsholmen Limestone; Ottenby, Öland; RM Br 20733; ×7. □J. Ventral internal mould; Bjørkåsholmen Limestone; Ottenby, Öland; specimen lost; ×10. □K. Exfoliated dorsal valve; Bjørkåsholmen Limestone; Ottenby, Öland; RM Br 20792; ×10. □L. Exfoliated ventral exterior; lectotype (here selected) of *Acrothele barbata* Moberg & Segerberg (1906, Pl. 3:7); Bjørkåsholmen Limestone; Ottenby, Öland; LO 1786T; ×10. □M. Ventral interior (same as Fig. 113D); Bjørkåsholmen Limestone; Ottenby, Öland; RM Br 20793; ×10. □N. Ventral exterior; Kidryas Formation (sample B-768-1); Tyrmantau Ridge, South Urals; RM Br 136292; ×4. □O. Ventral exterior; Kidryas Formation (sample B-768-1); Tyrmantau Ridge, South Urals; RM Br 136293; ×4. □P. Partially exfoliated ventral valve; Bjørkåsholmen Limestone; Ottenby, Öland; RM Br 20707; ×8.

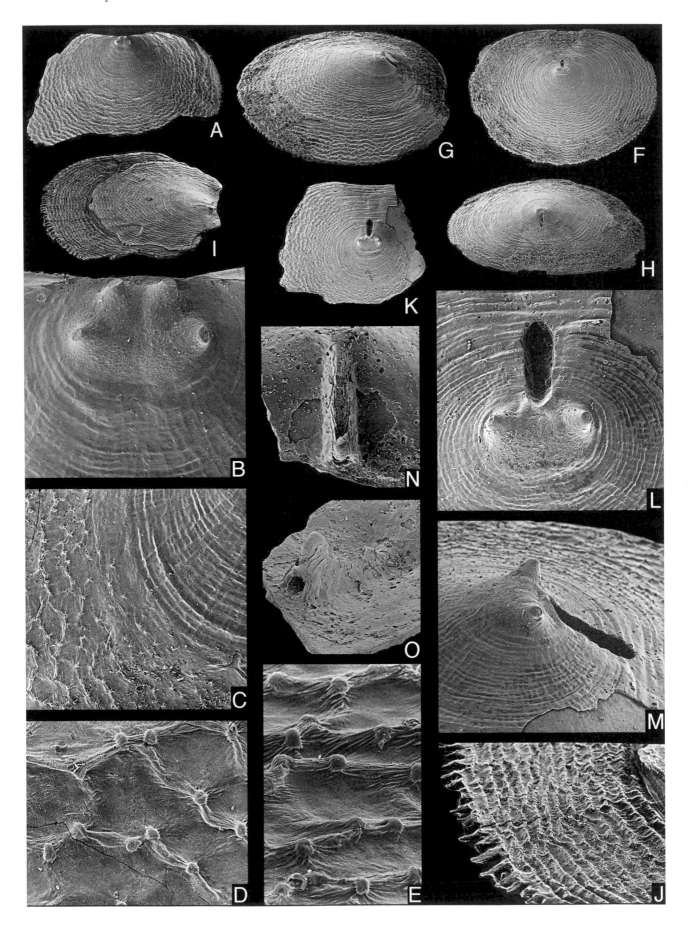

Ventral apex about 34% of the total valve length from the posterior margin. The ventral pseudointerarea is broadly triangular and relatively poorly defined laterally; it invariably lacks the ornamentation of the irregular rows of knob-like pustules; the intertrough is shallow and narrowly triangular in outline (Fig. 115F, H, L). The pedicle foramen is placed at the end of a listrium-like pedicle track that is partly outside the larval shell; it is lenticular, about 0.3 mm long, extending between points 31% and 24%, respectively, of the valve length from the posterior margin; the pedicle track is almost completely covered by a concave plate (Figs. 113L, 115L, M). The interior pedicle tube is long and extends along the entire posterior slope of the valve; it is continued as a short knob-like median septum (Figs. 113D, M, 115N, O). The ventral muscle scars are usually not particularly well defined, but they all appear to be placed on the posterior slope; they are closely spaced on either side of the pedicle tube and septum (Fig. 114C, M). The ventral *vascula lateralia* are usually extremely well defined and show secondary bifurcation, with one long, anteriorly directed trunk and up to 7–8 minor canals (Fig. 114B, C, G, H, J, P).

The dorsal valve is flattened in lateral view (Figs. 113B, 115I), on average 81% as long as wide (Table 53); in plane view it is slightly indented medially, with lateral flanks deflected somewhat upwards (Figs. 113A, 115A). The dorsal pseudointeraea forms a poorly defined, undivided, wide and short shelf along the posterior margin, on average occupying 54% of the total valve width (Table 53), flattened in lateral view, almost orthocline, becoming somewhat anacline laterally (Fig. 114E, I). The cardinal buttress is small, triangular, continuing into an extremely faint median ridge, which extends as a median tongue to the anterior lateral muscle scars (Fig. 114D, I). The dorsal cardinal muscles are small, elongate oval, and rather closely spaced, occupying only 35% of the valve width and extending for 27% of the length (Table 53). The scars of the dorsal central and anterior lateral muscles are also well defined and clearly separated, albeit closely spaced; the central scars are placed on average at 44% (OR 37–47%; N3) of the valve length, and the anterior lateral at 57% (OR 53–63%; N3) of the length (Fig. 114D, E, I). The dorsal *vascula lateralia* are not as well defined as in the ventral valve, but a pattern with minor canals, similar to that of the ventral trunks, can be observed in some specimens; the

*vascula media* are rather indistinct, but are apparently present directly anterior to the median tongue (Fig. 114K).

The larval shells of both valves are well defined, transversely oval, around 0.4–0.5 mm wide and 0.3–0.4 mm long, and have larval spines, up to about 70 µm long (Figs. 113C, L, 115B, L, M); the larval pits are of two sizes, but the pattern is not well preserved. Outside of the larval shell, the juvenile shell (up to a width of somewhat more than 1 mm) is ornamented by fine costellae, up to about 70 µm apart, and rugellae, up to about 40 µm apart (Fig. 115B, C); this area is well separated from the adult shell by a disturbance in growth, and outside of this area, the adult shell has the usual ornamentation of irregularly spaced knob-shaped pustules, up to 10 µm high, that form high, irregular, wavy rows, up to about 0.15 mm apart (Fig. 115D, E). At about the stage when the valve was 4–5 mm in width, the growth of the adults were disturbed (related to spawning?) to form distinct growth-lamellae (up to 2 in a shell 7 mm wide), with rows of closely spaced, short spines, up to 0.15 mm long, along the margin (Figs. 113N, 115J). In one nearrly mid-sagittal section through a ventral valve (Fig. 113H), the shell is generally up to 50 µm thick and consists of baculate laminae, up to 20 µm thick (Fig. 113I); the primary layer is extremely thin. The baculae are very thin and long, less than 500 nm across (Fig. 113E–G).

*Discussion.* – As noted by Walcott (1912, p. 640), *O. barbata*, from the Bjørkåsholmen Limestone of Öland, is a junior synonym of the coeval *O. ceratopygarum* from the Oslo region; the only specimen of the latter species illustrated by Brøgger (1882) is unfortunately lost, but topotypes from the Bjørkåsholmen Limestone in the Oslo region (Fig. 114) clearly show no morphological difference from *O. barbata*. Walcott (1908) described *Acrothele borgholmensis* (Fig. 114A) from the Ceratopyge Shale at Borgholm, island of Öland; the dorsal valve of this species is poorly known, but topotypes of the ventral valve (Fig. 114C) are identical to those of *O. ceratopygarum*. Gorjansky (1969) recorded *O. barbata* and an unnamed *Orbithele* from the Leetse beds of Latvia and Estonia; the detailed morphology of these is poorly known, and only the ventral valves are known, but they may possibly be synonymous with *O. ceratopygarum*. An equally poorly known species, *O. bicornis*, was described by Biernat (1973), and as far as can be judged it is very close to the Scandinavian form.

The specimens referred to *O. ceratopygarum* from the South Urals show no difference from the Scandinavian specimens in the convexity and outline of the ventral valve and the morphology of the internal pedicle tube; however, the dorsal valve of the South Ural material is not well known.

As noted above, the type species is very poorly known; it is considerably larger than *O. ceratopygarum*, and according to Sdzuy (1955), the apex of the latter species is more centrally placed. The Bohemian species discussed by Mergl (1981) are also generally larger than the Scandinavian form but other-

---

*Fig. 115.* *Orbithele ceratopygarum* (Brøgger); Bjørkåsholmen Limestone (sample Öl-1); Ottenby, Öland. □A. Dorsal exterior; LO 6564t; ×11. □B. Larval shell of A; ×95. □C. Detail of A, showing boundary between juvenile and adult shells; ×75. □D. Detail of A, showing adult pustulose ornamentation; ×157. □E. Lateral view of pustules of A; ×650. □F. Ventral exterior; RM Br 20791a; ×9. □G. Lateral view of F; ×12. □H. Posterior view of F; ×9. □I. Lateral view of dorsal exterior; RM Br 207091b; ×9. □J. Detail of spinose margin of I; ×40. □K. Ventral exterior; LO 6565t; ×15. □L. Larval shell of K; ×75. □M. Lateral view of larval shell of K; ×75. □N. Ventral interior with pedicle tube and septum; LO 6566t (same as Fig. 113E); ×45. □O. Lateral view of N; ×50.

wise very similar. The dorsal *vascula media* (?) in *O. discontinua* [may be a junior synonym to *O. secedens* (Barrande)] of Mergl (1981, Fig. 1) seems to be better developed than in *O. ceratopygarum*, and, moreover, the dorsal umbonal muscle scar is very large in the Bohemian species. However, a ventral valve of the same species (the holotype) illustrated by Mergl (1981, Pl. 1:7) clearly shows the same type of change in ornamentation between the juvenile and adult shell, as described above from *O. ceratopygarum*; the juvenile shell in the Bohemian species seems to be around 1.3 mm wide, which is close to that recorded from *O. ceratopygarum*; the valve of *O. discontinua* also has a spinose lamellae marking an interruption in the growth, at around 4 mm in width. *Orbithele maior* Mergl, 1981 is also very close in morphology to *O. discontinua*, whereas *Orbithele undulosa* (Barrande) [=?*O. rimosa* Mergl, 1981] has a completely different type of ornamentation from any other species of *Orbithele*.

*Occurrence.* – *O. ceratopygarum* occurs in the Ceratopyge Shale and Bjørkåsholmen Limestone in most parts of Scandinavia, including Öland, Västergötland, and the Oslo region (Figs. 5–6; Appendix 1K), but it has not been recorded at these levels in Scania. In the South Urals, it occurs at the following localities: Tyrmantau Ridge; Bolshaya and Malaya Kayala rivers; and the Koagash River (Appendix 1A).

# References

Ancygin, N.Y., Nassedkina, V.A. & Rozov, S.N. 1977: Pogranichnye sloi kembriya i ordovika levobererzh'ya reki Medes (Juzhnyi Ural). [Cambrian–Ordovician boundary beds of the left side of the Medes River (South Urals).] *In* Zhuravleva, I.T. & Rozova, A.V. (eds.): Biostratigrafiya i fauna verkhnego kembriya i pogranichnykh s nim sloev (Novyje dannye po aziatskoj chasti SSSR). *Trudy Instituta geologii i geofiziki Akademii Nauk SSSR, Sibirskoe otdelenie 313*, 184–197.

Andersson, A., Dahlman, B., Gee, D.G. & Snäll, S. 1985: The Scandinavian alum shales. *Sveriges Geologiska Undersökning Ca 56*, 1–50.

Apollonov, M.K. 1991: Cambrian–Ordovician boundary beds in the U.S.S.R. *In* Barnes, C.R. & Williams, S.H. (eds.): *Advances in Ordovician Geology, Geological Survey of Canada, Paper 90–9*, 33–45.

Bagnoli, G., Stouge, S. & Tongiorgi, M. 1988: Acritarchs and conodonts from the Cambro–Ordovician Furuhäll (Köpingsklint) Section. *Rivista Italiana Paleontologia Stratigrafia 94*, 163–248.

Balashova, E.A. 1961: Nekotorye tremadokskie trilobity Aktjubinskoj oblasti. [Some Tremadocian trilobites of the Aktyubinsk district.] *Trudy Geologicheskogo Instituta Akademii Nauk SSSR 18*, 102–145.

Barrande, J. 1868: Silurische Fauna aus der Umgebung de Hof in Bayern. *Neues Jahrbuch für Mineralogie*, 641–696.

Barrande, J. 1879: *Systême Silurien du centre de la Bohême, Pt. 1, Recherches paléontologiques, vol. 5, Classe des Mollusques. Ordre des Brachiopodes.* 226 pp. Praha.

Bednarczyk, W. 1959a: On the Genus *Conotreta* from the Lower Ordovician of the Holy Cross Mts. *Bulletin de l'Academie Polonaise des Sciences 7*, 463–468.

Bednarczyk, W. 1959b: Four new species of *Conotreta* from the Upper Tremadocian of the Holy Cross Mts. *Bulletin de l'Academie Polonaise des Sciences 7*, 509–513.

Bednarczyk, W. 1964: Stratigrafia i fauna tremadoku i arenigu (Oelandianu) regionu kieleckiego Gór Swiętokrzyskich. *Biuletyn Geologiczny Universytetu Warszawskiego 4*, 1–216.

Bednarczyk, W. 1971: Stratigraphy and palaeogeography of the Ordovician in the Holy Cross Mts. *Acta Geologica Polonica 21*, 573–616.

Bednarczyk, W. 1986: Inarticulate brachiopods from the Lower Ordovician in northern Poland. *Annales Societatis Geologorum Poloniae 56*, 409–418.

Bednarczyk, W. & Biernat, G. 1978: Inarticulate brachiopods from the Lower Ordovician of the Holy Cross Mountains, Poland. *Acta Paleontologica Polonica 23*, 293–316.

Bell, W.A. 1944: Early Upper Cambrian brachiopods. *In* Lochman, C. & Duncan, D.: Early Upper Cambrian faunas of central Montana. *Geological Society of America, Special Paper 54*, 144–155

Bergström, J. 1982: Scania. *In* D.L. Bruton (ed.): Field excursion guide. IV International Symposium on the Ordovician System. *Paleontological Contributions from the University of Oslo 279*, 184–197.

Biernat, G. 1973: Ordovician brachiopods from the Poland and Estonia. *Paleontologia Polonica 28*, 1–116.

Brøgger, W.C. 1882: Die Silurischen Etagen 2 and 3 im Kristianiagebiet und auf Eker. *Universitätsprogramm*, 1–376.

Bruton, D.L., Erdtmann, B.–D. & Koch, L. 1982: The Naersnes section, Oslo Region, Norway: a candidate for the Cambrian–Ordovician boundary stratotype at the base of the Tremadoc Series. *In* Bassett, M.G. & Dean, W.T. (eds.): *The Cambrian – Ordovician Boundary: Sections, Fossil Distributions and Correlations*. pp 61–68. National Museum of Wales, Geology Series 3, Cardiff.

Bruton, D.L., Koch, L. & Repetski, J.E. 1988: The Naersnes section, Oslo Region, Norway: trilobite, graptolite and conodont fossils reviewed. *Geological Magazine 125*, 45–455.

Bruton, D.L. & Owen, A.W. 1982: The Ordovician of Norway. *In* D.L. Bruton (ed.): Field excursion guide. IV International Symposium on the Ordovician System. *Paleontological Contributions from the University of Oslo 279*, 10–14.

Bulman, O.M.B. & Rushton, A.W.A. 1973: Tremadoc faunas from boreholes in Central England. *Geological Survey of Great Britain, Bulletin 43*, 1–40.

Callaway, C. 1877: On a new area of Upper Cambrian rocks in south Shropshire, with description of a new fauna. *Geological Society of London, Quarterly Journal 33*, 652–672.

Cockerell, T.D.A. 1911: The name *Glossina*. *Nautilus 25*, 96.

Cocks, L.R.M. 1978: A review of British Lower Palaeozoic brachiopods, including a synoptic revision of Davidson´s. *Palaeontographical Society, Monograph 549*, 1–256.

Cooper, G.A. 1956: Chazyan and related brachiopods. *Smithsonian Miscellaneous Collection 127*, 1–1245.

Cope, J.C.W., Fortey, R.A. & Owens, R.M. 1978: Newly discovered Tremadoc rocks in the Carmarthen district, South Wales. *Geological Magazine 115*, 195–198.

Curry, G.B. & Williams, A. 1983: Epithelial moulds on the shells of the early Palaeozoic brachiopod *Lingulella*. *Lethaia 16*, 111–118.

Dall, W.H. 1877: Index to the names which have been applied to the subdivisions of the Class Brachiopoda. *United States National Museum, Bulletin 8*, 1–88.

Eichwald, E. 1840: *Ueber das silurische Schichten-System von Esthland.* 210 pp. St. Petersburg.

Eichwald, E. 1843: Ueber die Obolen und den silurischen Sandstein von Esthland und Schweden. *Beiträge zur Kenntnis des Russischen Reiches und der angränzenden Länder Asiens*. 139–156. St. Petersburg.

Eichwald, E. 1859: *Lethaea rossica ou Paléontologie de la Russie*. Atlas. 62 plates. Stuttgart.

Eichwald, E. 1860: *Lethaea rossica ou Paléontologie de la Russie. Vol. 1:2*, 681–929. Stuttgart.

Fischer, P.H. 1887: *Manuel de conchyliologie et de paléontologie conchyliologique, ou histoire naturelle des mollusques vivants et fossiles. Part 11*, 1189–1334. Paris.

[Fjelldal, Ø. 1966: The Ceratopyge Limestone (3aγ) and limestone facies in the Lower Didymograptus Shale (3b) in the Oslo Region and adjacent districts. Unpublished B.Sc. Thesis, University of Oslo. 129 pp.]

Fortey, R.A., Bassett, M.G., Harper, D.A.T., Hughes, R.A., Ingham, J.K., Molyneux, S.G., Owen, A.W., Owens, R.M., Rushton, A.W.A. & Sheldon, P.R. 1991: Progress and problems in the selection of stratotypes for the bases of series in the Ordovician System of the historical type area in the U.K. *In* Barnes, C.R. & Williams, S.H. (eds.): *Advances in Ordovician Geology, Geological Survey of Canada, Paper 90:9*, 5–25.

[Gjessing, J. 1976: Tremadocian stratigraphy and fauna in the Oslo Region, Norway. Unpublished B.Sc. Thesis, University of Oslo. 136 pp.]

Gorjansky, V.J. 1969. Bezzamkovye brakhiopody kembrijskikh i ordovikskikh otlozhenij severo-zapada Russkoj platformy. [Inarticulate brachiopods of the Cambrian and Ordovician of the northwest Russian Platform.] *Ministerstvo Geologii RSFSR, Severo-Zapadnoe Territorial'noe Geologicheskoe Upravlenie 6*, 1–173.

Harrington, H.J. 1937: On some Ordovician fossils from Northern Argentina. *Geological Magazine 73*, 97–124.

Harrington, H.J. 1938: Sobre las faunas del Ordoviciano inferior del norte arentino. *Museo de La Plata (Buenos Aires), Revista, N.S. 1*, 109–289.

Havlíček, V. 1980. *Conotreta* Walcott (Brachiopoda) in the Lower Ordovician of Bohemia. *Vestník Ústředního Ústavu Geologického 55*, 297–299.

Havlíček, V. 1982: Lingulacea, Paterinacea, and Siphonotretacea (Brachiopoda) in the Lower Ordovician sequence of Bohemia. *Sborník Geologickych Ved. Paleontologie 25*, 9–82.

Hede, J.E. 1951: Boring through Middle Ordovician – Upper Cambrian Strata in the Fågelsång district, Scania. 1. Succession encountered in the Boring. *Lunds Universitets Årsskrift, N.F. 2*, 1–80.

Henderson, R.A. 1974: Shell adaption in acrothelid brachiopods to settlement on a soft substrate. *Lethaia 7*, 57–61.

Henderson, R.A, Debrenne, F., Rowell, A.J. & Webers, G.F. 1992: Brachiopods, archaeocyathids, and Pelmatozoa from the Minaret Formation of the Ellsworth Mountains, West Antarctica. *Geological Society of America, Memoire 170*, 249–267.

Henderson, R.A. & MacKinnon, D.I. 1981: New Cambrian inarticulate Brachiopoda from Australia and the age of the Tasman Formation. *Alcheringa 5*, 289–309.

Henningsmoen, G. 1982: The Ordovician of the Oslo Region – A short history of research. *In* D.L. Bruton (ed.): Field excursion guide. IV International Symposium on the Ordovician System. *Paleontological Contributions from the University of Oslo 279*, 92–98.

Holl, H.B. 1865: On the geological structure of the Malvern Hills and adjacent districts. *Geological Society of London, Quarterly Journal 21*, 72–102.

Holmer, L.E. 1986: Inarticulate brachiopods around the Middle–Upper Ordovician boundary in Västergötland. *Geologiska Föreningens i Stockholm Förhandlingar 108*, 97–126.

Holmer, L.E. 1989: Middle Ordovician phosphatic inarticulate brachiopods from Västergötland and Dalarna, Sweden. *Fossils and Strata 26*, 1–172.

Holmer, L.E. 1993: The Lower Ordovician elkaniid brachiopod *Lamanskya* and the Family Elkaniidae. *Transactions of the Royal Society of Edinburgh: Earth Sciences 84*, 151–160.

Holmer, L.E. & Popov, L.E. 1990: The acrotretacean brachiopod *Ceratreta tanneri* (Metzger) from the Upper Cambrian of Baltoscandia. *Geologiska Föreningens i Stockholm Förhandlingar 112*, 249–263.

Holmer, L.E. & Popov, L.E. 1994: Revision of the type species of *Acrotreta* and related lingulate brachiopods. *Journal of Paleontology 68*, 433–450.

Ivanov, K.S. & Puchkov, V.N. 1984: *Geologia Sakmarskoi zony Urala (novye dannye)* [*Geology of the Sakmara Zone of Urals (New Data)*]. 86 pp. Institut geologii i geokhimii, Sverdlovsk.

Ivshin, N.K. 1956: *Verkhnekembrijskie trilobity Kazakhstana. Chast' 1. (Kuyandinskij faunistieskij gorizont mezhdurechiya Olenty–Shyderty)* [*Upper Cambrian trilobites of Kazakhstan. Part 1. (Kujandinian faunal horizon of the Olenty–Shyderty)*]. 98 pp. Akademia Nauk Kazakhskoj SSR, Alma-Ata.

Ivshin, N.K. 1962: *Verkhnekembrijskie trilobity Kazakhstana. Chast' 2. Seletinskij Gorizont Kuyandinskogo Jarusa Tsentral'nogo Kazakhstana* [*Upper Cambrian Trilobites of Kazakhstan. Part 2. Seletinian Horizon of the Kujandinian Stage of Central Kazakhstan*]. 412 pp. Akademia Nauk Kazakhskoj SSR, Alma-Ata.

Ivshin, N.K. 1972. Kembrijskaya sistema. [Cambrian System] *In* Esenov, S.E. & Shlygin E.D. (eds.): *Tsentral'nyj Kazakhstan, Geologicheskoe opisanie. Geologiya SSSR 20:1*, 93–153.

Jaanusson, V. 1982a: Introduction to the Ordovician of Sweden. *In* Bruton, D.L. (ed.): Field excursion guide. IV International Symposium on the Ordovician System. *Paleontological Contributions from the University of Oslo 279*, 1–9.

Jaanusson, V. 1982b: Ordovician in Västergötland. *In* D.L. Bruton (ed.): Field excursion guide. IV International Symposium on the Ordovician System. *Paleontological Contributions from the University of Oslo 279*, 164–183.

Jaanusson, V. & Mutvei, H. 1982: *Ordovician on Öland. Guide to Excursion 3. IV International Symposium on the Ordovician System, Oslo 1982.* 23 pp. Section of Palaeozoology, Swedish Museum of Natural History, Stockholm.

Jeremejew, P. 1856: Geognostische Beobachtungen an den Ufern des Wolchow. *Russisch-Kaiserliche Mineralogische Gesellschaft Verhandlungen 1855–1856*, 63–84.

Kaljo, D., Heinsalu, H., Mens, K., Puura, I. & Viira, V. 1988: Cambrian–Ordovician boundary beds at Tonismägi, Tallinn, North Estonia. *Geological Magazine 125*, 457–463.

Kaljo, D., Rõõmusoks, A. & Männil, R. 1958: O seriyakh pribaltijskogo ordovika i ikh znachenii. [On the Series of the Baltic Ordovician and their significance.] *Eesti NSV Teaduste Akadeemia Toimetised. Tehniliste ja Füüsikalis–Matemaatiliste Teaduste Seeria 7*, 71–74.

Keller, B.M. & Rozman, H.S. 1961: Otlozheniya nizhnego ordovika Aktyubinskoj oblasti Kazakhstana i smezhnykh rayonov zapadnogo sklona Urala. [The Lower Ordovician strata of the Aktyubinsk district of Kazakhstan and surrounding areas on the western slope of the South Urals.] *Trudy Geologicheskogo Instituta Akademii Nauk SSSR 18*, 93–101.

Keller, B.M. & Rukavishnikova, T.B. 1961: Tremadokskie i smezhnye s nimi otlozheniya khrebta Kendyktas. [Tremadocian and related strata in the Kendyktas Range.] *Trudy Geologicheskogo Instituta Akademii Nauk SSSR 18*, 22–28.

Koliha, J. 1924: Atremata z krušnohorsk ch vrstev. *Palaeontographica Bohemiae 10*, 5–61.

Koneva, S.P. 1986. Novoe semejstvo kembrijskikh bezzamkovykh brakhiopod. [A new family of the Cambrian inarticulate brachiopods.] *Paleontologicheskij Zhurnal*, 49–55.

Koneva, S.P. & Popov, L.E. 1983: Nekotorye novye lingulidy iz verkhnego kembriya i nizhnego ordovika Malogo Karatau. [On some new lingulids from the Upper Cambrian and Lower Ordovician of the Malyj Karatau Range.] *In* Apollonov, M.K., Bandaletov, S.M. & Ivshin N.K. (eds.): *Stratigrafiya i paleontologiya nizhnego paleozoya Kazakhstana*, 112–124. Nauka, Alma–Ata.

Koneva, S.P. & Popov, L.E. 1988: Akrotretidy (bezzamkovye brakhiopody) iz pogranichnykh otlozhenij kembrij–ordovik khrebta Malyj Karatau (Yuzhnyj Kazakhstan). *Ezhegodnik Vsesoyuznogo Paleontologicheskogo Obshchestvo 31*, 52–72.

Koneva, S.P., Popov, L.E., Ushatinskaya, G.T. & Esakova, N.V. 1990: Bezzamkovye brakhiopody (akrotretidy) i mikroproblematiki iz verkhnego kembriya severo-vostochnogo Kazakhstana. [Inarticulate brachiopods and microproblematica from the Upper Cambrian of northeastern Kazakhstan.] *In* Repina, L.N. (ed.): Biostratigrafiya i paleontologiya kembriya Severnoj Azii. *Trudy Instituta Geologii i Geofiziki Akademii Nauk SSSR, Sibirskoe Otdelenie 765*, 158–170.

Korynevskij, V.G. 1983. Stratotipicheskij razrez kidrjasovskogo gorizonta nizhnego ordovika Urala. [The stratotype section of the Lower Ordovician Kidryas Stage of the Urals.] *Izvestiya Akademii Nauk SSSR, Seriya Geologicheskaya 8*, 46–54.

Korynevskij, V.G. 1989. Opornye razrezy nizhnego ordovika Yuzhnogo Urala (terrigennye facii). [Key sections of the Lower Ordovician of South Urals (clastic facies).] *Uralskoe Otdelenie Akademii Nauk SSSR*, 1–67.

Korynevskij, V.G. 1992. Khmelevskij gorizont verkhnego kembriya na Urale – stratigraficheskaya oshibka. [The Upper Cambrian Khmelievian Stage at the South Urals as a stratigraphic error.] *Isvestiya Akademii Nauk 2*, 131–135.

Krause, F.F. & Rowell, A.J. 1975: Distribution and systematics of the inarticulate brachiopods of the Ordovician carbonate mud mound of

Meiklejohn Peak, Nevada. *The University of Kansas Paleontological Contributions 61*, 1–74.

Kutorga, S.S. 1848: Über die Brachiopoden–Familie der Siphonotretacea. *Russich–Kaiserliche Mineralogische Gesellschaft Verhandlungen 1847*, 250–286.

Leonenok, N.I. 1955: Silurijskie otlozheniya Kos–Istekskogo rajona (Severnye Mugodzhary). [Silurian strata of the Kos–Istek area (northern Mugodzhary Range).] *Trudy Laboratorii Uglja Akademii Nauk SSSR 3*, 116–225.

Lermontova, E.V. 1951: *Verkhnekembrijskie trilobity i brachiopody Bostchekulja (Severo–Vostochnyj Kazakhstan).* [Upper Cambrian trilobites and brachiopods from the Bostchekul (northeastern Kazakhstan).] 49 pp. VSEGEI, Leningrad.

Lermontova, E.V. & Razumovskij, N.K. 1933: O drevnejshikh otlozheniyakh Urala (nizhnij silur i kembrij v okrestnostyakh derevni Kidryasovo na yuzhnom Urale). [On the ancient strata of the Urals (Lower Silurian and Cambrian at the outskirts of the Kidryasovo Village in the South Urals.] *Zapiski Rossijskogo Mineralogicheskogo Obshchestva 62:1*, 185–217.

Lissogor, K.A. 1961: Trilobity tremadokskikh i smezhnykh s nimi otlozhenij Kendyktasa. [Trilobites from the Tremadocian and related strata of the Kendyktas Range.] *Trudy Geologicheskogo Instituta Akademii Nauk SSSR 18*, 55–92.

McClean, A.E. 1988: Epithelial moulds from some Upper Ordovician acrotretide brachiopods of Ireland. *Lethaia 21*, 43–50.

Mägi, S. & Viira, V. 1976: Rasprostranenie konodontov i bezzamkovykh brakhiopod v Tseratopigevom i Latorpskom gorizontakh severnoj Estonii. [On the distribution of conodonts and inarticulate brachiopods in Ceratopyge and Latorp stages.] *Eesti NSV Teaduste Akadeemia Toimetised, Keemia–Geoloogia 25*, 312–318.

Markovskij, B.P. (ed.) 1960: *Novye vidy drevnikh rastenij i Bezpozvonochnykh S.S.S.R.* [*New Species of Ancient Plants And Invertebrates of the USSR.*] 612 pp. VSEGEI, Moscow.

Martinsson, A. 1974: The Cambrian of Norden. *In* Holland, C.H. (ed): *Cambrian of the British Isles, Norden, and Spitsbergen*, 185–283. Wiley, London.

Matthew, G.F. 1901: New species of Cambrian fossils from Cape Breton. *New Brunswick Natural History Society, Bulletin 4*, 269–286.

Mei, S. 1993: Middle and Upper Cambrian inarticulate brachiopods from Wanxian, Hebei, North China. *Acta Palaeontologica Sinica 32*, 400–429.

Melnikova, L.M. 1990: Ranne- i pozdnekembrijskie Bradoriida (ostrakody) severo-vostoka Tsentral'nogo Kazakhstana. [Early and Late Cambrian Bradoriida (ostracodes) from the northeast of Central Kazakhstan.] *In* Repina, L.N. (ed.): Biostratigrafiya i paleontologiya kembriya Severnoj Azii. *Trudy Instituta Geologii i Geofiziki Akademii Nauk SSSR, Sibirskoe Otdelenie 765*, 170–176.

Mens, K., Bergström, J. & Lendzion, K. 1987: *Kembrij vostochno-evropejskoj platformy (korrelyatsionnaya skhema i ob"yasnitel'naya zapiska).* [The Cambrian System on the East European Platform (Correlation chart and explanatory notes).]. 118 pp. Valgus, Tallinn.

Mergl, M. 1981: The genus *Orbithele* (Brachiopoda, Inarticulata) from the Lower Ordovician of Bohemia and Morocco. *Vestník Ústředního Ústavu Geologického 56*, 287–292.

Mergl, M. 1986: The Lower Ordovician (Tremadoc–Arenig) *Leptembolon* Community in the Komárov area (SW part of the Prague Basin; Bohemia). *Folia Musei Rerum Naturalium Bohemiae Occidentalis, Plzen, Geologica 24*, 3–34.

Mickwitz, A. 1896: Über die Brachiopodengattung *Obolus* Eichwald. *Mémoires de l'Académie Impériale des Sciences de St.Pétersbourg 4*, 1–215.

Moberg, J.C. 1910: Guide for the principal Silurian districts of Scania (with notes on some localities of Mesozoic beds). *Geologiska Föreningens i Stockholm Förhandlingar 32*, 45–194.

Moberg, J.C. & Segerberg, C.O. 1906: Bidrag till kännedomen om ceratopygeregionen med särskild hänsyn till dess utveckling i Fogelsångstrakten. *Lunds Universitets Årsskrift N.F. 2:2*, 1–116.

Nazarov, B.B. & Popov, L.E. 1980: Stratigraphy and fauna of Ordovician siliceous–carbonate deposits of Kazakhstan [Stratigrafiya i fauna krem-nisto-karbonatnykh otlozhenij Kazakhstana.] *Trudy Geologicheskogo Instituta Akademii Nauk SSSR 331*, 1–190.

Nikitin, I.F. 1956: *Brachiopody kembriya i nizhnego ordovika severo-vostoka Tsentral'nogo Kazakhstana.* [*Cambrian and Lower Ordovician Brachiopods from the Northeastern Part of Central Kazakhstan.*] 141 pp. Nauka, Alma–Ata.

Nikitin, I.F. 1972: *Ordovik Kazakhstana. Chast' 1, stratigrafiya.* [*Ordovician of Kazakhstan. Part 1, Stratigraphy.*] 242 pp. Nauka, Alma–Ata.

Norford, B.S. 1991: The international working group on the Cambrian–Ordovician boundary: report of progress. *In* Barnes, C.R. & Williams, S.H. (eds.): Advances in Ordovician Geology. *Geological Survey of Canada, Paper 90:9*, 27–32.

Owen, A.W., Bruton, D.L., Bockelie, J.F. & Bockelie, T.G. 1990: The Ordovician successions of the Oslo Region, Norway. *Norges Geologiske Undersøkelse, Special Publication 4*, 3–54.

Owens, R.M., Fortey, R.A., Cope, J.C.W., Rushton, A.W.A. & Bassett, M.G. 1982: Tremadoc faunas from the Carmarthen district, South Wales. *Geological Magazine 119*, 1–112.

Pander, C.H., 1830: *Beiträge zur Geognosie des Russichen Reiches.* 165 pp. St. Petersburg.

Pelman, Yu.L. 1977: Ranne- i srednekembrijskie bezzamkovye brakhiopody Sibirskoj platformy. [Early and Middle Cambrian inarticulate brachiopods of the Siberian Platform.] *Trudy Instituta Geologii i Geofiziki Akademiya Nauk SSSR, Sibirskoe Otdelenie 316*, 1–168.

Percival, I.G. 1978: Inarticulate brachiopods from the Late Ordovician of New South Wales, and their palaeoecological significance. *Alcheringa 2*, 117–141.

Popov, L.E., Khazanovitch, K.K., Borovko, N.G., Sergeeva, S.P. & Sobolevskaya, R.F. 1989: Opornye razrezy i stratigrafiya kembro-ordovikskoj fosforitonosnoj obolovoj tolshchi na severo-zapade Russkoj platformy. [The key sections and stratigraphy of the Cambrian–Ordovician phosphate-bearing Obolus beds on the north-eastern Russian platform.] *AN SSSR, Ministerstvo Geologii SSSR, Mezhvedomstvennyj stratigraficheskij komitet SSSR, Trudy 18*, 1–222. Nauka, Leningrad.

Popov, L.E. & Ushatinskaya, G.T. 1992. Lingulidy, proizkhozhdenie discinid, sistematika vysokikh taksonov. [Lingulids, the origin of discinids, systematics of higher taxa]. *In* Repina, L.N. & Rozanov, A.U. (eds.): *Drevnejshie brakhiopody territorii Severnoi Evrazii*, 59–67.Novosibirsk.

Poulsen, C. 1922: Om Dictyograptusskiferen paa Bornholm. *Danmarks Geologiske Undersøgelse, Ser. 4:16*, 3–28.

Poulsen, C. 1923: Bornholms Olenuslag og deres fauna. *Danmarks Geologiske Undersøgelse, Ser. 2:20*, 54–56.

Poulsen, C. 1960: Notes on some Lower Cambrian fossils from French West Africa. *Matematisk–fysiske Meddelelser, Det kongelige Danske Videnskabernes Selskap 32*, 1–12.

Poulsen, V. 1966: Cambro-Silurian Stratigraphy of Bornholm. *Geological Society of Denmark, Bulletin 16*, 118–137.

Poulsen, V. 1971: Notes on an Ordovician acrotretacean brachiopod from the Oslo region. *Geological Society of Denmark, Bulletin 20*, 265–278.

Puura, I. & Holmer, L.E. 1993: Lingulate brachiopods from the Cambrian–Ordovician boundary beds in Sweden. *Geologiska Föreningens i Stockholm Förhandlingar 115*, 215–237.

Ramsay, A.C. 1866: The geology of North Wales. *Geological Survey of Great Britain, Memoire 3*, 1–381.

Regnéll, G. 1960: The Lower Palaeozoic of Scania. *In* Regnéll, G. & Hede, I.E. (eds.): The Lower Palaeozoic of Scania. The Silurian of Gotland. *International Geological Congress 21 Session Norden. Guidebook*, 1–43. Stockholm.

Rowell, A.J. 1962: The genera of the brachiopods superfamilies Obolellacea and Siphonotretacea. *Journal of Paleontology 36*, 132–152.

Rowell, A.J. 1965: Inarticulata. *In* Moore, R.C. (ed.): *Treatise on Invertebrate Palaeontology, Part H. Brachiopoda 1(2)*, H260–H296. Geological Society of America and University of Kansas Press, Boulder, Colorado, and Lawrence, Kansas.

Rowell, A.J. 1966: Revision of some Cambrian and Ordovician inarticulate brachiopods. *The University of Kansas Paleontological Contributions 7*, 1–36.

Rowell, A.J. 1980: Inarticulate brachiopods of the Pioche Shale of the Pioche District, Nevada. *The University of Kansas Paleontological Contributions 98*, 1–36.

Rowell, A.J. & Henderson, R.A. 1978: New genera of acrotretids from the Cambrian of Australia and United States. *The University of Kansas Paleontological Contributions 93*, 1–12.

Rukavishnikova, T.B. 1961: Brachiopody nizhnego ordovika khrebta Kendyktas. [The Lower Ordovician brachiopods of the Kendyktas Range.] *Trudy Geologicheskogo instituta Akademii Nauk SSSR 18*, 29–54.

Savazzi, E. 1986: Burrowing sculptures and life habits in Paleozoic lingulacean brachiopods. *Paleobiology 12*, 46–63.

Schindewolf, O.H. 1955: Über einige kambrische Gattungen inartikulater Brachiopoden. *Neues Jahrbuch für Mineralogie, Geologie und Paläontologie 1*, 1–15.

Schmidt, W. & Geukens, F. 1958: Nouveaux gîtes à brachiopodes dans le Salmien inférieur du Massif de Stavelot. *Societé Belge de Geologie, de Paleontologie et d'Hydrologie, Bulletin 67*, 159–161.

Scotese, C.R. & McKerrow, W.S. 1991: Ordovician plate tectonic reconstructions. *In* Barnes, C.R. & Williams, S.H. (eds.): Advances in Ordovician Geology. *Geological Survey of Canada, Paper 90:9*, 271–282.

Sdzuy, K. 1955: Die Fauna der Leimitz–Schiefer (Tremadoc). *Senckenbergische Naturforschende Gesellschaft, Abhandlungen 492*, 1–73.

Shergold, J.H. 1988: Review of trilobite biofacies distributions at the Cambrian–Ordovician Boundary. *Geological Magazine 125*, 363–380.

Tjernvik, T.E. 1956: On the Early Ordovician of Sweden. *Bulletin of the Geological Institutions of the University of Uppsala 36*, 107–284.

Tjernvik, T.E. 1958: The Tremadocian beds at Flagabro in south-eastern Scania (Sweden). *Geologiska Föreningens i Stockholm Förhandlingar 80*, 259–276.

Tsay, D.T. 1983: Tremadokskie graptolity severo–vostoka Tsentral'nogo Kazakhstana. [Tremadocian graptolites from the north–east of Central Kazakhstan.] *In* Apollonov, M.K., Bandaletov, S.M. & Ivshin N.K. (eds.): *Stratigrafiya i paleontologiya nizhnego paleozoya Kazakhstana*. 97–104. Nauka, Alma–Ata.

Tullberg, S.A. 1882: Förelöpande redogörelse för geologiska resor på Öland. *Sveriges Geologiska Undersökning, Ser. C 53*, 220–236.

Ushatinskaya, G.T. 1992. Novye srednekembrijskie lingulaty iz Batenevskogo kryazha (Altae–Sayanskaya skladchataya oblast'). [New Middle Cambrian lingulates from Batenev Ridge (Altai–Sajan Mountain Area]. *In* Repina, L.N. & Rozanov, A.Yu. (eds.): *Drevnejshie brakhiopody territorii Severnoj Evrazii*. Novosibirsk. 80–88.

Vagranov, V.G., Ancygin, N.Ia., Nassedkina V.A., Milicina, V.S. & Shurygina, M.V. 1973: *Stratigrafiya i fauna ordovika Srednego Urala*. [Stratigraphy and faunas of the Ordovician of Central Urals.]. 228 pp. Nedra, Moskow.

Verneuil, E. de 1845: Paléontologie, mollusques, brachiopodes. *In* Murchison, R.I., Verneuil, E. de & Keyserling, A. de: *Géologie de la Russie d'Europe et des Montagnes de l'Oural 2:3*, 17–395. John Murray, London, and Bertrand, Paris.

Viira, V. 1966: Rasprostranenie konodontov v nizhneordovikskikh otlozheniyakh razreza Sukhrumägi (g. Tallinn). [Distribution of conodonts in the Lower Ordovician sequence of Sukhrumägi (Tallinn). *Eesti NSV Teaduste Akadeemia Toimetised. Füüsika–Matemaatika*, 150–155.

Voinova E.V., Kirichenko G.I. & Konstantinova, L.I. 1941: *Geologicheskoe stroenie Orsko-Khalilovskogo rajona*. [*Geological Structure of the Orsk–Khalilovo District.*] 123 pp. Gosgeolizdat, Moscow.

Waern, B. 1952: Palaeontology and stratigraphy of the Cambrian and Lowermost Ordovician of the Bödahamn core. *Bulletin of the Geological Institutions of the University of Uppsala 34*, 224–250.

Walcott, C.D. 1889: Description of a new genus and species of inarticulate brachiopod from the Trenton Limestone. *Proceedings of the United States Museum 12*, 365–366.

Walcott, C.D. 1902: Cambrian Brachiopoda: *Acrotreta*; *Linnarssonnella*; *Obolus*; with descriptions of new species. *United States National Museum, Proceedings 25*, 577–612.

Walcott, C.D. 1908: Cambrian Geology and Paleontology. No. 3 – Cambrian Brachiopoda, descriptions of new genera and species. *Smithsonian Miscellaneous Collections 53*, 53–137.

Walcott, C.D. 1912. Cambrian Brachiopda. *Monograph of the U.S. Geological Survey 51*, 1–812.

Westergård, A.H. 1909: Studier öfver Dictyograptusskifferen och dess gränslager med särskild hänsyn till i Skåne förekommande bildningar. *Lunds Universitets Årsskrift 5:3*, 1–79.

Westergård, A.H. 1944: Borrningar genom alunskifferlagret på Öland och i Östergötland 1943. *Sveriges Geologiska Undersökning C 463*, 1–22.

Williams, A. 1974: Ordovician Brachiopoda from the Shelve District, Shropshire. *Bulletin of the British Museum (Natural History) Geology 11*, 1–163.

Williams, A. & Holmer, L.E. 1992: Ornamentation and shell structure of acrotretoid brachiopods. *Palaeontology 35*, 657–692.

Wright, A.D. 1963: The fauna of the Portrane Limestone. 1. The inarticulate brachiopods. *Bulletin of the British Museum (Natural History) Geology 8*, 223–254.

Wright, A.D. & MacClean, A.E. 1991: Microbrachiopods and the end-Ordovician event. *Historical Biology 5*, 123–129.

Zell, M.G. & Rowell, A.J. 1988: Brachiopods of the Holm Dal Formation (late Middle Cambrian), central North Greenland. *In* Peel, J.S. (ed.): Stratigraphy and palaeontology of the Holm Dal Formation (late Middle Cambrian), central North Greenland. *Meddelelser om Grønland, Geoscience 20*, 119–44.

Zima, M.B. 1976: Rod *Aletograptus* Obut et Sobolevskaya, 1962 v Tien–Shane. [On the genus *Aletograptus* Obut et Sobolevskaya, 1962 from the Tien–Shan Range.] *In* Kaljo D. & Koren, T.N. (eds.): *Graptolites and Stratigraphy*. 40–43. Academy of Sciences of Estonian SSR, Institute of Geology, Tallinn.

# Appendix 1

Distribution of lingulates in examined samples (complete shells : ventral valves : dorsal valves); + present; ? questionable; * dorsal/ventral ratio indeterminable.

## A

Tyrmantau Ridge

| | B-768-1 | B-768-3 | G-135-1/G-141-1 |
|---|---|---|---|
| *Broeggeria salteri* | 2:2:2 | | |
| *Thysanotos siluricus* | | | 0:7:4 |
| *Leptembolon lingulaeformis* | | | 0:3:3 |
| *Palaeoglossa? razumovskii* | | 0:6:2 | |
| *Siphonobolus uralensis* | | | 0:1:1 |
| *Eurytreta* cf. *sabrinae* | 0:0:10 | 0:6:22 | |
| *Dactylotreta batkanensis* | 0:13:18 | 0:1:0 | |
| *Orbithele ceratopygarum* | 0:4:2 | | |

## B

Alimbet Farm

| | B-578-1 | B-578-2 | 381-N | B-578-4 | B-578-5 | B-676 |
|---|---|---|---|---|---|---|
| *Thysanotos siluricus* | 0:2:0 | 1:0:0 | | | | |
| *Leptembolon linulaeformis* | | 0:1:0 | 5:2:0 | 0:4:1 | 0:3:2 | 0:1:1 |
| *Siphonobolus uralensis* | 0:7:13 | 0:108:14 | 0:104:29 | | 0:0:1 | |
| *Eurytreta chabakovi* | 0:5:2 | 0:124:50 | | 0:16:12 | | 0:7:4 |
| *Semitreta?* aff. *magna* | | | | 0:2:0 | | |

## C

Alimbet River

| | B-606-D | B-606-1 | B-619 | B-778-1 | B-778-2 | B-780-1 | B-603-1 | B-605-1 | B-605-5 | B-606 | B-606-C |
|---|---|---|---|---|---|---|---|---|---|---|---|
| *Thysanotos siluricus* | | | | | | 0:1:2 | | | | | |
| *Leptembolon lingulaeformis* | 0:4:4 | 0:1:1 | 0:10:6 | 0:2:0 | 0:1:1 | 0:8:4 | 0:21:24 | 0:12:8 | 0:2:0 | 0:1:0 | 0:1:0 |
| *Siphonobolus uralensis* | | | | 0:1:1 | | 0:18:16 | 0:1:0 | | | | |
| *Eurytreta chabakovi* | | | | 0:1:1 | | 0:21:32 | | | | | |
| *Semitreta?* aff. *magna* | | | | 0:9:0 | | 0:1:0 | | | | | |

## D

Akbulaksai River

| | B-607-1 | B-607-6 | B-607-8 | B-607-9 | B-607-15 | B-607-17 | G-42-3 | B-607-6-1 |
|---|---|---|---|---|---|---|---|---|
| *Thysanotos siluricus* | | | | | | 0:3:4 | 1:0:0 | |
| *Leptembolon lingulaeformis* | 0:30:31 | 0:1:1 | 0:5:6 | 0:5:2 | | 0:0:1 | | 0:2:1 |
| *Hyperobolus andreevae* | | | | 0:12:13 | 4:8:5 | | | |
| *Acrotreta korynevskii* | | | | | 0:18:20 | | | |
| *Eurytreta chabakovi* | | | | | 0:7:7 | | | 0:5:5 |

## E

| | Karabutak River | | | | Bolshaya–Malaya Kajala rivers | | Kosistek River |
|---|---|---|---|---|---|---|---|
| | B-236 | B-236-5 | B-236-5-1 | 1163 | G-104-1 | G-113 | G-30-18 |
| *Thysanotos siluricus* | | 0:1:0 | 4:3:4 | + | | | 0:0:2 |
| *Leptembolon lingulaeformis* | 0:1:2 | | | | | 0:2:3 | |
| *Siphonobolus uralensis* | | | | | | 0:9:7 | 0:1:1 |
| *Siphonotretella* sp. | | | | 0:1:0 | | | |
| *Acrotreta korynevskii* | | | | | | | 0:0:3 |
| *Eurytreta chabakovi* | | | | | | 0:30:31 | |
| *Eurytreta* cf. *sabrinae* | | | | | 0:7:1 | | |
| *Mamatia retracta* | | | | 0:15:9 | | | |
| *Otariella intermedia* | | | | 0:0:5 | | | 0:0:1 |
| *Orbithele ceratopygarum* | | | | | | 0:1:0 | 0:1:0 |

## F

| | Koagash River Akbulaksai Formation | | | | | | | | Koagash Formation | |
|---|---|---|---|---|---|---|---|---|---|---|
| | 360 | B-517-2 | B-517-5 | B-518-1 | B-518-5 | B-569-22 | B-786 | G-196-1 | B-523 | K-458 |
| *Thysanotos siluricus* | | | | | | | 0:1:1 | | | |
| *Leptembolon lingulaeformis* | | 0:2:1 | 0:1:0 | 0:0:2 | 0:0:1 | 0:2:0 | 0:5:1 | | | |
| *Siphonobolus uralensis* | | | | | | 1:7:7 | | 0:1:0 | | |
| *Ferrobolus fragilis* | | | | | | | | | | 0:1:0 |
| *Siphonotretella* sp. | | | | | | | | | | 0:0:2 |
| *Acrotreta korynevskii* | | | | 0:1:0 | | | 0:3:2 | | 0:14:95 | |
| *Eurytreta chabakovi* | | | | | | 0:1:1 | | | | |
| *Mamatia retracta* | + | | | | | | | | | 0:9:7 |
| *Otariella intermedia* | | | | | | | | | | 0:0:6 |
| *Eoconulus primus* | | | | | | | | | 0:0:38 | |
| *Orbithele ceratopygarum* | | | | | | | | 0:2:0 | + | |
| *Lacunites alimbeticus* | | | | 0:1:0 | | | | | | |
| *Semitreta?* aff. *magna* | | | | | | | 0:9:0 | | | |
| *Lamanskya splendens* | 0:1:0 | | | | | | | | | |

## G

| | Kujandy section Kujandy Formation | | Olistoliths of Kujandy Fm. within Satpak Formation | | | | | | Satpak Fm. |
|---|---|---|---|---|---|---|---|---|---|
| | 7823-4 | 7823-6 | 7822-4 | 7822-5 | 7822-6 | 7825-12.5 | 7825-13.5 | 7825-13 | C-35 |
| *Experilingua* cf. *divulgata* | 0:4:7 | | | | | | | | |
| *Broeggeria salteri* | | | | | | | | | 0:2:1 |
| *Broeggeria* sp. | | | 0:13:14 | 0:2:8 | 0:19:23 | 0:8:13 | 0:1:0 | 0:1:1 | |
| *Zhanatella rotunda* | | | | | 0:1:1 | | | | |
| *Fossuliella konevae* | 0:2:6 | | | | | | | | |
| *Dysoristus orientalis* | 0:2:1 | | | | | | | | |
| *Ferrobolus concavus* | | | 0:1:3 | | 0:0:1 | 0:2:2 | | | |
| *Galinella retrorsa* | | | 0:56:32 | 0:24:8 | 0:106:45 | 0:35:9 | | 0:2:2 | |
| *Quadrisonia simplex* | | 0:1:2 | | | | | | | |
| *Quadrisonia declivis* | | | 0:31:15 | 0:1:1 | 0:14:14 | 0:5:0 | 0:1:1 | 0:2:2 | |
| *Treptotreta bella* | 0:1:0 | 0:0:1 | | | | | | | |
| *Eoscaphelasma satpakensis* | | 0:3:1 | | | | | | | |

## H

| | Aksak–Kujandy Section Kujandy Formation | | Satpak Formation | | | | Olistoliths within Satpak Formation | | Ol. within Erzhan Fm. | Selety River | Sasyksor Lake |
|---|---|---|---|---|---|---|---|---|---|---|---|
| | 7827 | 7827-1 | 7844-5 | 8119 | 79106 | 79109 | 7844-2 | 79108 | 79101 | 325 | 601 |
| *Experilingula* cf. *divulgata* | | | | | | | | | 0:0:1 | | |
| *Broeggeria salteri* | | | 0:2:2 | | | | | | | 0:5:1 | |
| *Fossuliella konevae* | 9:2:0 | 12:3:1 | | | | | 0:0:1 | | 0:4:64 | | |
| *Zhanatella rotunda* | 0:1:1 | 0:1:0 | | | | | 0:9:13 | | 0:0:2 | | |
| *Dysoristus orientalis* | | | | | | | | | 0:14:31 | | |
| *Ferrobolus fragilis* | | | | | | | | | | | 0:4:4 |
| *Mirilingula* sp. | | | | | | | | | | 0:9:10 | |
| *Siphonotretella* sp. | | | | | | | | | | | 0:0:1 |
| *Conotreta shidertensis* | | | | | | | | | | | 0:11:21 |
| *Quadrisonia declivis* | 0:23:14 | 0:8:8 | | | | | | | | 0:7:7 | |
| *Quadrisonia simplex* | | | | | | | 0:10:26 | | | | |
| *Galinella retrorsa* | 0:27:13 | 0:4:4 | | | | | | | | | |
| *Treptotreta bella* | | | | | | | | 0:4:3 | 0:95:162 | | |
| *Akmolina olentensis* | 0:3:25 | 2:10:60 | | | | | | | | | |
| *Marmatia retracta* | | | | | | | | | | | 0:32:84 |
| *Sasyksoria rugosa* | | | | | | | | | | | 0:5:2 |
| *Eoscaphelasma satpakensis* | | | | | | | 0:5:4 | | 0:3:6 | | |
| *Otariella prisca* | | | | 0:1:23 | 0:1:2 | 0:1:1 | | | | | |
| *Eoconolus primus* | | | | | | | | | | | 0:4:9 |
| *Broeggeria* sp. | 0:3:1 | | | | | | | | | | |
| *Eurytreta* sp. | | | 0:0:2 | | | 0:0:1 | | | | | |
| *Lingulella* sp. | | | 0:1:3 | | | | | | | | |

**I**

| | Satpak Syncline | | | Satpak-2 | | | Olistolith | |
| | Satpak-1 | | | | | | | |
| | 7835 | 7843 | | 7836a | 7863b | 7863c | 79110 | |
|---|---|---|---|---|---|---|---|---|
| *Zhanatella rotunda* | 0:14:29 | 1:11:15 | | 0:8:0 | | | | |
| *Fossuliella konevae* | | | | 0:0:8 | 0:0:4 | 0:0:1 | | |
| *Dysoristus orientalis* | 0:0:1 | | | 0:2:13 | | | | |
| *Quadrisonia simplex* | 0:2:5 | 0:5:6 | | | | | | |
| *Treptotreta bella* | 0:2::3 | | | 0:28:33 | | 0:2:5 | 0:2:1 | |
| *Eoscaphelasma satpakensis* | 0:5:11 | 0:3:11 | | 0:11:327 | 0:19:127 | 0:5:57 | 0:0:1 | |

**J**

| | Southern Kendyktas Range | | | | | | | | Agalatas Fm. | | Kurdai |
| | Kendyktas Formation | | | | | | | | | | |
| | 551 | 552 | 553 | 554 | 556 | 556-1 | 558 | 563 | 560 | 564 | 559 |
|---|---|---|---|---|---|---|---|---|---|---|---|
| *Lingulella antiquissima* | 0:8:9 | | 0:10:11 | 0:5:9 | 0:0:2 | | | 0:2:4 | + | | |
| *Elliptoglossa linguae* | 1:10* | 0:1:0 | 0:12* | 2:42* | 2:2* | | | 0:2* | | | |
| *Agalatassia triangularis* | | | | | | | | | | 0:1:2 | 0:34:22 |
| *Broeggeria salteri* | 5:8:11 | 0:1:1 | 0:6:8 | 6:1:5 | 2:27:22 | 0:6:7 | 0:1:0 | 3:11:19 | | 0:8:10 | |
| *Keskentassia multispinulosa* | | | | | | | | | | 0:24:17 | |
| *Eurytreta sabrinae* | 0:22:20 | | 0:4:4 | 0:8:22 | 0:2:1 | 0:1:1 | 0:0:2 | 0:1:7 | | | |
| *Eurytreta* aff. *curvata* | | | | | | | | | | 0:24:39 | |
| *Mamatia* cf. *retracta* | | | | | | | | | | 0:0:8 | |
| *Cristicoma? keskentassica* | | | | | | | | | | 0:1:5 | |
| *Otariella intermedia* | | | | | | | | | | 0:0:6 | |
| *Schizambon* sp. | | | | | | | | | | 0:1:0 | |
| *Lacunites* cf. *alimbeticus* | | | | | | | | | | 0:0:1 | |

**K**

| | Öland | Scania | | | Västergötland | | Oslo |
| | Öl-1 | Sk-1 | Sk-2 | Sk-3 | Vg-1 | Vg-2 | Ng-1 |
|---|---|---|---|---|---|---|---|
| *Broeggeria salteria* | 0:29:17 | | | | + | | 0:1:5 |
| *Lingulella antiquissima* | 0:39:24 | | | + | + | + | 0:8:3 |
| *Lamanskya splendens* | + | | | | + | + | |
| *Elliptoglossa linguae* | | | | 0:28* | | | |
| *Rowellella* sp. | 0:1* | | | + | 0:3* | | |
| *Siphonotretella jani* | 0:9:13 | 0:0:3 | + | + | 0:0:5 | + | |
| *Ottenbyella carinata* | 0:75:63 | 0:2:0 | | | 0:?:4 | + | 0:76:42 |
| Acrotretidae gen. et sp. nov. a | | | 0:1:1 | | | 0:3:1 | |
| *Longipegma tulensis* | 0:6:14 | | | + | 0:?:14 | | |
| *Dactylotreta pharus* | 0:25:62 | 0:1:2 | | | 0:3:14 | | |
| *Eurytreta minor* | 0:1:3 | 0:40:9 | 0:27:19 | + | 0:?:7 | 0:30:16 | 0:148:90 |
| *Eurytreta* cf. *sabrinae* | 0:25:20 | 0:0:2 | | | 0:?:? | | 0:0:2 |
| *Eurytreta* sp. a | | | 0:2:2 | | | | |
| *Pomeraniotreta biernatae* | 0:157:122 | 0:168:89 | | + | 0:98:45 | 0:40:80 | 0:273:158 |
| *Otariella* sp. | | | | | 0:0:3 | | |
| *Eoconulus* sp. | | | | | 0:0:4 | | |
| *Orbithele ceratopygarum* | 0:16:6 | | | | | | |